Elements of Dual Scaling:

An Introduction to Practical Data Analysis

Shizuhiko Nishisato
The Ontario Institute for Studies in Education,
and University of Toronto

LEA
LAWRENCE ERLBAUM ASSOCIATES, PUBLISHERS
1994 Hillsdale, New Jersey Hove and London

Lawrence Erlbaum Associates, Inc., Publishers
365 Broadway
Hillsdale, New Jersey 07642

Library of Congress Cataloging-in-Publication Data

Nishisato, Shizuhiko, 1935-
Elements of dual scaling : an introduction to practical data analysis / Shizuhiko Nishisato.
p. cm.
Includes bibliographical references and index.
ISBN 0-8058-1209-1
1. Correspondence analysis (Statistics) I. Title.
QA278.5.N57 1994
519.5'354--dc20 93-25276
CIP

Books published by Lawrence Erlbaum Associates are printed on acid-free paper, and their bindings are chosen for strength and durability.

Printed in the United States of America

10 9 8 7 6 5 4 3 2 1

To

My wife Lorraine and son Ira in Canada

My mother Fusako,

Michiko, Tsunehiko, and Akihiko in Japan

George and Rosetta in South Africa

Contents

Preface

This book is devoted to applications of a technique for data analysis. The technique is called *dual scaling*. As some may know, dual scaling has a plethora of other names.[1] Proponents of these names and their collaborators have developed their own methods, within their own framework, and with specific orientations. Thus, in spite of the fact that all of them have a common starting point, some of them have passed the stage of "basic formulation," moving into the phase of their own unique advancement. Therefore, the current methodology of *analyse des correspondances* in France, for example, is to a great extent different from that of *homogeneity analysis* in the Netherlands. The same applies to *dual scaling* in Canada. This phenomenon of branching out will continue, and we may see the day when the name dual scaling will no longer be used synonymously with correspondence analysis.

I received a frank suggestion from a colleague that this volume's title should include dual scaling and correspondence analysis, instead of simply dual scaling to encourage sales and promotion of the book. Although correspondence analysis is better known among some quarters of researchers, I decided to use the original title because it would be impossible to cover recent developments in correspondence analysis adequately enough to satisfy French statisticians, given the branching-out phenomenon already mentioned. This inadequacy of knowledge about developments in other countries became more obvious while I was compiling a bibliography of materials dealing with quantification of categorical data in 1986.

From the outset of dual scaling research, the main orientation has been to extend its applicability to a wider variety of categorical data. As a result, dual scaling can handle not only contingency tables and multiple--choice data (i.e., the main data types for correspondence analysis), but also paired comparison, rank order, successive categories, sorting, and multiway

---

[1]For instance, correspondence analysis, homogeneity analysis, quantification theory, optimal scaling, biplot, reciprocal averaging, centroid scaling, basic structure content scaling, simultaneous linear regression, additive scoring, nonlinear multivariate analysis, and multivariate descriptive analysis.

data matrices. In addition to this wider applicability, an analogue of discriminant analysis for categorical data has been developed under the name *forced classification*, and the procedure of generalized forced classification is also available. Other computational procedures were also developed, including the method of successive data modification (SDM) for quantification of ordered categories, partially optimal scaling for data with preassigned weights or mixed (categorical plus continuous) data, and the piecewise method of reciprocal averages for handling a large number of multiple-choice items.[2] The introduction and discussion of many of these procedures would certainly be enough for one book—another reason for staying within the familiar territory of dual scaling. Consequently, there may be an excessive number of references to those studies conducted in Toronto.

Until recently, there were four standard English books on this topic of quantification methodology. They are Gifi (1990), Greenacre (1984), Lebart, Morineau, and Warwick (1984), and Nishisato (1980a), representing, respectively, *homogeneity analysis*, *correspondence analysis*, *analyse des correspondances*, and *dual scaling*. Each of these books is very informative; they will continue to be excellent sources of reference. However, they are too advanced for many people who are primarily interested in applications. Since 1980, a number of specialized books have been published in English by Dutch psychologists.[3] These books are again too advanced for many researchers in applied fields. There are many other books in other languages as well.[4] All of these assume that the readers have some training in statistics. At the opposite end of the spectrum, there is an introductory, totally application-oriented book written by Nishisato and Nishisato (1984). It is particularly suitable for a short-term workshop, or as an introductory textbook for an application-oriented course. But the book is not geared for those who are also interested in derivation of formulas. Looking at the current availability of English books and continued interest in this type of technique, there is a need for a "medium-level" book which bridges the two

---

[2] See all the relevant publications listed in the References.

[3] For instance, Heiser (1981), Kiers (1989), Koster (1989), Meulman (1982, 1986), van der Heijden (1987), van Rijckevorsel (1987), van der Burg (1988), and van Rijckevorsel and de Leeuw (1988).

[4] To name a few, in Japanese, Hayashi (1977), Iwatsubo (1987), Komazawa (1978, 1982), Nishisato (1975, 1982), Yanai and Takane (1977); in French, Benzécri (1992), Benzécri et al. (1973), Benzécri and Benzécri (1980), Bouroche (1977), Bouroche and Saporta (1980), Cailliez and Pages (1976), Lebart, Morineau, and Tabard (1977), and Saporta (1979).

extremes, that is, a book that is application-oriented, easy to follow with many examples, and equipped with at least a minimal package of technical information, hence the main motivation for this book. I hope that it will appeal to a wider range of readers and serve their needs well.

The book begins with a nontechnical introduction to data analysis and various characterizations of dual scaling with illustrative examples (chapters 1 and 2). The reader is then asked if the problem at hand is suitable for analysis by dual scaling; if not, the book offers a few possible alternatives (chapter 3). After this general introduction, the reader is given a first look at some of the fundamental ideas behind dual scaling (chapter 4). The next two chapters are written for the reader who wishes to look at mathematics of dual scaling, starting from a minimal package of basic quantitative skills (chapter 5) to a formal derivation of dual scaling (chapter 6). Although these two chapters are not essential for the ensuing applications, they can help one to understand how the basic ideas are put into practice. By this time, the reader is fully prepared to appreciate applications of dual scaling to six different types of categorical data (chapters 7 to 12). The final five chapters are devoted to specific, advanced applications of dual scaling to different problems, and these chapters represent comparatively unique developments, carried out under the theme *extensions of dual scaling* in Toronto, Canada. The entire book is organized in a gradual progression from the simple and general to the advanced and specific. The reader should find this sequence easy to follow.

I am indebted to many people, but only a few names will be mentioned here. First, Dr. R. Darrell Bock—a mentor who introduced me to his *optimal scaling*; Dr. Lyle V. Jones and Dr. Masanao Toda, who gave me invaluable opportunities for study in the United States and Japan, respectively; Mrs. Mildred Ringwalt, her late husband Mr. Arthur Ringwalt, and their family for their warmth, encouragement, and affection during my Chapel Hill days; Prof. Dr. Wolfgang Gaul, who invited me as *Gast Professor* at Institut für Entscheidungstheorie und Unternehmensforschung, der Universität Karlsruhe, Germany, in 1987, and gave me an impetus to write this book; Dr. Ross E. Traub, with whom I have been involved since 1967 in a large number of MA, PhD, and EdD thesis committees; Dr. Merlin W. Wahlstrom, whose expertise has been a great contribution to OISE's annual workshop on dual scaling; many other supportive colleagues at OISE, and fellow researchers throughout the world. Among our graduates, the following have contributed through their theses [5] to the development of dual scaling: Hyung Ahn (Im), Diana Chan, Heather M. Chipuer, Deborah Day, Marc E. Gessaroli, Lindsay Gibson, Stuart Halpine, Yukio

[5] See their theses listed in the References.

Inukai, Daniel R. Lawrence, Oscar Millones, Wang-Pui Poon, John Sachs, Wen-Jenn Sheu, Yukihiko Torii, Margaret Willett, and Liqun Xu. My cordial appreciation also goes to co-authors Wen-Jenn Sheu, Hyung Ahn (Im), Prince S. Arri, Dalbir Bindra, Donald C. Donderi, Wolfgang Gaul, Yukio Inukai, Takashige Iwamoto, Daniel R. Lawrence, Kuo-Sing Leong, H. Edward Massengill, Roderick P. McDonald, Junichi Nakahara, Ira Nishisato, Tadasu Oyama, Emir H. Shuford, Yoshio Sugiyama, Yukihiko Torii, Philip Weingarden, Raymond A. Wiesen, Jack S. Wise, Fumiyasu Yamada, and Hirotsugu Yamauchi.

In preparing the manuscript, I must thank first of all Mrs. Patricia O'Connor for her painstaking and expert work, without which the manuscript would not have been completed; Ms. Patricia J. Probert for her initial work of the manuscript conversion; Ms. Mary Kolić and Liqun Xu for their work on graphs and examples. My special thanks to my son Ira who edited the manuscript and offered numerous suggestions to ensure that my writing was not too technical for nonspecialists, and Dr. Judith Amsel, Vice President, Editorial, LEA, Mr. Art Lizza, Vice President, Production, LEA, and Ms. Linda Eisenberg, Production Editor, LEA, for their advice, and two copy-editors. My research on quantification of categorical data started in 1969, the year in which my son Ira was born, and has continued to date, always with financial assistance from the federal and provincial governments, in particular, the Natural Sciences and Engineering Research Council of Canada (NSERC) for the past 15 years. I would like to thank all those mentioned here and others who have contributed much to the publication of this book in one way or another.

— Shizuhiko Nishisato

PART I

Background

Part 1 contains six chapters to introduce basic ideas on data analysis, data types and dual scaling. If you are already familiar with dual scaling, correspondence analysis, Hayashi's theory of quantification, optimal scaling, or homogeneity analysis, you may skip Part 1, and start with Part 2. However, if you are not familiar with any one of these terms, try to digest the material covered here before you engage yourself in applications of dual scaling in Part 2. Those topics in Part 1 will prove important for subsequent chapters even if they may look elementary.

CHAPTER 1
To Begin With

There are a few things you might want to look at before you get involved in dual scaling. They are concerned with the nature of data analysis. As you will see, most points to be mentioned here are what your common sense would dictate. Nevertheless, putting those points in one place may be of some use as an introduction to data analysis.

Data collection is your first topic. If you are a teacher, you may give several tests to a group of students, and their scores on the tests will be your data. If your work is in marketing research, your data may consist of consumers' responses to a set of so-called multiple-choice questions plus their biographical information (e.g., gender, age, profession). If you are a clinical psychologist, patients' responses to an inkblot test may be your data. If you are a public relations officer of a company, a list of complaints from the public would constitute a data set. Whatever your task, you must collect "valid" information, valid in the sense that it is worth analyzing and can be analyzed. This is a very important point with many relevant problems, yet frequently tends to be overlooked or ignored completely.

First of all, you must consider the *task of the respondents* or subjects: for instance, to answer a set of questions, to compare the taste of Coca Cola and Pepsi, or to rank five candidates for a committee of three. Are these tasks simple and clear enough for your subjects? It is easy to assume, because you are familiar with the area under investigation, that your subjects would be able to answer all the questions, say about pollution problems, social welfare problems, or mandatory retirement issues. If you want to solicit reactions of people in your community to the government's recent tax reform proposal, make the questionnaire short and easy to answer.

Once you know what you want to find out from data analysis, you must collect data suitable for your purposes. If two classes of students are to be compared on their performance in mathematics, for example, make sure that the same mathematics test is given to both classes, and avoid such a day for testing when many students in one class are absent from school. Would you not also wish to discover what other information might be related to their differences in achievement, for instance, aptitude, career orientation, curricula, use of computers, and so on? If you do, collect the information from the students.

In many situations, you want to find out whatever your data can tell you. As compared with data collection for the purpose of research, it might be more difficult to collect data for so-called exploratory work, where you want to gather not only specific but also general information. In an attempt to collect as much information as possible, however, don't be too ambitious. There are many studies in which the investigator employs so-called open-ended questionnaires consisting of such a question as "What do you think of capital punishment? State your view in the following blank space." There is nothing wrong with this kind of open-ended question if you want to collect opinions about an issue. However, from the data analysis point of view, you would definitely prefer hearing from everyone in your study sample whether he or she stands for or against capital punishment, whether or not there should be a referendum on capital punishment, whether he or she thinks the death penalty serves as an effective deterrent to stamp out serious crimes, whether or not capital punishment is morally wrong, or whether he or she thinks the current criminal justice system leaves too much risk of an innocent person being executed. Then you can expect answers from everyone to specific issues, and you will be able to analyze the data in a presentable way. In contrast, when you collect data with open-ended questions, all you can do, despite your good intentions, is list those opinions, with your comments, and most likely stash the results away in an obscure place in your report, such as the appendix. The time you spend in developing a data-collection instrument such as a survey questionnaire, then, will amply reward your efforts. It is certainly worth taking the additional time. Try by all means to avoid open-ended questions.

An example of an ambitious effort with relatively little benefit is the case of what you might call conditional questions. "Do you agree with the proposal for banning cigarette smoking in your work place? If you agree, skip the remaining questions; if you disagree, proceed to the next question." It is clear that you want to find out more about reasons for disagreement in this example. But this type of question often restricts the kind of data analysis you can do once data are collected. You would probably tabulate proportions of agreement and disagreement, and proportions of heavy smokers, medium smokers, light smokers, and nonsmokers, given that they disagreed with the proposal. The results look interesting, and you may be satisfied with them. However, conditional questions create a big blank of missing responses in your data set, which would not allow you to investigate, for example, other variables such as different lifestyles and personality traits which distinguish between proponents and opponents of cigarette banning. With additional effort, you could eliminate and modify those questions into a form that everyone could answer. Whether you are a heavy smoker or not can certainly be asked even if you agree with the banning of cigarette smoking. Consider the question: "Do you agree that the administration should respect nonsmokers' rights? Agree( ), Disagree( )." Even though most smokers might not agree with this question, there is nothing wrong in soliciting their views. Similarly, the same applies to the nonsmokers when the question is "Do you agree that the administration should respect smokers' rights? Agree( ), Disagree( )." As you will see later, it is often very important to have complete data from all the respondents. Try by all means to avoid conditional questions.

Another problem with conditional questions lies in the possibility of *incorrect attribution* of data analysis results. For instance, consider the instruction: "Are you allergic to animal hair, ragweed, house dust, Brazil nuts, or shellfish? If your response is yes, answer the next 20 questions; if your response is no, skip to Question 21." Suppose that the research analyzes responses to those 20 questions and finds out that sufferers of those allergies are mostly those in the middle class, born in this country, non-vegetarians, and employed. One of the pitfalls in data analysis is then to attribute these characteristics to allergy sufferers. This is wrong, for nonsufferers may also be characterized in the same way as allergy sufferers in the aspects mentioned here. In other words, those characteristics are not peculiar to the allergy sufferers, hence the researcher cannot attribute either them to allergy, or allergy to them. This point of incorrect attribution may appear obvious to you, but it is often overlooked. We should collect data from both allergy sufferers and nonsufferers, and compare the two groups to identify unique characteristics of allergy sufferers.

The kind of data you collect often determines which methods of data analysis you can use. While you might find several ways to analyze your data, you should know that *techniques of data analysis are designed for specific types of data*. On the one hand, therefore, you must be clear about the purpose of your study and collect your data accordingly. On the other hand, you must know what a particular method of analysis can do for you. You must consider these two sides before the collection of your data. You may find it easier, though, to first find a method of data analysis which appears suitable for your research, and then determine the method of data collection in line with the format required. Later on, you will see several data types which are appropriate for dual scaling. Examine and study how these types of data are collected, so that at the time of data analysis you will not have to worry about the appropriateness of dual scaling for your data.

Data analysis can mean many things, from a simple counting of children in each district to hypothesis testing with a complex statistical model. Whatever you decide to do, you would no doubt like to gain a better understanding of your data as a result of data analysis, as well as communicate the results to other people who are interested in the problem you are investigating. One popular way of summarizing data is averaging—for example, test scores in a class. If you are interested in comparing mathematics achievement of several classes, these class averages (means) would be informative enough for most people. Your data contain, however, more information than that, and you should ask yourself if you wish to make better use of the data.

Proportions and means are familiar statistics, and we are used to considering them as always meaningful. Both are easily calculated from all the numbers in the data set, and are thus, you would think, representative of the group. Yet an average no longer reflects individual differences. For example, consider an average mark. Suppose that you are excellent in English, French, History, and Economics, but that you are poor in Physics and Chemistry. It is obvious that your overall average wouldn't tell us anything about such an uneven distribution of marks. Should there not be two averages calculated, one for English, French, History, and Economics, and the other for Physics and Chemistry? You would then have one high average and one low average, rather than one only slightly better than the class average. In terms of information about yourself, two averages in this example are much better than a single overall average. The same reasoning applies to the class average as well. In one class, all the students may do equally well, while in another half the students are excellent and the remaining half very poor academically. Let's stretch your imagination a little further. In the year end test in mathematics, students are examined about all the topics discussed during the year. If two students receive very similar final marks, most of us tend to think that they are equally competent in mathematics. This is not necessarily the case. One student may have collected high scores from a few topics, and the second student may have done equally well on all the topics, indicating different kinds of competence. Comparisons of subject averages of the two students would be more meaningful than comparing two overall averages for describing the two students.

Proportion is also widely used. Yet it has a problem similar to the mean. For instance, we often compute the proportion of correct answers out of n questions for each student, ignoring completely whether some questions are more difficult than others, or if some questions are more relevant to what we want to assess than others. Similarly, we often calculate the proportion of correct or "yes" responses to each question, without taking into consideration if high scorers contributed more to the proportion, or if correct responses have come from the entire spectrum of distribution of scores in a random fashion. In other words, the proportion is calculated by weighting all the questions (first case) and the subjects (second case) equally. Would it make sense? Yes, it does in some cases. From the data analysis point of view, however, we would prefer some differential weighting of questions (difficult questions are worth more than easier ones) and subjects to generate such summary statistics as mean and proportion.

This discussion tells us that even the most commonly used statistics may not be appropriate for interpretation or understanding of your data. Simplifying data often results in a useful result, but simplification should not be done at the expense of valuable information. You need a simple, understandable description which reflects the characteristics of the data. In this regard, you will soon see the amazing ability of dual scaling. It simplifies complex data remarkably well, and provides a detailed description of practically every bit of information in the data, yielding a simple yet exhaustive analysis. Dual scaling is also said to be a technique of *multivariate descriptive analysis* (Lebart, Morineau, and Warwick, 1984), as opposed to *multivariate inferential analysis*. What this means is that it is basically a method to describe only a given data set, rather than to draw inferences from given data about a more general set, called population. "Multivariate" means data on more than one measure (e.g., several personality tests). This distinction between the two approaches has many implications: advantages and disadvantages, possibilities and limitations. In contrast to inferential statistics, which confirms or refutes "hypotheses," dual scaling belongs to so-called exploratory data analysis, which specializes in "discovering" through detailed analysis of a given data set. As the adage holds, "To discover is more exciting than to confirm"—a view which you may agree with as you go through later chapters. Rigorous comparisons of the two approaches, however, are beyond the scope of this introductory text, and must be left to your own investigation. The point here is that because it is a *descriptive* method, dual scaling can prove extremely useful in elucidating an otherwise complex set of data.

Another way of looking at dual scaling is that it is a *nonlinear multidimensional* method of quantification. Both "nonlinear" and "multidimensional" stem from its characteristic to analyze data exhaustively. For instance, low blood pressure (x_1), medium blood pressure (x_2), and high blood pressure (x_3) may be quantified as $x_1 < x_2 < x_3$ to capture symptomes of an illness that are linearly related to the blood pressure (i.e., the higher the blood pressure, the more serious the illness), and as $x_2 < x_1 < x_3$, or $x_2 <$

$x_3 < x_1$, to identify symptoms that are nonlinearly related to the illness (i.e., if the blood pressure is very high or very low, the illness is serious). In elucidating the behavior of the blood pressure linearly and nonlinearly, dual scaling analyzes data into mathematically distinct patterns of relations, which is called multidimensional analysis. Let us now look at what dual scaling can do for you.

CHAPTER 2
What Can Dual Scaling Do For You?

You will see many applications of dual scaling to data analysis later in this book. However, this section provides a preview through a few simple examples before discussing technical aspects of the method. We hope that this chapter will give you some ideas about the application of dual scaling, as well as sufficient motivation to reach beyond this chapter. Let us start with a very small example.

2.1 EVALUATING TEACHERS' PERFORMANCE

Suppose that three teachers, White, Brown, and Green, were evaluated on their teaching performance in terms of "good," "average," and "poor." Ten students evaluated White. Table 2.1 shows that, out of the 10, 1 person placed the teacher in the category "good," 3 the category "average," and 6 in the category "poor." Likewise, you can interpret the data in Table 2.1 for Brown and Green. Those numbers are all that you are given as summary data. This table offers a neat summary of the evaluation, and you, or your colleague, may be satisfied with it for a final report. But is this format the "best" way to show how good these teachers are? Perhaps not; or, to be bold, definitely not. There are other ways to summarize this type of data in a slightly more useful fashion than Table 2.1. Dual scaling is one of those alternatives. Whether or not dual scaling can do a better job than Table 2.1 is a matter for your judgment, but you will likely be impressed with the ability of dual scaling to simplify the job of data interpretation.

Let us see what dual scaling would do with Table 2.1. Figure 2.1(a) and (b) present what you might call "dually scaled" data. Overlaying (a) and (b) shows that each teacher is close to one of the evaluation categories. Therefore, you may conclude that White, Brown, and Green are respectively poor, average, and good teachers. Your reaction is likely one of either bewilderment at the simplicity of this approach, or concern that dual

TABLE 2.1
Evaluation of Teaching Performance

Teacher	Good	Average	Poor	Total
White	1	3	6	10
Brown	3	5	2	10
Green	6	3	0	9
Total	10	11	8	29

scaling may have oversimplified the data. For peace of mind, remember that (a) and (b) of Figure 2.1 contain enough information to reproduce fairly accurately the original data in Table 2.1. In other words, the information contained in (a) and (b) is not much different from the information in Table 2.1, except that Figure 2.1 is more readily interpretable. Later on, you will see that Figure 2.1 explains 92.1% of what we call the *total association* in the data (for now, don't worry about the meaning of this statement) You may wonder why (a) and (b) are not combined into a single graph. This too will be explained shortly.

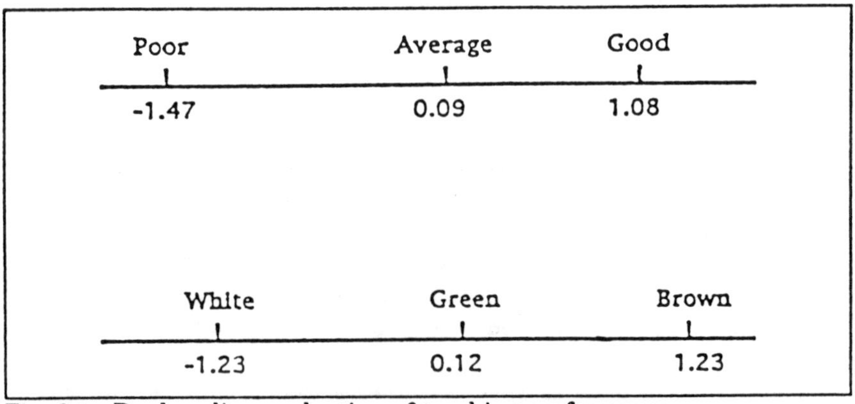

Fig. 2.1. Dual scaling evaluation of teaching performance

This example shows that dual scaling can simplify your job of extracting easily interpretable information from the data. But what will happen if your data are much more complex than this basic example? Let us now look at a more typical example.

2.2 GENDER, AGE, AND SUICIDES

Table 2.2 presents a data set extracted from a German study on suicide.[1] There are 2 "gender" groups, 17 age groups, and 9 means of suicide. The numbers in the table are totals of four years, 1974 to 1977. This is again a summary table. However, unlike Table 2.1, there are too many entries to grasp what these numbers tell us. We often say that the human brain is an incredible information processor, but can you extract anything at all from this table that would make sense? What would you tell yourself, or your colleague, about the outcome of your survey or assignment? You may, of course, further analyze Table 2.2 in anyway you like. But how? There are several methods to analyze data of this type, and dual scaling is one of them and possibly one of the best for data interpretation.

Analysis of Table 2.2 indicates that you would need at least a *two-dimensional space* to capture some characteristics of the data, that is, a graph with not only a horizontal axis (Solution 1), but also a vertical axis (Solution 2). In other words, the data contain at least two distinct characteristics.[2] Let us find out what this means in terms of data analysis. Parts (a) and (b) of Figure 2.2 summarize the results of dual scaling. These graphs may not be easy to decipher, but they provide you with the following conclusions:

1. There seems to be a clear difference in the mode of suicides between females and males. You see all the data points of the female group in (a) are on the left-hand side on axis I, and the data points of the male group on the other side.
2. Age also seems to be a key factor in determining the modes of suicides. We see a steady move of age points from the group 20 to 90+, although the youngest group (10-15) in both genders appears to behave more like an old-age group.

You will of course want to know the relation between these two key variables (i.e., gender and age) and other modes of suicides. Two graphs (a) and (b) are constructed in as much a comparable way as possible. Therefore, by overlaying the two graphs, you obtain "approximate" locations of the modes of suicides in relation to the points of age groups for each

---

[1] J. Heuer (1979). The original table contained 18 groups, but for reasons of simplicity 2 groups have been merged.

[2] In multidimensional data analysis, we often use words "solution," "dimension," "component," and "factor" in an interchangeable way. In this book, the word "solution" is used in most instances.

TABLE 2.2
Gender, Age and Suicides (Heuer, 1979)*

	Age	Mat	Gas-h	Gas-o	Hang	Drwn	Guns	Stab	Jump	Others
	10-15	28	0	3	20	0	1	0	10	6
	16-20	353	2	11	81	6	15	2	43	47
	21-25	540	4	20	111	24	9	9	78	67
	26-30	454	6	27	125	33	26	7	86	75
	31-35	530	2	29	178	42	14	20	92	78
F	36-40	688	5	44	272	64	24	14	98	110
e	41-45	566	4	24	343	76	18	22	103	86
m	46-50	716	6	24	447	94	13	21	95	88
a	51-55	942	7	26	691	184	21	37	129	131
l	56-60	723	3	14	527	163	14	30	92	92
e	61-65	820	8	8	702	245	11	35	140	114
s	66-70	740	8	4	785	271	4	38	156	90
	71-75	624	6	4	610	244	1	27	129	46
	76-80	495	8	1	420	161	2	29	129	35
	81-85	292	3	2	223	78	0	10	84	23
	86-90	113	4	0	83	14	0	6	34	2
	90+	24	1	0	14	4	0	2	7	0
	10-15	4	0	0	247	1	17	1	6	9
	16-20	348	7	67	578	22	179	11	74	175
	21-25	808	32	229	699	44	316	35	109	226
	26-30	789	26	243	648	52	268	38	109	226
	31-35	916	17	257	825	74	291	52	123	281
	36-40	1118	27	313	1278	87	293	49	134	268
M	41-45	926	13	250	1273	89	299	53	78	198
a	46-50	855	9	203	1381	71	347	68	103	190
l	51-55	684	14	136	1282	87	229	62	63	146
e	56-60	502	6	77	972	49	151	46	66	77
s	61-65	516	5	74	1249	83	162	52	92	122
	66-70	513	8	31	1360	75	164	50	115	95
	71-75	425	5	21	1268	90	121	44	119	82
	76-80	266	4	9	806	63	78	30	79	34
	81-85	159	2	2	479	39	18	18	46	19
	86-90	70	1	0	259	16	10	9	18	10
	90+	18	0	1	76	4	2	4	6	2

*Notes: Mat = drugs, poison; Gas-h = gas at home; Gas-o = gas, others; Hang = hanging; Drwn = drowning; Guns = gunshot; Stab = stabbing; Jump = jumping; Others = other means

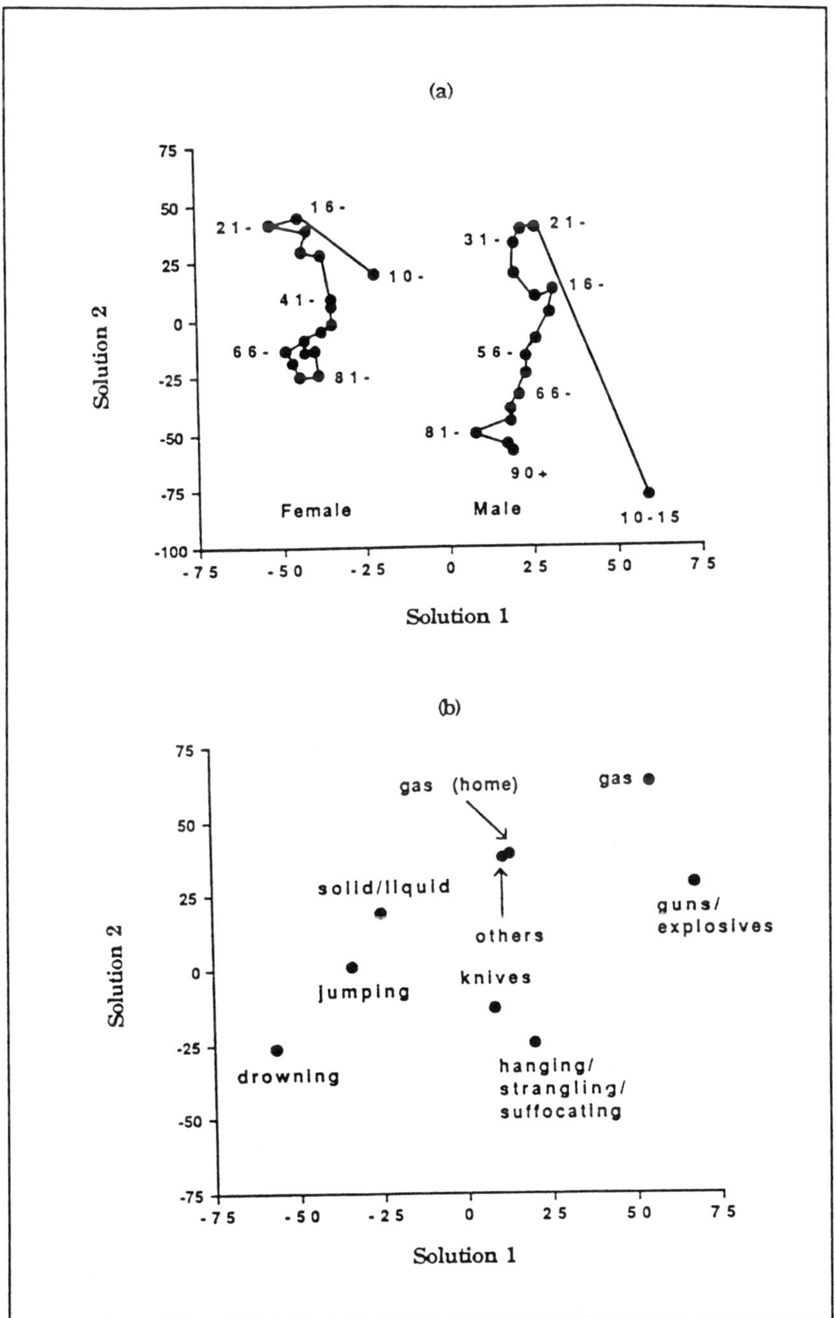

Fig. 2.2. Two-dimensional interpretation of data in Table 2.2: (a) row weights and (b) column weights.

gender. (Note that "approximate" is used here to avoid any suggestion that these are geometrically exact locations.) You can now see some interesting associations as shown in Table 2.3.

TABLE 2.3
Summary of Associations

Drowning	Female	Older group
Jumping	Female	Around 50-60
Hanging Strangling Suffocating	Male	Around 60-70
Knifing	Male	Around 50-60
Guns	Male	Around 15-40
Gas	Male	Around 20-30

Don't you agree that these graphs tell you much more about suicides by the two genders and age groups than the original summary table? The graphs also tell you that the gender difference (Solution 1) explains approximately 52% of "association" (this will be explained later), and that age differences (Solution 2) account for roughly 38%. Thus, fully 90% of the data are explained by the combination of the two solutions. You can see in Figure 2.2 the *clustering* of data points, that is, identifiable and interpretable patterns. Such clusters can be used to summarize your analysis. This example indicates that your data may require more than one solution to accommodate some hidden patterns of responses. If so, the first solution tells you about only one part of the data, leaving the remainder unanalyzed. What is desirable then is to extract the second solution to see what the remainder of the data can tell you. Of course, it is possible that in some more complicated cases the first two solutions could leave a substantial amount of information in the data unaccounted for. To carry out an exhaustive analysis, you would have to extract more solutions. This is exactly what dual scaling will do.

These two examples have demonstrated the ability of dual scaling to reorganize given data into a simpler and more easily interpretable form. This may be all that you want of data analysis. If so, that is a very reasonable expectation. You will be pleased to know, however, that dual scaling can offer another useful way of looking at data, namely by allowing you to focus your analysis on a specific aspect of the data, rather than on the entire data set. This procedure of focusing is called *forced classification*.

2.3 HIGHLY OPINIONATED, AND WHAT ELSE?

Table 2.4 presents a questionnaire to collect information on various aspects of personality relative to the average person. Once data are collected from a group of people, you can analyze the data by the standard method of dual scaling as in the previous two examples. It would reveal some patterns of responses, which would undoubtedly be very interesting to look at. But suppose that you are interested in the personality trait of being "highly opinionated," and that you would like to find its *correlates*, that is, other traits which seem to go together with it. This is exactly the task for which the procedure of forced classification was developed.

Consider the application of forced classification to data collected from 24 people. Table 2.5 lists some of the traits which are correlated with "highly opinionated." This table says that if you are highly opinionated you tend to be less quiet, more outspoken, more presumptuous, more creative, more flamboyant, less timid, and less courteous. The results make sense, don't they? It seems that the relations among these correlated traits confirm more or less our expectations. So this procedure of forced classification can offer many possible answers to such questions as:

- What personal background is characteristic of users of Clinique cosmetics?
- What characteristics of Oil of Olay appeal to its users?
- What personality makeup and family structure are correlated with high-school drop-outs?
- What philosophy of life and personality traits distinguish between hard-core criminals and others?
- What combinations of personality traits make bankers successful?

You can no doubt think of many more examples for which the procedure of forced classification may effectively be used. It is unquestionably one of the most useful options of dual scaling.

After having seen three different examples, you must have some ideas of your own about dual scaling. These examples reveal only the tip of the iceberg, and you will see many more interesting examples later on in this book. It is essential for you to use the technique to appreciate its wide applicability. Only then will you see the iceberg itself, an image of the versatility of dual scaling. In the meantime, you would probably like to know what more dual scaling can do for you. Thus, without technical discussions as yet, let us list some characteristics of dual scaling. Do not worry about the details. This is only for your information, not for your data analysis work.

TABLE 2.4
"As Compared with an Average Person. . ." (Nishisato, 1984c)

PLEASE CHECK ONE OF THE TWO ANSWERS TO EACH STATEMENT.

As compared with an average person:

1.	I am more friendly	( ) Yes ( ) No
2.	I am quieter	( ) Yes ( ) No
3.	I am more mature	( ) Yes ( ) No
4.	I am more clumsy	( ) Yes ( ) No
5.	I am more outspoken	( ) Yes ( ) No
6.	I am more miserly	( ) Yes ( ) No
7.	I am more self-conscious about my behavior	( ) Yes ( ) No
8.	I am more high-handed and opinionated	( ) Yes ( ) No
9.	I spend more time chatting	( ) Yes ( ) No
10.	I am more courteous	( ) Yes ( ) No
11.	I am more impatient	( ) Yes ( ) No
12.	I am more optimistic about life	( ) Yes ( ) No
13.	I am more meticulous	( ) Yes ( ) No
14.	I am more inquisitive	( ) Yes ( ) No
15.	I am more presumptuous	( ) Yes ( ) No
16.	I am more creative	( ) Yes ( ) No
17.	I am more timid	( ) Yes ( ) No
18.	I am more inflexible	( ) Yes ( ) No
19.	I eat more	( ) Yes ( ) No
20.	I am more ambitious	( ) Yes ( ) No
21.	I am more humorous	( ) Yes ( ) No
22.	I am more cheerful	( ) Yes ( ) No
23.	I am more tasteful	( ) Yes ( ) No
24.	I am more pleasant in dealing with others	( ) Yes ( ) No
25.	I am more flamboyant	( ) Yes ( ) No
26.	I am more of a spendthrift (waster)	( ) Yes ( ) No
27.	I am more generous	( ) Yes ( ) No
28.	I am more articulate (vocal)	( ) Yes ( ) No
29.	I am more fortunate and lucky	( ) Yes ( ) No
30.	My feelings are more easily hurt	( ) Yes ( ) No

TABLE 2.5
Forced Classification Toward "Being Opinionated"

"As compared with an average person. . ."

More opinionated then I am	If I am r*	Less opinionated then I am
Less quiet	0.44	Quieter
More outspoken	0.44	Less outspoken
More presumptuous	0.40	Less presumptuous
More creative	0.37	Less creative
More Flamboyant	0.27	Less Flamboyant
Less timid	0.27	More timid
Less Courteous	0.26	More courteous

*r indicates *product-moment correlation* (to be explained in chapter 4) between the trait "opinionated" and each trait listed.

2.4 SOME CHARACTERISTICS OF DUAL SCALING

The following is a preview of some of the characteristics of dual scaling, which will make sense once you have completed this book. You will no doubt find other characteristics as well which are not listed here:

1. Dual scaling provides a simpler, often clearer, description of data, thus serving as a technique to form a useful summary of otherwise complex data.
2. It derives a numeric (quantitative) description from such non-numeric (qualitative) data as "female vs. male," "manic vs. depressive," "urban vs. rural," and "users vs. non-users."
3. It handles analysis of a variety of so-called categorical data, which you will see throughout this book.
4. It offers an exhaustive analysis of information in the data, often through *multidimensional analysis*, as we saw in the second example.
5. It serves as a technique for *discriminant analysis* of categorical data.
6. It extracts information from data in optimal ways (e.g., derives test scores which have maximal reliability).

7. It uses individual differences in judgment to explore the data, rather than averaging them out as in most statistical analyses. Individual differences are often more interesting than average responses.
8. It can quantify qualitative information so that traditional analysis (e.g., analysis of variance) for quantitative data may be carried out.

This list can continue, but it looks as though these points are sufficient for now. After all, it is only for your information, and it is not our intention to get you involved in a list of technical matters. Indeed, the list does include topics which require a great deal of knowledge about data analysis. Therefore, have no fear if some of the points sound rather theoretical or esoteric. It would be useful, however, to try to understand why such characteristics as those listed above may be desirable for a method of data analysis. A glimpse at such reasons might at least generate some interest to find out what dual scaling is all about. Some of these reasons stem from the nature of response data. So let us look at the very basics of human judgment.

2.5 SOME OBSERVATIONS ON HUMAN JUDGMENT

In spite of the title of this section, be advised that dual scaling is not restricted to handling only human judgment. It can analyze such nonhuman data as ecological distributions of animal species over different terrains, or data on air pollution and climatic conditions. Human judgment, however, presents several interesting problems which tend to favor such a technique for data analysis as dual scaling over some others. Only a few aspects are presented here. All of them indicate how important it is to examine individual differences.

(a) Judgmental Criteria. Each person has his own unique record of experiences and mode of thought. So, if you ask a group of people which one of 10 listed cars they would like to purchase, some people may look at only attribute (such as price, fuel economy, power, style, and manufacturer), while others may consider specific combinations of these attributes. Some may look at only affordable cars, while others may ignore affordability. Some may consider a car only for themselves, while others may think of their regular passengers. Under these circumstances, how can you expect to obtain meaningful results by averaging responses from different respondents?

(b) Shift in Judgmental Criteria. It is typical that people have relatively fixed and stable sets of judgmental criteria. But you can also observe a

dynamic shift of judgmental criteria within a person. Takane (1980) mentioned the following example. In a similarity judgment on Jamaica, Cuba, and Russia, people would say:

1. Jamaica is similar to Cuba, and
2. Cuba is similar to Russia. However,
3. Russia is *not* similar to Jamaica.

Note that (1) is based on "geographical" proximity, and that (2) is based on "political" proximity. Jamaica and Russia, however, share neither geographical nor political similarity. This kind of lack of transitivity is not rare at all. Consider, for example, major league baseball teams such as the Toronto Blue Jays, the New York Yankees, and the Detroit Tigers. If the Blue Jays beat the Yankees and the Yankees beat the Tigers, this does not mean that the Blue Jays will beat the Tigers. Although this does not offer an immediate example of the shift in criteria, it shows a more general situation in which response-generating conditions may change during the course of data collection.

(c) Directionality. There are many examples where symmetry or reciprocity does not exist in judgment. Suppose in a car-switching study 20% of owners of the Ford cars switched to General Motors cars at the time of their second purchase. This does not mean that 20% of owners of GM cars switched to Ford cars. In a more mundane example, Person A may like Person B, but B may not like A. Or, in the previous example, people might say that Cuba is similar to Russia, but it would be rather unlikely that they would say Russia is similar to Cuba, because of the size difference. Likewise, it is more likely for respondents to say North Korea is similar to China, rather than China is similar to North Korea. In case these examples are not convincing enough, how about the following: "Dog bites man—not news; man bites dog—news!"

(d) Active versus Passive or "Finder's versus Loser's" Psychology. Depending on whether a person belongs to the "finder's circle" or the "loser's circle," the person's judgment of an essentially symmetric situation can be drastically different. Look at the following example:

(Finder) "If you find a $50 bill in a public place, would you report it to the police?"
38% (yes), 35% (I don't know), 29% (no)

(Loser) "If you drop a $50 bill on your way to work and report it to the police, do you think you would get it back?"
2% (yes), 12% (maybe), 86% (no).

Without scrutiny, you can get the message, can't you?

(e) Hidden Constraints on Judgment. No one can be free from the effects in judgment of one's upbringing, experiences, or knowledge. Thus, adults are not always better judges than children. Consider the following

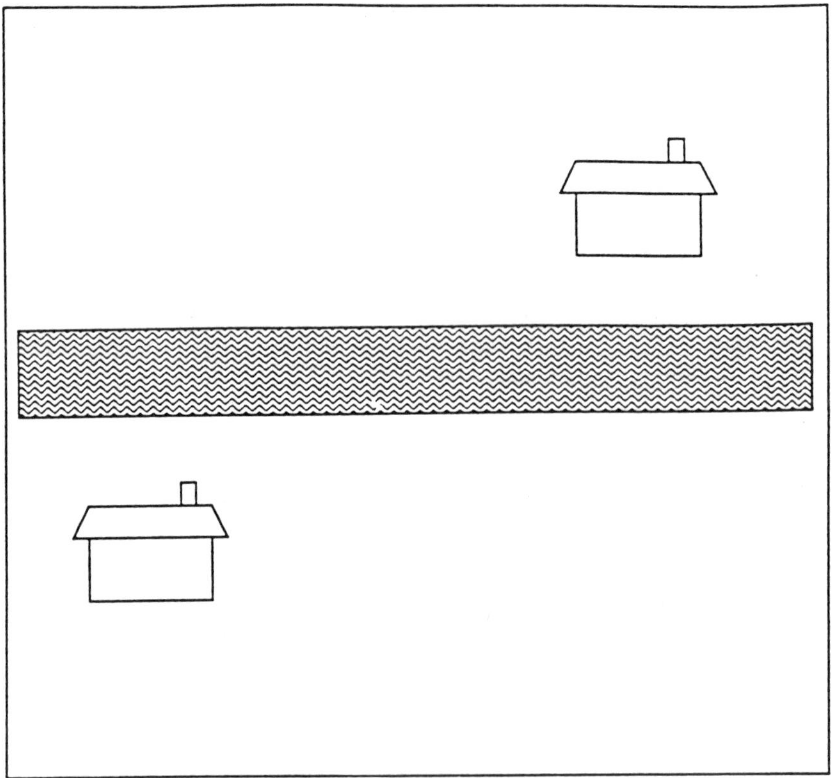

Fig. 2.3. Where should we build a bridge?

example, shown in Figure 2.3. We want to build a bridge over the river to get to the warehouse from the office. The bridge must be perpendicular to the river. The main consideration here should be that the person wants to get to the warehouse in the quickest possible way. Where do you think the bridge should be built? It is extremely difficult for an adult to come up with the right answer, because his common sense would not allow him to think of alternative approaches to the "usual" way. However, a precocious 7-year old could probably give you the optimal answer. (If you are truly unable to solve the problem, you can find the solution in Figure 2.4!)[3]

These are some of the aspects peculiar to human judgment. You can see that individual differences are a legitimate and fundamental topic for

[3] This example is from "Atama no Taiso [Exercises for the Brain] by A. Tago, 1966, *1*, pp. 63-64. In the original quiz, Tago specified the river being 100 meters wide and the office 300 meters away from the warehouse. He allowed 3 minutes to solve the quiz.

investigation. Averaging out all the individual differences might serve the purpose of identifying an "average person," but such a person may not resemble anyone in the same group, or for that matter anyone at all! One of the primary characteristics of dual scaling is that it is capable of identifying patterns of *individual differences* in the data. This characteristic alone might be interesting enough to learn more about dual scaling.

This chapter has presented a nontechnical introduction to dual scaling. By now, we hope that you have a basic understanding of the technique. As you will see, it is a very versatile method which can be applied to many different problems. To get you started, chapter 3 is devoted to a nontechnical examination of whether or not your data set is appropriate for dual scaling. This is always the most important first step.

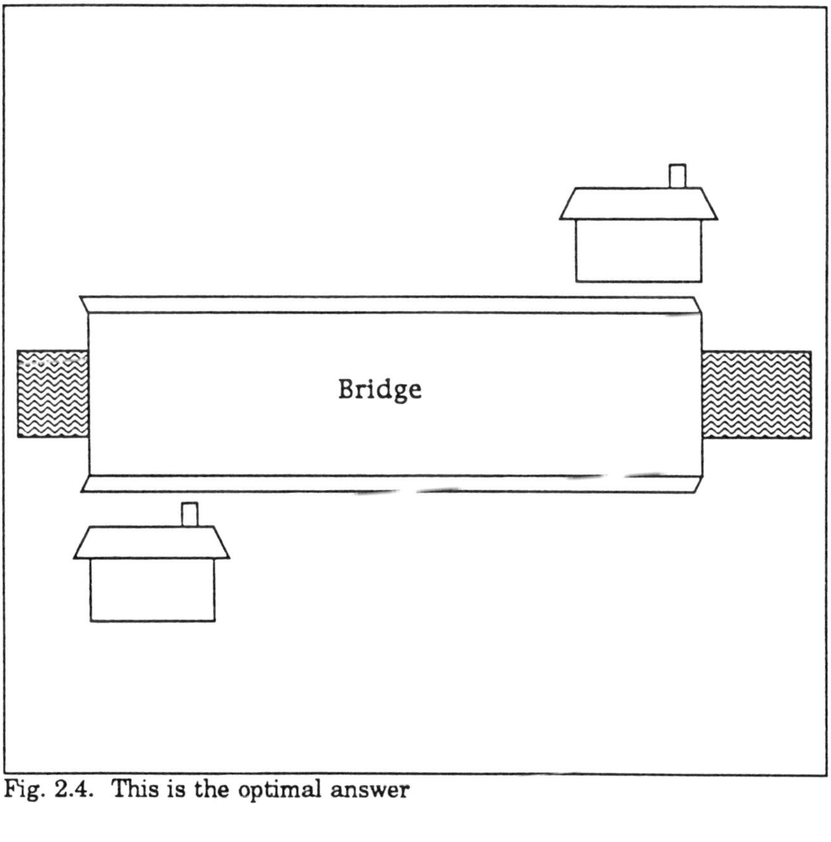

Fig. 2.4. This is the optimal answer

CHAPTER 3
Is Your Data Set Appropriate for Dual Scaling?

3.1 EXPLORATORY ANALYSIS AND DATA TYPES

Once you have collected data, it is important to establish how best to analyze them. For any given data set, it is common to find several alternative methods which are appropriate. People often say that the choice of a method for analysis depends on the purpose of the study. True, this is usually the case. But there are other occasions in which a specific purpose does not exist, in which you wish to find out whatever your data can tell you. This is so-called *exploratory analysis*, and the most frequent use of dual scaling.

Even if your interest is not in exploratory analysis, you may still need dual scaling. A more direct way of telling whether or not your data set is appropriate is to look at the "type" of data. The identification of the data type is not always easy. If you find the task difficult, do not despair. We will present examples of different data types which can be analyzed by dual scaling, to allow you to identify to which type your data set belongs. Dual scaling can currently handle the following types of data:

1. Contingency/frequency data.
2. Multiple-choice data.
3. Paired comparison data.
4. Rank-order data.
5. Successive categories (rating) data.
6. Sorting data.
7. Categorized continuous data and other types of categorical data.
8. Multiway data.

If you are familiar with these data types, you may proceed to Section 3.10. Otherwise, start with Section 3.2, and see if your data set is of the same type as any of the examples given. Note that the same type of data may exist under different names.

3.2 CONTINGENCY/FREQUENCY DATA

Table 2.1 in Chapter 2 is an example of this data type. Does "3" in your data set, for example, mean "three responses"? If so, it is likely that your data set is a contingency/frequency table. In this data type, we include all kinds of so-called two-way tables of response occurrences, or "count data." The following are two more examples of this data type. The first example (Table 3.1) is from Hollingshead (1949), who found that the members of a small Midwestern community in the United States divided themselves into five social classes. He investigated his prediction that adolescents in different social classes would enroll in different curricula at a high school. Social classes I and II were merged because the frequencies were small. The second example (Table 3.2) is a part of Table 8 from R. M. Reitman and L. A. Davison's *Clinical Neuropsychology* (1974), in which frequencies of complaints and types of complaints following head injury are listed.

TABLE 3.1
Different Curricula and Social Classes (Hollingshead, 1949)

	Social Classes				
Curricula	I&II	III	IV	V	Total
College preparatory	23	40	16	2	81
General	11	75	107	14	207
Commercial	1	31	60	10	102
Total	35	146	183	26	390

We hope that these examples are sufficient for you to identify the contingency and frequency tables. If not, look at some more examples in chapter 7.

3.3 MULTIPLE-CHOICE DATA

Multiple-choice data are one of the most popular types. Although it is common to ask respondents to choose one option (alternative) per question, you will occasionally find data in which respondents are allowed to choose more than one option per question. This so-called multiple response case is very cumbersome, and cannot be recommended. However, this section discusses both types of multiple-choice data in dual scaling.

TABLE 3.2
Incidence of Complaints and Types of Complaints
(Reitman and Davison, 1974)

Types of complaints	1 Year after trauma				2 Years after trauma			
	YB	YG	OB	OG	YB	YG	OB	OG
PM	6	5	6	2	5	2	8	1
LD	15	7	16	3	12	2	9	1
IC	1	2	8	1	1	0	1	0
IR	16	2	12	3	11	0	3	0
PC	23	13	17	8	15	5	10	2
FT	15	9	8	5	4	2	2	3
DS	2	2	5	2	6	0	5	0
SP	2	0	0	0	4	0	1	0
TI	1	0	0	0	1	0	0	0
DZ	3	4	15	7	3	0	7	2
HD	19	9	19	10	12	4	10	3
VA	6	5	9	4	7	4	6	1
IV	4	3	5	2	5	1	4	0
PL	2	1	3	1	2	1	3	0
SZ	4	2	3	2	2	2	0	1

Note. YB = younger boys; YG = younger girls; OB = older boys; OG = older girls; PM = poor memory and/or concentration; LD = learning difficulties; IC = intellectual changes; IR = irritability; PC = personality changes; FT = fatigability; DS = difficulty with sleep; SP = speech problems; TI = tics; DZ = dizzy spells (vertigo, giddiness); HD = headache; VA = visual or auditory defect; IV = impaired voluntary movement; PL = paralysis; SZ = seizures.

3.3.1 Standard Form: Choosing One Option per Question

In the collection of multiple-choice responses, a number of questions are usually asked, and each respondent must choose one of the response options (alternatives) for each question. For example, consider the following set of questions:

(1) Which do you like better, coffee or tea?
1. Coffee ( ) 2. Tea ( ) 3. Neither ( )

(2) Do you regularly drink milk?
1. Yes ( ) 2. No ( )

(3) When can you work best, morning, afternoon, or evening?
1. Morning ( ) 2. Afternoon ( ) 3. Evening ( )

The data may be presented in the format shown in Table 3.3. As you may have guessed, a "1" in this table indicates a choice, and a "0" a nonchoice. So, in this example, Subject 1 chose coffee, Yes, and Evening, respectively, for the three questions. This format is often called a *response-pattern table* (Nishisato, 1980a), an *indicator matrix* (Gifi, 1990), or an *incidence matrix* (Greenacre, 1984). In practice, this format tends to produce a large table—for example, if N subjects answered 100 questions, with 5 options per question, it would result in an N×500 table. For this reason, a preferred format is to construct a subjects-by-questions table of chosen option numbers. Table 3.4 shows the data in this format constructed from Table 3.3. Notice that the entries in the table are merely *codes*, not quantities. In other words, a "3" means the choice of the third option, and does not mean that it is greater than a "2," or that three responses were given.

TABLE 3.3
Multiple-Choice Data (Response Patterns)

Question		1			2		3		
Option		1	2	3	1	2	1	2	3
Subject	1	1	0	0	1	0	0	0	1
	2	0	1	0	1	0	1	0	0
	3	1	0	0	0	1	0	1	0
	.	.	.	.	.	.	.	.	.
	.	.	.	.	.	.	.	.	.
	.	.	.	.	.	.	.	.	.
	N	0	0	1	0	1	1	0	0

TABLE 3.4
Multiple-Choice Data (Condensed Format)

Question		1	2	3
Subject	1	1	1	3
	2	2	1	1
	3	1	2	2
	.	.	.	.
	.	.	.	.
	.	.	.	.
	N	3	2	1

We will now look at an example involving real data. Table 3.6 shows a 14-item questionnaire, and data obtained from 17 subjects are presented in Table 3.5. Please note that a "0" in Table 3.5 indicates a missing response (i.e., the subject left the question blank). For additional examples, refer to chapter 8.

TABLE 3.5
Multiple-Choice Data (Condensed Format)

	Items													
Subject	1	2	3	4	5	6	7	8	9	10	11	12	13	14
1	1	2	1	2	1	2	2	2	1	1	2	1	1	1
2	1	2	0	1	2	3	2	1	2	1	2	2	1	2
3	2	2	0	2	1	3	2	2	2	2	2	2	2	1
4	1	1	1	2	2	1	2	2	2	1	2	2	1	2
5	2	2	0	1	1	3	0	2	1	1	2	2	1	1
6	1	1	1	1	1	1	2	1	1	1	1	1	2	1
7	1	2	1	1	1	3	2	2	2	2	2	2	1	2
8	2	2	2	2	1	3	1	1	2	2	0	1	2	2
9	2	2	2	2	2	3	2	2	2	1	1	0	1	2
10	2	2	2	2	2	3	2	2	2	2	2	1	1	2
11	2	2	1	2	2	3	2	1	2	1	2	1	1	1
12	1	2	2	1	1	3	1	1	2	1	2	1	2	2
13	2	2	0	1	2	1	2	1	1	1	1	2	1	1
14	1	1	2	1	1	2	2	2	1	1	2	2	1	2
15	2	2	2	1	1	1	2	1	2	1	2	1	1	2
16	1	1	1	1	2	2	1	2	2	2	1	1	2	2
17	2	1	1	1	0	1	1	1	1	2	2	1	2	2

Note. From Nishisato and Nishisato (1984).

Occasionally you may encounter a case in which a respondent chooses more than one option per item, that is, the case of *multiple response data*. Let us look at such data.

3.3.2 Multiple-Response Form: Choosing More Than One Option per Question

Whenever it is reasonable to expect the respondents to choose only one option per question, do your best to ensure they do so. In other words, avoid, as much as possible, collecting multiple-response data. Look at Table 3.7, which corresponds to the response patterns of Table 3.3 except that a few multiple responses are involved. As you can see, a "1" is equally divided among the multiple choices: if the respondent chooses two options in a question, these options receive 0.5 each; if the respondent chooses three, four, or five options each choice receives 1/3, 0.25, or 0.2, respectively.

As this example illustrates, you must devise a way to simplify the data representation, as we did in Table 3.4. One example of such simplification is shown in Table 3.8. For dual scaling, this simplified table must be transformed back to the form of Table 3.7, which may not be as easy as transforming Table 3.4 to Table 3.3. Multiple responses for dual scaling

TABLE 3.6
A 14-Item Questionnaire

1. Were you born in Canada?
(1 = yes, 2 = no)

2. Which do you like better, coffee or tea?
(1 = coffee, 2 = tea)

3. Which do you like better, whisky or beer?
(1 = whisky, 2 = beer)

4. Which do you like better, baked potatoes or french fries?
(1 = baked potatoes, 2 = french fries)

5. Which do you like better, sour pickles or sweet pickles?
(1 = sour pickles, 2 = sweet pickles)

6. When can you work best?
(1 = morning, 2 = afternoon, 3 = evening)

7. Do you spend more time chatting with people than the average person?
(1 = yes, 2 = no)

8. Are you allergic to any food, animal hair, dust, pollen or anything else?
(1 = yes, 2 = no)

9. Do you often say things and later regret them?
(1 = yes, 2 = no)

10. Are your feelings easily hurt?
(1 = yes, 2 = no)

11. Are you easily startled by unexpected stimuli, noise, voices, etc.?
(1 = yes, 2 = no)

12. Do you like social events very much?
(1 = yes, 2 = no)

13. Are you bothered by people watching you while at work?
(1 = yes, 2 = no)

14. Do you keep worrying about humiliating experiences?
(1 = yes, 2 = no)

TABLE 3.7
Multiple-Choice Data with Multiple Responses

Questions	1			2		3		
Option	1	2	3	1	2	1	2	3
1	.5	.5	0	1	0	1/3	1/3	1/3
2	0	1	0	.5	.5	.5	0	.5
3	.5	0	.5	0	1	0	1	0
.	.	.	.	.	.	.	.	.
.	.	.	.	.	.	.	.	.
.	.	.	.	.	.	.	.	.
N	0	.5	.5	.5	.5	.5	.5	0

TABLE 3.8
Condensed Format?

Q	1	2	3
1	12	1	123
2	2	12	13
3	13	2	2
.	.	.	.
.	.	.	.
.	.	.	.
N	23	12	12

introduce data that are less discriminative and more complex to handle than the standard format of one choice per question.

There are occasions in which data in the multiple-response form can be changed to the standard form. For example, consider the following question:

Which of the following places have you been during the past ten years?

1. Joshua Tree ( ) 2. Sanibel Island ( ) 3. Soo Lookout ( )

It is quite probable that some subjects will check more than one of the three places, resulting in multiple responses. This situation, however, can be changed by dividing the above question into three dichotomous questions:

1. During the past ten years, have you visited Joshua Tree?
( ) yes ( ) no
2. During the past ten years, have you visited Sanibel Island?
( ) yes ( ) no
3. During the past ten years, have you visited Soo Lookout?
( ) yes ( ) no

In this way, we can often avoid the problem of multiple responses.

3.4 PAIRED COMPARISON DATA

Unlike the previous two data types, paired comparison data are generally not well known. So let us use a small example to illustrate how this type of data can be collected and presented. Suppose you want to assess *preference values* of ginger ale (A), Coca Cola (B), root beer (C), and 7-Up (D). You will have to conduct a series of taste experiments: consider all possible pairs (A,B), (A,C), (A,D), (B,C), (B,D), and (C,D); present each pair to each subject; and ask the subject which of the two in the pair he or she likes better. It looks easy, doesn't it? Once all the pairs are judged in this way, the responses, *paired comparisons*, are presented in the form shown in Table 3.9. A "1" indicates the choice of the first drink over the second, a "2" indicates the choice of the second over the first, and a "0" represents a tied judgment. Hence, subject 1 preferred A to B, C to A, D to A, C to B, D to B, and C to D. Likewise, subject N preferred A to B, C to A, A to D, C to B, C to D, and showed an equal preference for B and D. Is this example clear? For additional examples, see chapter 10.

TABLE 3.9
Paired Comparisons of Ginger Ale (A), Coca Cola (B), Root Beer (C), and 7-Up (D)

Pair	A B	A C	A D	B C	B D	C D
Subject 1	1	2	2	2	2	1
2	2	1	2	1	2	2
3	1	1	1	1	1	1
.	.	.	.	.	.	.
.	.	.	.	.	.	.
.	.	.	.	.	.	.
N	1	2	1	2	0	1

There are occasions in which *order effect* or *position effect* may be present. For instance, your preference for food A and food B in a taste exprериment may change if you taste A first and then B, as compared to B first and then A (order effect), or your judgment of relative weights of objects X and Y in a psychological experiment on size and weight may change if you are presented X and Y to your right and left hands, respectively, as opposed to your left and right hands (position effect). In these cases, we should use all possible permutations of two stimuli: for instance, in addition to (A,B), (A,C), (A,D), (B,C), (B,D), and (C,D), use also (B,A), (C,A), (D,A), (C,B), (D,B), and (D,C).

3.5 RANK-ORDER DATA

Rank-order data are relatively easy to identify. Given seven political candidates A, B, C, D, E, F, and G, each subject is asked to rank them in order of preference, starting with 1 for the most preferred, 2 for the second most preferred, and so on. The data can be presented in a subjects-by-candidates table with rank numbers as its entries (see Table 3.10). In this example, Subject 1, for instance, ranked the candidates in the order C, E, A, G, F, B and D.

TABLE 3.10
Ranking of Seven Political Candidates

Candidate		A	B	C	D	E	F	G
Subjects	1	3	6	1	7	2	5	4
	2	4	1	7	2	3	6	5
	3	3	1	7	2	3	6	5
	.	.	.	.	.	.	.	.
	.	.	.	.	.	.	.	.
	.	.	.	.	.	.	.	.
	N	6	4	5	2	7	1	3

You will sometimes see genuine *tied ranks* and tied ranks generated by missing responses. This latter case will happen if, for example, a subject ranks the first three candidates out of seven, but cannot rank the rest of the four candidates. In this case, assuming that these last four candidates would have occupied rank positions 4, 5, 6 and 7, you may assign the average rank of these four numbers, that is (4+5+6+7)/4 = 5.5, to *each* of the four candidates. Similarly, if another subject ranked the three best (1, 2, 3) and the worst (7) out of seven, but could not distinguish between the remaining three, you would assign the average rank that they would have obtained to each of the three, that is, (4+5+6)/3 = 5. Thus, the seven candidates are ranked as 1, 2, 3, 5, 5, 5, 7. This kind of adjustment becomes necessary when data are collected by such a question as "Rank as many candidates as you like."

We do not wish to complicate the matter unnecessarily by introducing tied ranks. However, we hope that the above explanation is helpful in handling rank-order data with or without tied ranks. Can you now distinguish rank-order data from the other three types so far discussed? We hope so, but in case you are not completely certain, take a look at a few more examples given in Chapter 11.

3.6 SUCCESSIVE CATEGORIES (RATING) DATA

We will begin by asking how rating data are collected. Prepare a set of successively ordered response categories: for instance, "never, sometimes, often, always," "poor, fair, good, excellent,", and "low, medium, high." Using *only one set* of these categories, ask each subject to classify each *stimulus* into the most appropriate category. In essence, this is not much different from multiple-choice data that are collected from questions answered in terms of categories such as strongly disagree, disagree, neutral, agree, strongly agree. Indeed, rating or successive categories data can be analyzed as multiple-choice data by dual scaling. However, the reverse is not always possible, even if those multiple-choice options are successively ordered. The basic distinctions between the two types of data are that (a) multiple-choice options need not be ordered (e.g., "male, female"; "yes, no, no comment"), while categories of rating data must be ordered (e.g., "never, sometimes, often, always"; "strongly disagree, moderately disagree, neutral, moderately agree, strongly agree"), and (b) different multiple-choice items can have different options (e.g., "yes, no" for Question 1, "agree, disagree, no comment" for Question 2), while rating data require all the stimuli to be evaluated in terms of a *common* set of successive categories.

TABLE 3.11
Rating (Successive Categories) Data

Speaker		1	2	3	4	5
Judge	1	2	3	4	1	4
	2	3	2	3	2	3
	3	1	4	4	2	3
	.	.	.	.	.	.
	.	.	.	.	.	.
	.	.	.	.	.	.
	N	2	3	4	1	4

Let us look at one example of rating data (see Table 3.11). In a speech contest, five speakers were rated by N judges as: 1. Poor; 2. Fair; 3. Good; and 4. Excellent. The entries for this table show the category numbers chosen. Thus, Judge 1 rated speakers 1, 2, 3, 4, and 5, respectively, as fair, good, excellent, poor, and excellent. You can see that all the speakers were rated by a *single* set of successive categories. If more than one set of categories are introduced in data collection, the data must be treated as multiple-choice data. Is this example good enough to serve as a prototype for rating data? Please see the other examples of this data type in chapter 12.

3.7 SORTING DATA

While you may not be familiar with this data type, it can be particularly useful. The data are collected by preparing a set of stimuli (e.g., photos of movie actors), and asking the subject to classify the stimuli into as many piles as he or she likes, under the condition that stimuli within each pile are similar in some sense and different from those in other piles. This data type is also very similar to multiple-choice data, except that multiple-choice items are replaced with sorting subjects, and that subjects answering multiple-choice questions are replaced with stimuli to be sorted into piles. The following brief example will clarify this method.

In sorting political leaders, Subject 1 placed Churchill and Roosevelt in one pile, Stalin in one pile, and Hitler and Mussolini in one pile, thus using three piles to sort the names. Subject 2 used two piles, the first pile containing Churchill and Roosevelt, and the second pile Stalin, Hitler, and Mussolini. Subject 3 used four piles, Churchill, Roosevelt, and Stalin being sorted separately, and Hitler and Mussolini together into one pile. Table 3.12 is essentially a table of response patterns, with the role of subjects and items being interchanged. Thus, as with multiple-choice data, it is more practical to present Table 3.12 in a condensed format as shown in Table 3.13

TABLE 3.12
Sorting Data (Response Patterns)

Subject		1			2		3			
Pile		1	2	3	1	2	1	2	3	4
Stimulus:	Churchill	1	0	0	1	0	1	0	0	0
	Roosevelt	1	0	0	1	0	0	1	0	0
	Stalin	0	1	0	0	1	0	0	1	0
	Hitler	0	0	1	0	1	0	0	0	1
	Mussolini	0	0	1	0	1	0	0	0	1

Although this may be a data type new to you, the instructions for collection of the data are usually so distinct as to make it very obvious. If you are still concerned about your task of data identification, examine the

TABLE 3.13
Sorting Data (Condensed Format)

Subject		1	2	3
Stimulus:	Churchill	1	1	1
	Roosevelt	1	1	1
	Stalin	2	2	3
	Hitler	3	2	4
	Mussolini	3	2	4

additional examples in chapter 9 to isolate those characteristics peculiar to sorting data.

3.8 CATEGORIZED CONTINUOUS DATA AND OTHER TYPES OF CATEGORICAL DATA

Continuous data include such measurements as blood pressure, income, and age. They can usually be categorized into several classes (e.g., low, medium, and high). Thus, they, too, can be subjected to dual scaling. The converted data are treated as the multiple-choice type. You might think that this kind of categorization and reorganization of quantitative data would result in a substantial loss of information. In a certain sense, you are right. But you would equally be amazed at the very revealing results which dual scaling can yield from such categorized data. It's worth a try. Categorization of continuous data into the multiple-choice type, however, is not without problems. You should be concerned with the number of categories (or options) of each variable and the distribution of responses over the categories (see the discussion on "over-quantification" and "expanded dimensionality" problems in Part 4).

You can also generate paired comparison data and rank-order data from continuous data. For instance, you can rank 10 baseball players with respect to batting averages, the number of stolen bases, runs batted in, and salaries, all of which are continuous measurements.

There are a number of other forms of categorical data, that is, observations recorded in terms of predetermined or postspecified classes (e.g., options, discrete rating categories, absence or presence). Try to reorganize such data into one of these data types, and see if dual scaling can analyze your data. It is often the case that you will be able to find a way of making effective use of dual scaling for a given data set.

3.9 MULTIWAY DATA

If you can identify your data as one of those types discussed so far, dual scaling is an appropriate technique for your analysis. A slight complication arises when those data types, which are all in the format of the so-called "two-way" tables, acquire an additional set of classification variables. For instance, if you suddenly remember that those subjects who answered a large number of multiple-choice questions can actually be divided into boys and girls (one classification variable), into the enrichment class and the regular class (another variable), and that the items are divided into personality items and aptitude items (third variable), this is an example of "multiway data." Since this type encompasses a large number of possible examples, we will not look at examples here. After you have examined standard applications of dual scaling to two-way data, we will return to a few examples of multiway data matrices and their analyses in Part 4.

3.10 WHAT IF YOUR DATA SET IS NOT APPROPRIATE FOR DUAL SCALING?

If your data cannot be handled by dual scaling, you need another method. What follows is a very sketchy description of each of a few selected alternatives. Please note that each method would require at least a single book. Therefore, the description given here will be helpful only for possible identification of a technique you might consider using for your data.

3.10.1 Principal Component Analysis (PCA) and Factor Analysis (FA)

When your data are *quantitative* and are obtained from a number of subjects on a set of measures (e.g., students' marks on several tests), or when the *correlation* or *covariance* matrix on these measures is available, you may want to find out how many "factors" are involved in these measures, or how many "dimensions" are needed to accommodate the relations between the measures. PCA will provide you with the most dominant component or factor (i.e., a linear combination of the measures, containing a maximal amount of information) first, then the second most dominant component, and so on. These components or factors are uniquely determined, since PCA amounts to interpreting data in terms of so-called "principal planes" (see chapter 5). In other words, it is a technique of unique mathematical decomposition of data. In contrast, FA is based on a statistical model with uncertainty in it. Although the aim of FA is not much different from that of PCA, FA decomposition of data is not unique. We have the problem of so-called rotation of axes for interpretation of the results. In both PCA and FA, analysis reveals not only hidden "factors" or "components" among variables (e.g., measures, tests), but also the differential contributions of scores of subjects to each factor. As you will see later, dual scaling is essentially the same as PCA, except that dual scaling enjoys the freedom of differentially weighting response options in arriving at multidimensional decomposition of data. Notice that PCA deals, for example, with a uniquely determined, thus fixed, correlation matrix for decomposition, while dual scaling generates through weighting a "most informative" correlation matrix for decomposition.

3.10.2 Cluster Analysis

When your data can be converted to proximity measurements (e.g., correlation coefficients, association coefficients) or dissimilarity measurements (e.g., Euclidean distance), and you are interested in finding groupings (clusters) of respondents or objects, you may consider cluster analysis. There are, however, many proximity measures and dissimilarity measures one can choose from for a given data set, which may make it difficult to identify one that is most appropriate for your data. In addition to this problem, there are two main families of clustering methods, hierarchical (nested) and nonhierarchical (nonnested) clustering. Within each of the two, a number of clustering methods are available, depending how you define clusters and intercluster distances. Therefore, it is important to gain knowledge about different measures of proximity and dissimilarity, and clutering methods before you engage yourself in cluster analysis. How to use cluster analysis appropriately is not an easy matter, and it is advisable to seek guidance from experienced researchers.

3.10.3 Nonmetric Multidimensional Scaling (MDS)

When your data are at least *ordinal measurement*,[1] nonmetric multidimensional scaling offers another way of finding components as in PCA. As long as you can derive a "reasonably" accurate estimate of similarity between a

---

[1] Numbers are used to reflect order relations, such as 3, 2, and 1 given to blood pressure 180, 130, and 125. Ordinal measurement is not precise enough, however, for the operation of addition or subtraction. In this example, 1 + 2 would not produce 3, which is blood pressure 180.

pair of stimuli, MDS offers a way to arrive at multidimensional decomposition of your data. When the number of stimuli is larger than 10, results of nonmetric MDS based on ordinal measurements are very similar to PCA results based on the corresponding continuous measurements.

3.10.4 Analysis of Variance (ANOVA)

If your data are quantitative in the sense that arithmetic operations of addition and subtraction result in meaningful quantities, and if your subjects are classified into subgroups, for instance, by age and profession, then analysis of variance may be appropriate. If each subject has more than one measurement, the analysis is called multivariate analysis of variance (MANOVA). You can find from ANOVA or MANOVA whether the distribution of data is significantly influenced by the classification variables (i.e., age and profession). The data matrix is a multiway table, its only difference from that for dual scaling being that the elements of the table are quantitative (e.g., scores).

3.10.5 Regression Analysis (RA)

When your data contain two subsets of continuous variables, and when you are interested in predicting one set from the other, this is the problem of (multivariate) regression analysis. For instance, suppose that you have collected from the same group of students two sets of data on academic achievement, one in the last year of high school and the other in the second year of university. You want to investigate whether or not their high school achievement can indeed predict their university performance. This is an important problem for the validity of using high school records for university admission purposes, and regression analysis provides an answer to the problem.

3.10.6 Canonical Analysis (CA)

As in regression analysis, you have two sets of continuous data collected from the same group of subjects. Unlike RA, however, the main interest of your canonical analysis lies in the mutual relationship, rather than RA's causal relationship, between the two sets of variables. Suppose that you have a set of measures on aptitudes for several professions and a set of measures on personality traits, and that you wish to study possible relations between aptitude and personality. CA will identify both the relative importance of each variable in maximally relating the two sets of variables, and the number of dimensions required to accommodate various patterns of relationships (e.g., personality traits for a specific profession).

3.10.7 Loglinear Analysis (LA)

When you have multiway contingency data, you can analyze them with dual scaling. As an alternative to multidimensional analysis of dual scaling, you may want to handle specific statistical hypotheses (e.g., Are the rows and the columns statistically independent?). If so, LA offers you an ideal method for data analysis. LA is in essence the analysis of variance of logarithmic transformations of frequency data. LA also has something in common with dual scaling, and both methods can be used in a complementary fashion to capitalize on their respective unique strengths (e.g., van der Heijden and de Leeuw, 1985).

This section could go on, but we must get back to the main theme of this book, dual scaling and its applications. We hope that you are now ready to become involved in explorations of categorical data with dual scaling. As a final step of this introduction, chapters 4 and 5 will present respectively some fundamental ideas underlying dual scaling and relevant quantitative background. Chapter 6 will then present a mathematical description of dual scaling, and the ensuing chapters will be devoted to dual scaling of different types of data, with emphases on both applications and the acquisition of technical knowledge. Although a single book could not hope to completely cover dual scaling, we endeavor to present as much useful and helpful information as possible.

CHAPTER 4
Some Fundamentals for Dual Scaling

There are several good references on the history and prehistory of the general methodology behind dual scaling.[1] In following its historical development, you will see a number of key concepts that form its backbone. It seems essential to look at some of those basic concepts and ideas in preparation for a close examination of dual scaling and its applications. This chapter takes a quick glance at them so that dual scaling of different types of categorical data may be viewed in terms of the same framework. If you are already familiar with basic ideas of scaling, you can skip this chapter.

4.1 SUM OF WEIGHTED RESPONSES (CONTINUOUS DATA)

Let us consider a small example to look at the important problem of combining responses. Suppose that six students wrote two tests, with the results shown in Table 4.1. We can show the positions of the students on a graph, using two axes X_1 and X_2 for the two tests, as shown in Figure 4.1. Student 1, for example, obtained scores of 2 and 3 on X_1 and X_2, respectively. Therefore, his position is located at the intersection of X_1 of 2 and X_2 of 3. Similarly, you can identify the position of Student 5, who obtained 9 and 8 on X_1 and X_2, respectively. Now look at the graph. You can clearly see the tendency that when the score of a student on X_1 is high (or low) his score on X_2 is also high (or low). In other words, there exists positive and high "correlation" between X_1 and X_2. In this case you can also say that X_1 and X_2 convey similar or redundant information about the students' performance. Can we not then combine X_1 and X_2 to come up with a single score, say Y? Of course we can. However, you would like to make sure that the combined score, Y (i.e., *composite score*), retains as much informa-

---

[1] See, for example, Benzécri (1982), de Leeuw (1973, 1983), Gifi (1990), Greenacre (1984), Nishisato (1979a, 1980), and Nishisato and Nishisato (1984).

TABLE 4.1
Scores of Six Students on Two Tests

Student	Test 1	Test 2
1	2	3
2	8	9
3	6	6
4	4	4
5	9	8
6	1	1

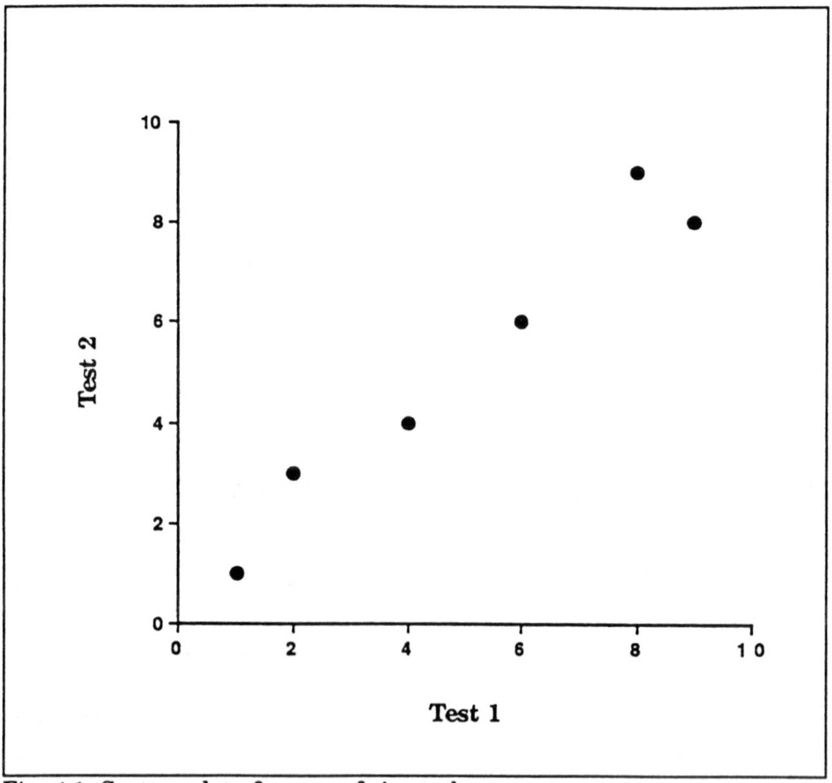

Fig. 4.1. Scatter plot of scores of six students on two tests.

tion on X_1 and X_2 as possible. How can you do it? The general expression for the composite score Y is written as

$$Y = c_1X_1 + c_2X_2 , \tag{4.1}$$

where $c_1^2 + c_2^2 = 1$. The terms c_1 and c_2 are *weights* for X_1 and X_2, respectively. The condition that $c_1^2 + c_2^2 = 1$ is used to preserve the unit of test scores. In other words, when you represent your composite score by another axis going through the origin of the graph, the composite score of each student can be expressed as the projection of the student's position on the new axis, provided that $c_1^2 + c_2^2 = 1$. This point can be explained using Figure 4.2. Suppose that composite score Z is chosen and that we wish to calculate the projection of the data point P on this axis Z, that is the distance between the origin and the point P*, d(O,P*). The data point P has coordinates (x, y). Pick an arbitrary point on the composite axis Z and call it R, which has coordinates (a, b). Then, recall the following trigonometric relations that you must have studied at high school:

$$\cos\theta = \frac{a}{\sqrt{a^2+b^2}}, \quad \sin\theta = \frac{b}{\sqrt{a^2+b^2}}, \quad \cos^2\theta + \sin^2\theta = 1 .$$

From the first two definitions of cosine and sine, we obtain

$$a = \sqrt{a^2+b^2}\ \cos\theta, \quad b = \sqrt{a^2+b^2}\ \sin\theta .$$

Therefore, the projection of P onto Z, that is, d(O,P*), can be derived as

$$d(O,P^*) = d(O,Q^*) + d(Q^*,P^*) = d(O,Q^*) + d(Q,P) .$$

But d(O,Q*) and d(Q,P) can be expressed as

$$d(O,Q^*) = x\ \cos\theta = \left(\frac{a}{\sqrt{a^2+b^2}}\right)x,$$

$$d(Q,P) = \left(\frac{b}{\sqrt{a^2+b^2}}\right)y .$$

Thus, for any arbitrary point P(x, y), its projection on the axis Z, Z_p, can be expressed as

$$Z_p = \left(\frac{a}{\sqrt{a^2+b^2}}\right)x + \left(\frac{b}{\sqrt{a^2+b^2}}\right)y .$$

Thus, if we indicate those weights for x and y respectively by c_1 and c_2, we obtain the results we wanted to examine; that is, the condition that the

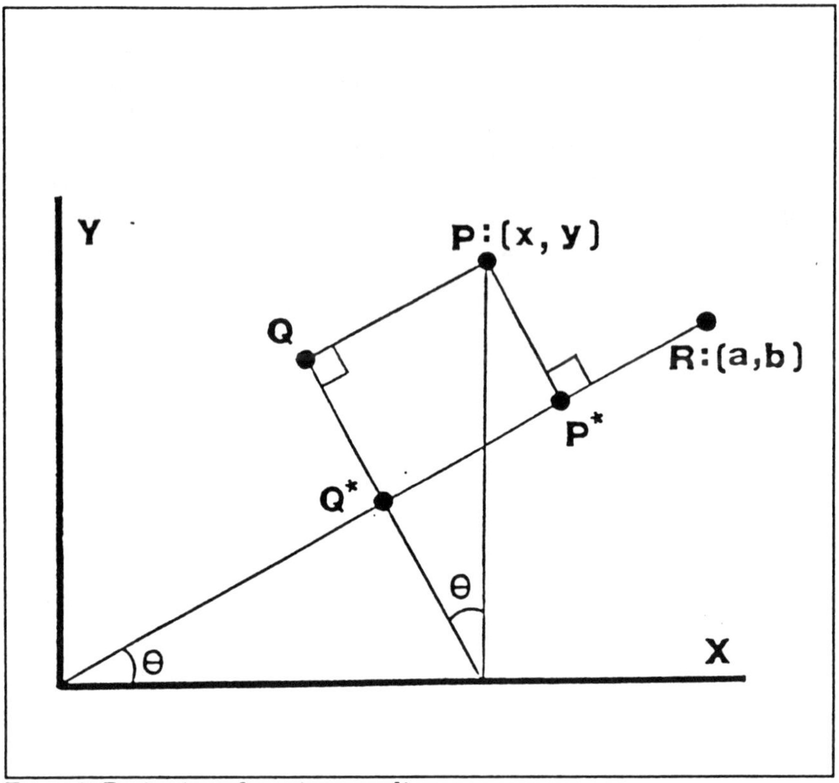

Fig. 4.2. Projection of a point on a line.

sum of squares of weights is 1 leads to the formula to calculate the projection of any data point on the chosen axis. Table 4.2 shows two sets of composite scores, Y_1 and Y_2, calculated by the expressions

$$Y_1 = \frac{3}{\sqrt{13}}X_1 + \frac{2}{\sqrt{13}}X_2 \quad \text{and} \quad Y_2 = \frac{-1}{\sqrt{17}}X_1 + \frac{4}{\sqrt{17}}X_2 \,.$$

Figure 4.3 shows the composite axes and projections of the original students' positions onto the composite axes, that is, composite scores. Can you figure out that the slope of each composite axis is given by c_1/c_2? Examine the graph closely until you know how each composite axis is represented. Look at the expression of Y_1. The axis Y_1 is drawn by taking three steps to the right along X_1 and two steps up along X_2. Then, 3 and 2 for X_1 and X_2, respectively, are divided by the square root of the sum of squares of them, that is, the square root of (3×3)+(2×2), which is $\sqrt{13}$. Is this clear now? Apply the same reasoning to Y_2. Once the equation (4.1)

TABLE 4.2
Composite Scores

Student	Test 1	Test 2	Composite Score Y_1	Composite Score Y_2	Composite Score Y^*
1	2	3	3.33	2.43	3.54
2	8	9	11.65	6.79	12.02
3	6	6	8.32	4.37	8.49
4	4	3	4.99	1.94	4.95
5	9	8	11.93	5.58	12.02
6	1	1	1.39	0.73	1.41

is understood graphically as in Figures 4.2 and 4.3, then you know that you can create quite different kinds of composite scores by manipulating the coefficients (weights) c_1 and c_2. In particular, you can generate composite scores which have the maximum spread. In the above example, such a composite is obtained when the composite axis has the angle of 45 degrees, that is $c_1 = c_2 = 1\backslash\sqrt{2}$ (see composite Y^*, the last column of Table 4.2). In other words, you can derive the most informative, representative composite formula by determining coefficients c_1 and c_2 in such a way that the spread of Y is a maximum. In statistics, the spread of scores is usually measured by a statistic, called *variance*. Thus, the above problem can be stated and handled as the problem of "determining c_1 and c_2 in such a way that, subject to the condition $c_1^2 + c_2^2 = 1$, the variance of Y is a maximum," where Y is given by (4.1). In this regard, if you look at Figure 4.1, you can see that the most representative axis would have approximately the angle of 45 degrees. In Figure 4.2, Y_2 does not look bad so far as the spread of composite scores is concerned, but it is obviously not close to the original cloud of points. Y_2 definitely has a smaller spread of projected points than Y_1.

Note that the most representative composite axis is called the *principal axis*, and, as you may have guessed, the analysis to find composite scores with the maximum variance is called *principal component analysis* (PCA). In general, PCA considers a composite of n tests (variables),

$$Y = c_1X_1 + c_2X_2 + \dots + c_nX_n \ . \tag{4.2}$$

Although a graphical illustration of this case similar to that of equation (4.1) is difficult, you already have its mathematical statement: Determine $c_1, c_2, \dots, c_n$, subject to the condition $c_1^2 + c_2^2 + \dots + c_n^2 = 1$, so as

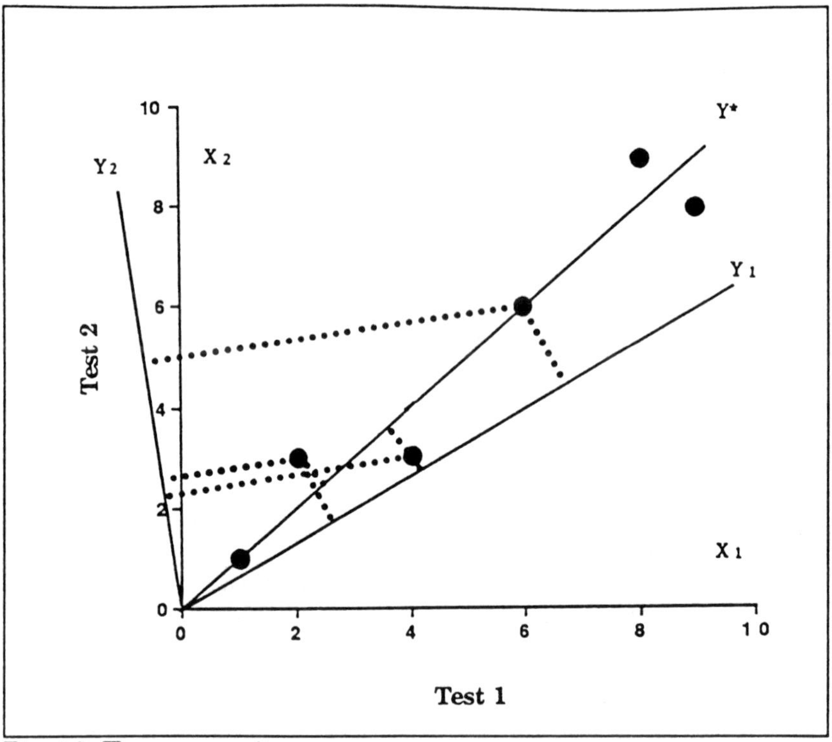

Fig. 4.3. Three composite scores

to maximize the variance of Y. Such a composite as this Y, called the *first principal component*, is very unlikely to represent the clouds of the subjects' points exactly. Therefore PCA examines the discrepancies between the original clouds of points and composite scores, and subject the discrepancies to the same analysis, resulting in the *second principal component*. Look at Figure 4.1 again. If the first principal component represents the projections of points on the axis with the angle of 45 degrees, then the second principal component shows the projections of points on the axis with the angle of 135 degrees. This may sound complicated, but the point to be made here is the fact that the two such composite axes are orthogonal (i.e., at an angle of 90 degrees). In other words, this means that the first principal component explains the maximum amount of information, and that the second principal component accounts for the maximum amount of information, under the condition that the information captured by the first component is not contained in the second component. In this way, you can extract n principal components, which are orthogonal to one another, from n tests (variables). Since you start with the most informative composite, you may be satisfied with only the first few major components to look at rather than all of them. Because of this convention, PCA serves as the most effective technique for "data reduction."

Within the context of PCA, the positive square root of the variance of each principal component is called a *singular value* of a data matrix such as Table 4.1 (see discussion in chapter 6).

4.2 SUM OF WEIGHTED RESPONSES (CATEGORICAL DATA)

The main reason why we looked at continuous data is to show you that dual scaling is indeed PCA as applied to categorical data. Let us consider a small example of multiple-choice data in Table 4.3. Six subjects answered three multiple-choice questions. Questions (items) 1 and 3 have two options, and question 2 has three options. Suppose we want to assign weights, x_1, x_2, . . . , x_7, to the seven options so that we may calculate scores for the six subjects, y_1, y_2, . . . , y_6. Notice that the data show only presence (1) or absence (0) of responses, or we may say a choice (1) or a nonchoice (0) of options. In other words, quantities such as test scores are not provided. Let us assume that those weights for options x_1, x_2, . . . , x_7, were given. Then six subjects' scores can be expressed as

$$
\begin{aligned}
y_1 &= x_1 + x_4 + x_7 \\
y_2 &= x_2 + x_3 + x_7 \\
y_3 &= x_1 + x_5 + x_7 \\
y_4 &= x_1 + x_3 + x_6 \\
y_5 &= x_2 + x_5 + x_6 \\
y_6 &= x_2 + x_4 + x_7
\end{aligned}
$$

Can you see how these expressions were obtained? See, for example, Subject 1, who chose Option 1 of Item 1, which is given weight x_1, Option 2 of Item 2, with weight x_4, and Option 2 of Item 3, with weight x_7, resulting in the first expression above. In case this is not clear, reexamine Table 4.3, where the option weights are indicated at the bottom margins. Once you can figure this out, the rest is easy. Rather than explaining the idea of the remaining procedure, therefore, let us state our problem for dual scaling: "Determine those seven option weights, x_1, x_2, . . . , x_7, subject to

TABLE 4.3
Multiple-Choice Data (Categorical Data)

	Item	1		2			3		
	Option	1	2	1	2	3	1	2	Score
Subject	1	1	0	0	1	0	0	1	y_1
	2	0	1	1	0	0	1	0	y_2
	3	1	0	0	0	1	0	1	y_3
	4	1	0	1	0	0	1	0	y_4
	5	0	1	0	0	1	1	0	y_5
	6	0	1	0	1	0	0	1	y_6
Weights		x_1	x_2	x_3	x_4	x_5	x_6	x_7	

an appropriate constraint,[2] in such a way that the variance of the composite scores, $y_1, y_2, \ldots, y_6$ will be a maximum." Is this not the same as PCA? Yes, the idea is the same, and this is important. Dual scaling is indeed PCA for categorical data. There are some differences, of course, between the traditional PCA and dual scaling in their computational processes, but such computational discrepancies cannot be equated to the importance of the common idea behind them.[3] When this mathematical decomposition procedure is presented, you will see many apparently different approaches converge to the same principle. In the meantime, however, we will look at a few more approaches to dual scaling.

4.3 WITHIN AND BETWEEN

As you saw in the previous two sections, continuous data have quantities to start with so that the data can be shown in a graph, no matter what kind of graph it may be. In contrast, categorical data such as those in Table 4.3 do not have quantities until weights are given to them. For this

[2] Since data consist of presence or absence of a response, the constraint used for continuous data in Section 4.1 no longer applies to the present case. However, we must still fix the unit of derived quantities, which now becomes the role of the constraint. Thus, the chosen constraint no longer serves the purpose of preserving the unit of measurement.

[3] Mathematically, both of them are based on the so-called *singular value decomposition*, or Eckart-Young decomposition of respective data matrices (see chapters 5 and 6 for discussion).

reason, we will again use continuous data to introduce some concepts in this section.

Let us consider another small example, in which five students wrote three tests (see Table 4.4). There are 15 (5×3) scores in total. The average of the 15 scores, called the *grand mean*, is therefore 180/15 = 12. If we indicate the score of student i on test j by X_{ij} and the grand mean by m, the quantity $(X_{ij} - m)$ indicates how different the score of student i on test j is from the grand mean. If we calculate $(X_{ij} - m)$ for all the students and all the tests, we can see variations among the 15 scores. To take this one step further, let's consider the average of Student i, which we indicate by m_i. Students' averages are listed in the last column of Table 4.4. The quantity $(X_{ij} - m_i)$ indicates how different each student's three scores are from his or her own average. In other words, $(X_{ij} - m_i)$ relfects the *within-student variation* of scores.

TABLE 4.4
Test Scores (Continous Data)

	Test				
Student	1	2	3	Total	Average
1	12	7	8	27	9
2	16	18	14	48	16
3	6	7	5	18	6
4	10	14	12	36	12
5	16	16	19	51	17
			Total	180	

We have so far looked at two expressions, $(X_{ij} - m)$ and $(X_{ij} - m_i)$. We now would like to relate them to show that the so-called within-student variations are a part of the total variation, that is, $(X_{ij} - m)$. We can rewrite the total variation in the following way,

$$(X_{ij} - m) = X_{ij} - m_i + m_i - m = (X_{ij} - m_i) + (m_i - m)$$

Look at the right-hand side of the expression, which consists of two parts, the within-student variation and $(m_i - m)$. This last term is the difference between the average of student i and the grand mean, and indicates how different each student is from the overall average. Therefore, $(m_i - m)$ is called the *between-student varation*. The left-hand side of the expression, $(X_{ij} - m)$, is called the *total variation*. The data in Table 4.4 therefore can be viewed in terms of three sources of variations. Let's subtract the grand mean of 12 from each of the 15 scores in the table and decompose the resultant numbers into the two kinds of variations:

TABLE 4.5
Within, Between, and Total Variations

$$\begin{bmatrix} 0 & -5 & -4 \\ 4 & 6 & 2 \\ -6 & -5 & -7 \\ -2 & 2 & 0 \\ 4 & 4 & 7 \end{bmatrix}_{(X_{ij} - m)} = \begin{bmatrix} 3 & -2 & -1 \\ 0 & 2 & -2 \\ 0 & 1 & -1 \\ -2 & 2 & 0 \\ -1 & -1 & 2 \end{bmatrix}_{(X_{ij} - m_i)} + \begin{bmatrix} -3 & -3 & -3 \\ 4 & 4 & 4 \\ -6 & -6 & -6 \\ 0 & 0 & 0 \\ 5 & 5 & 5 \end{bmatrix}_{(m_i - m)}$$

(Total) (Within) (Between)

Can you verify that the sum of the two corresponding elements on the right-hand side of the expression is equal to the corresponding element on the left? Futhermore, if you add 12, the grand mean, to each element on the left-hand side, you get the original set of scores in Table 4.4. You can see from the above decomposition how individual students are different (see Between in Table 4.5) and how one's performance fluctuates from test to test (see Within in Table 4.5). Note, however, that if you add all the 15 numbers in each of the three sets, Total, Within, and Between, each sum is zero. In statistics, therefore, we calculate not the sum of these numbers, but the sum of the squares of these numbers. The three corresponding sums of squares are called, respectively, the total sum of squares (SS_t), the within-student sum of squares (SS_w), and the between-student sum of squares (SS_b). These quantities are given by

(4.3) $$SS_t = \sum_{i=1}^{N}\sum_{j=1}^{n}(X_{ij}-m)^2 ,$$

(4.4) $$SS_w = \sum_{i=1}^{N}\sum_{j=1}^{n}(X_{ij}-m_i)^2 ,$$

(4.5) $$SS_b = n\sum_{i=1}^{N}(m_i-m)^2 ,$$

where N is the number of subjects (N=5), n the number of tests (n=3), and Σ is called the summation sign, indicating that the quantities following the sign are summed over the entire range of the subscript.[4]

It is well known in statistics that in terms of sums of squares too the following relation exists:

$$SS_t = SS_w + SS_b \tag{4.6}$$

In our example, you can certainly see this relation: 296 = 38 + 258. Each term of equation (4.6) is the sum of squares, hence nonnegative, but it does not have any upper bound. The more nonnegative numbers added, the greater the sum. It is therefore more convenient to divide both sides of equation (4.6) by SS_t to show relative variations within and between students,

$$\frac{SS_t}{SS_t} = 1 = \frac{SS_w}{SS_t} + \frac{SS_b}{SS_t} = (1-\eta^2) + \eta^2 , \tag{4.7}$$

where η^2, "eta square," is called the *squared correlation ratio* in statistics. It is the ratio of SS_b over SS_t. Since $SS_w = SS_t - SS_b$, it is obvious that $SS_w/SS_t = 1 - \eta^2$. In our example, $\eta^2 = 258/296 = 0.8716$. Since η^2 is between 0 and 1, the current result indicates that individual (interindividual) differences, as reflected on SS_b, are much greater than within-individual (intraindividual) differences, captured by SS_w.

Is our discussion clear so far? If not, please examine the example again until you feel comfortable with the current topic. Note at this point that once continuous data such as those in Table 4.4 are given, the squared correlation ratio is uniquely determined. In other words, there exists only one value of η^2 for a given data set. In contrast, when data are categorical such as those in Table 4.3, the value of η^2 will vary depending on those weights for the response options. To see this point, let us use as our example the data in Table 4.3. The weights (unknown) for the seven options are, as before, indicated $x_1, x_2, \ldots, x_7$ so that we obtain a table of

---

[4]For instance, $(m_1 - m) + (m_2 - m) + (m_3 - m) + (m_4 - m)$ can be expressed as $\Sigma(m_i - m)$. Thus, equation (4.3), (4.4), and (4.5) in our example are sums of squares of the respective sets of 15 numbers in Table 4.5. It is the general practice to indicate the range of the subscript, for instance, from $i = 1$ to n, the beginning value below and the last value above the sigma sign. But, when the range of summation is obvious like the present case, the range indication is typically omitted.

weighted responses (Table 4.6). The grand mean and subject's averages are defined as before,

$$m = \frac{\text{(Grand Total)}}{18}, \quad m_1 = \frac{(x_1 + x_4 + x_7)}{3}, \quad m_2 = \frac{(x_2 + x_3 + x_6)}{3},$$

and similarly, m_3, m_4, m_5, and m_6 can be calculated. Our task is to express η^2 for this data set. You can follow the same steps as used in the above example of continuous data, and you will eventually arrive at the expression you need. However, this would take many lines of formulas, and would be quite boring and confusing. Therefore, we will skip the lengthy derivation, and present here the final expressions of SS_t and SS_b (if you wish, verify these expressions):

$$SS_t = 3x_1^2 + 3x_2^2 + 2x_3^2 + 2x_5^2 + 3x_6^2 + 3x_7^2 - C,$$

$$SS_b = \frac{(A_1^2 + A_2^2 + A_3^2 + A_4^2 + A_5^2 + A_6^2)}{3} - C,$$

where

$$A_1 = x_1 + x_4 + x_7, \quad A_2 = x_2 + x_5 + x_6, \quad A_3 = x_1 + x_5 + x_7,$$

$$A_4 = x_1 + x_3 + x_6, \quad A_5 = x_2 + x_5 + x_6, \quad A_6 = x_2 + x_4 + x_7,$$

$$C = \frac{(3x_1 + 3x_2 + 2x_3 + 2x_4 + 2x_5 + 3x_6 + 3x_7)^2}{18}.$$

The last expression of C is referred to as the *correction term*, and it serves to adjust quantities involved to the zero origin. Can you now see that SS_t and SS_b, hence η^2 too, will change when the seven weights for options vary?

Since η^2 is bounded between 0 and 1, η^2 changes as a function of option weights within this range. In the previous section, you saw that dual scaling determines option weights so as to maximize the variance of composite scores. You are now ready to hear this: Dual scaling also determines the option weights so as to maximize η^2. In the previous section, you saw that PCA with continuous data and dual scaling (PCA for categorical data) both maximize the variance of composite scores, PCA in terms of weights for variables (e.g., tests) and dual scaling in terms of option weights. In this section, we have seen that η^2 is uniquely determined for a given set of continuous data, but that dual scaling determines option weights so as to maximize η^2. Somehow, these two cases, that is, PCA and

TABLE 4.6
Data in Terms of Weights for Options

Subject	Item			Total for each subject
1	x_1	x_4	x_7	$x_1 + x_4 + x_7$
2	x_2	x_3	x_6	$x_2 + x_3 + x_6$
3	x_1	x_5	x_7	$x_1 + x_5 + x_7$
4	x_1	x_3	x_6	$x_1 + x_3 + x_6$
5	x_2	x_5	x_6	$x_2 + x_5 + x_6$
6	x_2	x_4	x_7	$x_2 + x_4 + x_7$
Grand Total= $3x_1 + 3x_2 + 2x_3 + 2x_4 + 2x_5 + 3x_6 + 3x_7$				

maximization of η^2, do not seem to provide one-to-one correspondence. Yet you are told now, without proof, that these apparently different approaches produce identical option weights. Indeed, this method of optimal quantification has been used by many people (e.g., Guttman, 1941), that is, to determine option weights so as to maximize η^2, or, equivalently, to minimize $(1 - \eta^2)$.

4.4 DUALITY AND GUTTMAN'S PRINCIPLE OF INTERNAL CONSISTENCY

You have so far seen two ways of describing dual scaling, one as PCA for categorical data and the other as maximizing η^2. These are sometimes referred to as the PCA approach and the ANOVA (analysis of variance) approach. In these two formulations, we considered the problem of determining weights for response options, $x_1, x_2, \ldots, x_m$. The same methods are equally applicable to determining scores for subjects, $y_1, y_2, \ldots, y_N$. As you will see later, there exists perfect mathematical symmetry in the relation between the two sets, $x_1, x_2, \ldots, x_m$ and $y_1, y_2, \ldots, y_N$. This symmetry is referred to as *duality*, and served as a key concept in the adoption of the name "dual scaling" for the technique.[5] Franke (1985) stated that he "uses Nishisato's term for its generality and lack of ambiguity" (p. 63).

This dual aspect of the method is best described by Guttman's *principle of internal consistency* (Guttman, 1941, 1950), which can be summarized as follows: "Assign as similar scores as possible to those subjects who

[5] See Nishisato (1980a).

chose the same options of a question, and these scores should be as different as possible from the scores of those who chose other options," and "Assign as similar weights as possible to those options which were chosen by one subject, and those option weights should be as different as possible from the weights of options which were not chosen by this subject."

To illustrate this principle of deriving scores for the subjects, we will use the data in Table 4.3. Recall that we obtained Table 4.6 from Table 4.3 by replacing responses with corresponding option weights. Let us now construct another table, this time by replacing responses with corresponding scores for the subjects (Table 4.7). Can you follow how this table was obtained from Table 4.3? Compare Table 4.3 with Table 4.7. You can see, for example, that Option 1 of Item 1 was chosen by subjects 1, 3, and 4, whose scores are respectively y_1, y_3, and y_4, the three entries in the column for Option 1 of Item 1 in Table 4.7. You can now consider composites of subjects' scores for options, instead of composites of option weights for subjects. Similarly, you can define the *squared correlation ratio* as the ratio of the between-option sum of squares to the total sum of squares of responses weighted by $y_1, y_2, \ldots, y_6$. Notice that all we have done is to switch the role of rows and the role of columns of the table. An amazing aspect of duality is that the maximal value of η^2 in terms of $x_1, x_2, \ldots, x_7$ is exactly the same as the maximal value of η^2 obtained through manipulation of $y_1, y_2, \ldots, y_6$. In fact, those option weights generate scores for subjects that have the maximal variance, and those scores of subjects in turn produce weights for the options that have the maximal variance. This reciprocal relation between the two sets of weights (scores) provides the so-called *dual relations* (Nishisato, 1980a) or *transition formulas* (Benzécri et al., 1973). In the present case, the dual relations between x's and y's can be expressed as

$$\begin{aligned}
\eta y_1 &= (x_1 + x_4 + x_7)/3 & \qquad \eta x_1 &= (y_1 + y_3 + y_4)/3 \\
\eta y_2 &= (x_2 + x_3 + x_6)/3 & \qquad \eta x_2 &= (y_2 + y_5 + y_6)/3 \\
\eta y_3 &= (x_1 + x_5 + x_7)/3 & \qquad \eta x_3 &= (y_2 + y_4)/2 \\
\eta y_4 &= (x_1 + x_3 + x_6)/3 & \qquad \eta x_4 &= (y_1 + y_6)/2 \\
\eta y_5 &= (x_2 + x_5 + x_6)/3 & \qquad \eta x_5 &= (y_3 + y_5)/2 \\
\eta y_6 &= (x_2 + x_4 + x_7)/3 & \qquad \eta x_6 &= (y_2 + y_4 + y_5)/3 \\
& & \qquad \eta x_7 &= (y_1 + y_3 + y_6)/3
\end{aligned}$$

where η is the positive square root of the maximized squared correlation ratio. Notice that when $\eta^2 = 1$ (i.e., $\eta = 1$) there exists a perfect reciprocal averaging relation between x and y. When η^2 is smaller than 1 (the usual case), the composite score or weight is smaller than a simple average. The above two sets of expressions can be expressed more generally as

TABLE 4.7
Data in Terms of Scores for Subjects

Item	1		2			3	
Option	1	2	1	2	3	1	2
	y_1	y_2	y_2	y_1	y_3	y_2	y_1
	y_3	y_5	y_4	y_6	y_5	y_4	y_3
	y_4	y_6				y_5	y_6

(4.8) $$\eta y_i = \sum \frac{f_{ij}x_j}{f_{i\cdot}} \quad , \quad \eta x_j = \sum \frac{f_{ij}y_i}{f_{\cdot j}}$$

where, in the case of multiple-choice data such as in Table 4.3, f_{ij} is either 1 or 0, $f_{i.}$ is the number of responses of subject i (the sum of row i of the table) and $f_{.j}$ is the number of responses of option j (the sum of column j of the table). Formula 4.8 applies, of course, to the contingency table, in which case f_{ij} indicates the frequency of the i-th row and the j-th column, and $f_{i.}$ and $f_{.j}$ are respectively marginal frequencies of row i and column j.

4.5 CORRELATION AND SIMULTANEOUS LINEAR REGRESSION

Now that you have seen the *dual relations*, you are ready to look into a more direct relationship between x's and y's, that is, *correlation*. Let us again use the data set in Table 4.3. Suppose that those option weights and subject scores are simultaneously assigned to the responses (i.e., 1's) in the table, resulting in a table of weighted responses (Table 4.8). Can you figure out how this table is prepared? You can see the weights of the options of each item chosen by subjects in Table 4.6, which constitute the second term of each pair of Table 4.8. The left-hand side of each pair is nothing but the corresponding subject's score. As Guttman (1941) reasoned, you can say that the two unknown quantities in each pair in Table 4.8 are assigned to the same response, that is, they are the common descriptors of a single response, and therefore the two unknowns should be given as similar values as possible. One of the popular measures of relationship between a pair of variables is the so-called *product-moment correlation* or *Pearsonian correlation*. This measure indicates the degree of linear relationship, that is, the tendency that as one variable increases so too does the other. Let us

TABLE 4.8
Simultaneously Weighted Data

Subject	Item 1	Item 2	Item 3
1	(y_1,x_1)	(y_1,x_4)	(y_1,x_4)
2	(y_2,x_2)	(y_2,x_3)	(y_2,x_6)
3	(y_3,x_1)	(y_3,x_5)	(y_3,x_7)
4	(y_4,x_1)	(y_4,x_3)	(y_4,x_6)
5	(y_5,x_2)	(y_5,x_5)	(y_5,x_6)
6	(y_6,x_2)	(y_6,x_4)	(y_6,x_7)

indicate this correlation by ρ ("rho"). To simplify the expression for ρ, let's choose the units and the origins of y's and x's as follows: (The sum of squares of responses weighted by y_i) = (the sum of squares of responses weighted by x_j) = d, and (the sum of responses weighted by y_i) = (the sum of responses weighted by x_j) = 0. Don't worry about these conditions on y_i and x_j because they will not alter the value of ρ or η^2. Now, ρ can be expressed simply as

$$\rho = \frac{\text{(The sum of products of paired weights)}}{d} = \sum\sum \frac{f_{ij}y_i x_j}{d} \tag{4.9}$$

where f_{ij} = 1 or 0 as shown in Table 4.3.

Dual scaling is also a technique to determine y_i and x_j in such a way that ρ is a maximum. Note again that these subject scores, y_i, and option weights, x_j, are identical to those obtained by the methods discussed so far. In addition, you should note that

$$\rho = \eta, \quad \text{that is, } \eta^2 = \rho^2 \tag{4.10}$$

In statistics, the squared product-moment correlation is generally not equal to the squared correlation ratio. The equality between them, as shown in Equation 4.10, is strictly a result of duality of this scaling method.

It is important to look at this approach to dual scaling as applied to the contingency table, because it will offer you another opportunity to see the distinction between continuous data and categorical data in analysis. Let us consider a contingency table which is typically obtained by asking two multiple-choice questions. Consider the following questions:

Q.1: How do you feel about taking sleeping pills?
( ) strongly for ( ) for ( ) neutral ( ) against ( ) strongly against

Q.2: Do you sleep well every night?
( ) never ( ) rarely ( ) some nights ( ) usually ( ) always

Suppose you obtain the data from 140 subjects, as shown in Table 4.9. The important distinction between continuous and categorical data, referred to previously, can now be explained as follows. Suppose you assign weight y_1 to the "strongly for" option of Q.1. The element in the first row (strongly for) and the first column (never) in the table, that is, 15, is now given weight y_1 so that the weighted response is $15y_1$. So far, this is the same for both types of data. Once you consider the sum of squares of weighted responses, however, you will recognize the difference in meaning of the expression $15y_1$ between the two types. In continuous data, 15 is a single number, a quantity. Therefore, the square of this weighted response is $(15y_1)^2 = 225y_1^2$. In contrast, $15y_1$ in dual scaling means that each of 15 responses is given y_1, hence the sum of squared responses being equal to $y_1^2 + y_1^2 + \ldots + y_1^2 = 15y_1^2$. Do you see this distinction? When you derive formulas for categorical data, this is of the utmost importance, because it is one of the main distinctions in the formulation of categorical data analysis from that of continuous data.

Dual scaling of the data in Table 4.9 determines five weights y_i for the options of Q.1 and five weights x_j for the options of Q.2 in such a way that statistic ρ is a maximum. In Formula 4.8, f_{ij} is no longer 1 or 0 but the frequency of row i and column j of Table 4.9. You may wonder what the above operation of maximizing ρ really means. Let us start with a case of nonoptimal weights.

Suppose that you decide, as most people do, to use your subjective, or common-sense, weights of -2, -1, 0, 1, and 2 for the options "never, rarely, some nights, usually, always," respectively, for Q.2. Using these weights, calculate the mean weighted response for "strongly for" of Q.1. Thus,

$$m_1 \text{ (strongly for, Q.1)} = \frac{[15\text{x}(-2) + 8\text{x}(-1) + 3\text{x}0 + 2\text{x}1 + 0\text{x}2]}{28} = -1.3$$

Can you follow how m_1 is calculated? Look at the first row of Table 4.9, consisting of (15, 8, 3, 2, 0), a total of 28 responses. These responses are weighted by (-2, -1, 0, 1, 2) in the calculation of m_1. Similarly, you can calculate the means of weighted responses of the remaining four options, which are $m_2 = -0.8$, $m_3 = -0.6$, $m_4 = 0.6$, and $m_5 = 1.1$. If we now assign -2, -1, 0, 1, 2 to the five options of Q.1, we can calculate the means of weighted responses for the options of Q.2, $n_1 = 1.2$, $n_2 = 0.5$, $n_3 = 0.4$, $n_4 = 0.5$, $n_5 = 1.3$.

Table 4.9
Contingency Table

		Q.2				
		Never	Rarely	Some	Usually	Always
	Strongly for	15	8	3	2	0
	For	5	17	4	0	2
Q.1	Neutral	6	13	4	3	2
	Against	0	7	7	5	9
	Strongly against	1	2	6	3	16

How good are these common-sense weights in explaining the data? One way to check is to construct a graph where you plot these means, just calculated, against your subjective weights (Figure 4.4). Assuming as before that you assign weight y_i for row i (Q.1) and x_j for column j (Q.2), let us call the plot of m_i against the subjective column weights the "regression of y on x" and the plot of n_j against the row weights the "regression of x on y." This graph alone does not tell us much. So, just wait until you see the corresponding results when you use, instead of your subjective weights, *optimal weights* obtained from dual scaling, that is, those weights that maximize ρ. Since this example is given in Nishisato (1980a, pp. 65-68), we show you only the graph obtained without computations, but using dual scaling weights (Figure 4.5). Can you see that this is a remarkable plot? Both lines are straight and their slopes are identical! Those optimal weights had the effect of adjusting the spacings of rows and columns in such a way that the relation between rows and columns, after quantification, is linear, the condition under which ρ attains its maximum. This remarkable characteristic was termed *simultaneous linear regression* by Lingoes (1964). Indeed, it served as the criterion in Hirschfeld's 1935 formulation of this quantification method. Specifically, Hirschfeld posed the question: "Is it always possible to introduce new variates for the rows and the columns of a contingency table such that both regressions are linear?"

In retrospect, you can see Hirschfeld's results in the expressions of dual relations or transition Formulas 4.8. As you recall, η in Formula 4.8 is the same as ρ (see Formulas (4.9), and (4.10)), which is a key quantity, called the *parameter*, in linear regression. There is one more approach which is also obvious now that you know the dual relations. This approach provides a simple method of calculating "optimal weights," and should therefore be discussed before we move to the next chapter. It is called the *method of reciprocal averages*.

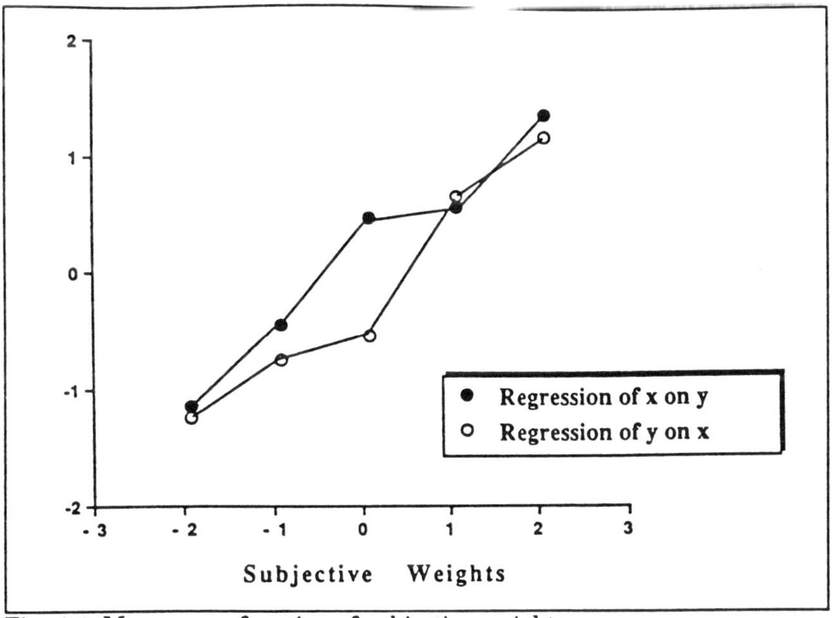

Fig. 4.4. Means as a function of subjective weights.

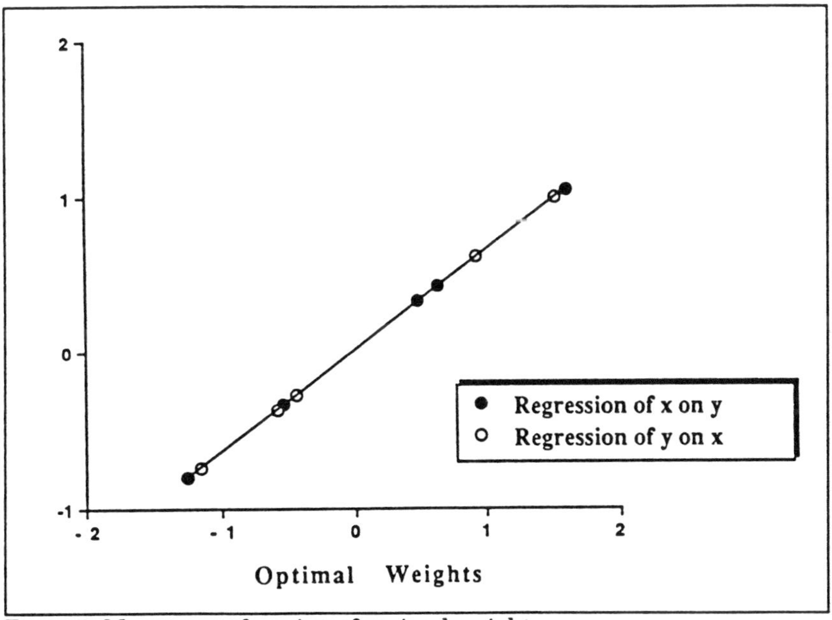

Fig. 4.5. Means as a function of optimal weights.

4.6 RECIPROCAL AVERAGING

Richardson and Kuder (1933) proposed the idea for a method which Horst (1935) termed the method of reciprocal averages (MRA). The same technique was also mentioned by Fisher (1940). Rather than discussing the idea, we present a numerical example, apply the method, and calculate optimal weights. The following description of the method and computational processes is from Nishisato and Nishisato (1984, pp. 115—120).

Suppose that three teachers (White, Green, and Brown) were rated with respect to their teaching performance by students, and that the results were as given in Table 4.10.

Table 4.10
Performance Assessment of Three Teachers

Teacher	Good	Average	Poor	Total
White	1	3	6	10
Green	3	5	2	10
Brown	6	3	0	9
Total	10	11	8	29

The MRA is carried out in the following way:

(1) The MRA starts with assigning arbitrary weights to columns (or rows, if preferred). Although such values are said to be arbitrary, one must avoid identical weights for all columns (or rows)—including zero. It is always a good strategy to use "reasonable" values. As an example of reasonable values, consider:

$$x_1 \text{ (good)} = 1,\ x_2 \text{ (average)} = 0,\ x_3 \text{ (poor)} = -1.$$

(2) Calculate the weighted averages of the rows. These are the first approximations for the optimal scores for the rows, that is, the teachers:

$$y_1(\text{White}) = \frac{1x_1 + 3x_2 + 6x_3}{10} = \frac{1\text{x}1 + 3\text{x}0 + 6\text{x}(-1)}{10} = -0.5000$$

Similarly,

$$y_2(\text{Green}) = \frac{3\text{x}1 + 5\text{x}0 + 2\text{x}(-1)}{10} = 0.1000$$

$$y_3(\text{Brown}) = \frac{6\times1 + 3\times0 + 0\times(-1)}{9} = -0.6667$$

(3) Calculate the mean responses weighted by y_1, y_2, and y_3, as shown below:

$$M(\text{mean}) = \frac{[10y_1 + 10y_2 + 9y_3]}{29}$$
$$= \frac{[10\times(-0.5000) + 10\times0.1000 + 9\times0.6667]}{29}$$
$$= 0.0690$$

(4) Subtract M from each of y_1, y_2, and y_3, and adjusted values should be indicated again by y_1, y_2, and y_3:

$$y_1 = -0.5000 - 0.0690 = -0.5690$$
$$y_2 = 0.1000 - 0.0690 = 0.0310$$
$$y_3 = 0.6667 - 0.0690 = 0.5977$$

(5) Divide y_1, y_2, and y_3 by the largest absolute value of y_1, y_2, and y_3. Adjusted values should again be indicated by y_1, y_2, and y_3:

$$y_1 = -0.5690 \,/\, 0.5977 = 0.9519$$
$$y_2 = 0.0310 \,/\, 0.5977 = 0.0519$$
$$y_3 = 0.5977 \,/\, 0.5977 = 1.0000$$

(6) Using these new values as weights, calculate the averages of the columns:

$$x_1 = [1y_1 + 3y_2 + 6y_3] \,/\, 10$$
$$= [1\times(-0.9519) + 3\times0.0519 + 6\times1.0000] \,/\, 10 = 0.5204$$
$$x_2 = [3\times(-0.9519) + 5\times0.0519 + 3\times1.0000] \,/\, 11 = 0.0367$$
$$x_3 = [6\times(-0.9519) + 2\times0.0519 + 0\times1.0000] \,/\, 8 = -0.7010$$

(7) Calculate the mean responses weighted by x_1, x_2, and x_3:

$$N(mean) = [10\times0.5204 + 11\times0.0367 + 8\times(-0.7010)] \div 29 = 0$$

(8) Subtract N from x_1, x_2, and x_3 (since $N = 0$, x_1, x_2, and x_3 remain the same).

(9) Divide each element of x_1, x_2, and x_3 by the largest absolute value of the three numbers (-0.7010 has the largest absolute value). Adjusted values should be indicated again by x_1, x_2, and x_3:

$$x_1 = 0.5204/0.7010 = 0.7424$$
$$x_2 = 0.0367/0.7010 = 0.0524$$
$$x_3 = -0.7010/0.7010 = -1.0000$$

Reciprocate the above averaging processes (step 2 through step 9) until all the six values are stabilized. If we continue the process, we arrive at the following results:

[Iteration 2]	$y_1 = -0.9954$	$x_1 = 0.7321$
	$y_2 = 0.0954$	$x_2 = 0.0617$
	$y_3 = 1.0000$	$x_3 = -1.0000$
[Iteration 3]	$y_1 = -0.9993$	$x_1 = 0.7321$
	$y_2 = 0.0993$	$x_2 = 0.0625$
	$y_3 = 1.0000$	$x_3 = -1.0000$
[Iteration 4]	$y_1 = -0.9996$	$x_1 = 0.7311$
	$y_2 = 0.0996$	$x_2 = 0.0625$
	$y_3 = 1.0000$	$x_3 = -1.0000$

Iteration 5 provides the identical set of numbers as iteration 4. Therefore it can be said that *the process has converged* to the optimal solution after four iterations.

(10) The unit for y_1, y_2, and y_3 now needs to be equated to the unit for x_1, x_2, and x_3. In the DUAL3 software series,[6] the unit is chosen in such a way that the sum of squares of weighted responses is equal to the number of responses. In this case, the constant multiplier for adjusting the unit of y_1, y_2, and y_3 is given by:

$$c_r = \sqrt{\frac{29}{10y_1^2 + 10y_2^2 + 9y_3^2}} = 1.2325$$

The constant multiplier for adjusting the unit of x_1, x_2, and x_3 is given by:

---

[6] See the appendix.

$$c_c = \sqrt{\frac{29}{10x_1^2 + 11x_2^2 + 8x_3^2}} = 1.4718$$

The final weights are obtained by multiplying y_1, y_2, and y_3 by c_r, and x_1, x_2, and x_3 by c_c, which are:

[Adjusted]	y_1 = -1.2320	x_1 = 1.0760
	y_2 = 0.1228	x_2 = 0.0920
	y_3 = 1.2325	x_3 = -1.4718

Since the origin and the unit of these weights are arbitrary, one can add a constant to them or multiply them by a constant without changing their property of "maximally discriminative weights." The two sets of weights y_i and x_j satisfy the dual relations with maximum correlation ρ of 0.6069. This correlation coefficient can be calculated by the following formula:

$$\rho_1 = \frac{\Sigma\Sigma f_{ij} y_{1i} x_{1i}}{\sqrt{\Sigma g_i y_{1i}^2}\sqrt{\Sigma f_j x_{1j}^2}}$$

Once the first solution $[\rho_1, y_{1i}, x_{1j}]$ is obtained, calculate the residual frequency, which we indicate by $f_{ij(1)}$, where

$$f_{ij(1)} = f_{ij}* - \rho_1 y_{1i} x_{1j} \left[\frac{g_i f_j}{f_t}\right]$$

and apply the MRA to $f_{ij(1)}$. When the convergence is obtained, calculate ρ_2. When the second solution $[\rho_2, y_{2i}, x_{2j}]$ is obtained, calculate the next residual frequency,

$$f_{ij(1)} - \rho_2 y_{2i} x_{2j} \left[\frac{g_i f_j}{f_t}\right] = f_{ij(2)}$$

Repeat the same process to obtain the other solutions. The maximum number of solutions from an Nxn table is restricted to the smaller value of N and n, -1. In our example, the table is 3x3. Therefore, we can obtain at most only two solutions. In other words, two solutions explain 100% of row—column association in the data.

When the **MRA** is applied to multiple-choice data, the format of the data table must be that of an *incidence matrix*, that is, a respondents-by-options table of 1s (choices) and 0s (nonchoices).[7]

Before leaving this section, let us look at a numerical example of the already mentioned decomposition of f_{ij}, using the 3x3 contingency table analyzed in Section 4.6. First, prior to dual scaling, we can decompose the observed contingency table into the matrix of frequencies expected under statistical independence of rows and columns, plus the residual matrix:

$$\underset{\mathbf{F}\text{(observed)}}{\begin{bmatrix} 1 & 3 & 6 \\ 3 & 5 & 2 \\ 6 & 3 & 0 \end{bmatrix}} = \underset{\mathbf{F}_0\text{(expected)}}{\begin{bmatrix} 3.25 & 3.79 & 2.76 \\ 3.45 & 3.79 & 2.76 \\ 3.1 & 3.41 & 2.48 \end{bmatrix}} + \underset{\mathbf{F}-\mathbf{F}_0\text{(residual)}}{\begin{bmatrix} -2.45 & -0.79 & 3.24 \\ -0.45 & 1.21 & -0.76 \\ 2.90 & -0.41 & -2.48 \end{bmatrix}}.$$

Recall that the matrix of frequencies expected when rows and columns are statistically independent can be calculated by "g_i times f_j divided by f_t," which is the element in row i and column j of $\mathbf{F}_0$. Dual scaling can now decompose the residual matrix, which is the matrix of associations between rows and columns. The "rank" (see chapter 5) of this matrix is 2, meaning that the matrix can be decomposed into two parts:

$$\underset{\mathbf{F}-\mathbf{F}_0}{\begin{bmatrix} -2.45 & -0.79 & 3.24 \\ -0.45 & 1.21 & -0.76 \\ 2.90 & -0.41 & -2.48 \end{bmatrix}} = \underset{\text{Solution 1}}{\begin{bmatrix} -2.78 & -0.26 & 3.03 \\ 0.27 & 0.03 & -0.30 \\ 2.50 & 0.24 & -2.73 \end{bmatrix}} + \underset{\text{Solution 2}}{\begin{bmatrix} 0.33 & -0.53 & 0.21 \\ -0.72 & 1.18 & -0.46 \\ 0.40 & -0.65 & 0.25 \end{bmatrix}}$$

Typical elements of solution 1 and solution 2 are calculated by Formula (4.11)

$$\left(\frac{g_i f_j}{f_t}\right)\rho_1 y_{1i} x_{1j}\,, \quad \left(\frac{g_i f_j}{f_t}\right)\rho_2 y_{2i} x_{2j}\,, \tag{4.11}$$

so that the above decomposition can be described in terms of a typical element by Formula (4.12).

---

[7] Regarding the other types of data in this book, the following articles describe the exact formats of the data for the MRA: For *paired comparison* and *rank order data*, see Nishisato (1978a); for *successive categories (rating) data*, see Nishisato and Sheu (1984).

$$f_{ij} - \left(\frac{g_i f_j}{f_t}\right) = \left(\frac{g_i f_j}{f_t}\right)[\rho_1 y_{1i} x_{ij} + \rho_2 y_{2i} x_{2j}] \ . \tag{4.12}$$

Chapter 4 should be sufficient for most practical purposes as an introduction to dual scaling. It has outlined the basic concepts and provided a practical method for computation. Should the reader wish to go a little further into a more mathematical treatment of dual scaling, the next two chapters will be especially useful.

CHAPTER 5
Useful Quantitative Tools

Because it contains a rather sudden introduction to technical information, this chapter might well be placed in the appendix. However, for those who want to look at the mathematical side of dual scaling, this seems to be the right place for these topics. If you are interested only in a working knowledge of dual scaling, you may skip this chapter and the next. However, these chapters will be useful to successively upgrade your technical knowledge and to help to attain smooth transition to Chapter 7 and to the advanced topics discussed in Part 4.

One of the main causes of difficulty in understanding textbooks is the notation. Different authors use their own symbols to describe the same idea or concept. Similarly, some authors employ the "scalar" notation, while others use the "matrix" notation, to present mathematical formulas. It is to our advantage to switch to the matrix notation from the familiar scalar notation in order to arrive at a description of dual scaling that is compact and simple enough to help us to further our understanding of its mathematical potential. We will therefore start with an introduction of the basic concepts used in the matrix notation.

5.1 SIMPLE VECTOR AND MATRIX OPERATIONS

Mathematical expressions are often an eloquent means of communication. This section is devoted to a brief introduction of very basic operations of vector and matrix calculus, which will be useful in understanding mathematical aspects of dual scaling. The topics covered here are by no means sufficient to read articles published in such a journal as *Psychometrika* or the *Journal of American Statistical Association*, but are relevant to the many discussions on dual scaling as presented in this book. Let us begin with some definitions and notation.

A *matrix* is a rectangular array of numbers, called elements. It is customary to indicate a matrix by a capital letter such as **A**, **B**, **C**, or **D**, and its elements by small letters with subscripts such as a_{ij}, b_{ij}, c_{ij}, or d_{ij}, where the first subscript i refers to the row (i.e., ith row) and the second subscript j to the column (i.e., jth column). For instance,

$$\mathbf{A} = \begin{bmatrix} a_{11} & a_{12} & a_{13} \\ a_{21} & a_{22} & a_{23} \end{bmatrix}.$$

The matrix with two rows and three columns is referred to as a 2×3 (two by-three) matrix. The expression 2×3 is called the *order* or *dimension* of the matrix.

The *transpose* of **A**, indicated by **A′**, is the matrix obtained from **A** by interchanging rows and columns. For instance, if

$$\mathbf{A} = \begin{bmatrix} 1 & 3 & 5 \\ 2 & 4 & 7 \\ 8 & 0 & 6 \end{bmatrix}, \text{ then } \mathbf{A}' = \begin{bmatrix} 1 & 2 & 8 \\ 3 & 4 & 0 \\ 5 & 7 & 6 \end{bmatrix}.$$

A *vector* is a single row or column of elements. Therefore, a *row vector* can be considered a 1×m matrix, and a *column vector* an n×1 matrix. For instance, the following are examples of a row vector and a column vector, respectively:

$$\begin{bmatrix} 3 & 2 & 5 \end{bmatrix}, \quad \begin{bmatrix} 5 \\ 2 \end{bmatrix}.$$

Column vectors are usually indicated by boldface small letters such as **a, b, c**, or **d**, and row vectors by **a′**, **b′**, **c′**, or **d′** as if these were transposes of column vectors. It is usually the case that matrices and vectors are also indicated by typical elements such as $\mathbf{A} = (a_{ij})$, $\mathbf{B} = (b_{ij})$, $\mathbf{a} = (a_i)$, $\mathbf{b}' = (b_j)$.

When two matrices are of the same order, that is, if **A** and **B** are of the same dimension, **A** and **B** are said to be *conformable* for addition and subtraction, which are defined as follows:

$$\begin{aligned} \mathbf{A} + \mathbf{B} &= (a_{ij}) + (b_{ij}) = (a_{ij} + b_{ij}) \\ \mathbf{A} - \mathbf{B} &= (a_{ij}) - (b_{ij}) = (a_{ij} - b_{ij}) \end{aligned}.$$

In other words, these operations are carried out with respect to the corresponding elements. For example,

$$\begin{bmatrix} 2 & 3 & 1 \\ 5 & 6 & 2 \\ 1 & 7 & 9 \end{bmatrix} + \begin{bmatrix} 0 & 2 & 3 \\ 4 & 1 & 1 \\ 2 & 4 & 7 \end{bmatrix} = \begin{bmatrix} 2 & 5 & 4 \\ 9 & 7 & 3 \\ 3 & 11 & 16 \end{bmatrix} , \quad \begin{bmatrix} 2 & 3 \\ 5 & 6 \\ 1 & 7 \end{bmatrix} - \begin{bmatrix} 0 & 1 \\ 4 & 1 \\ 2 & 4 \end{bmatrix} = \begin{bmatrix} 2 & 1 \\ 1 & 5 \\ -1 & 3 \end{bmatrix} .$$

The multiplication of a matrix by a constant k (scalar) is called *scalar multiplication* and is defined by

$$k\mathbf{A} = (ka_{ij}) .$$

For instance,

$$8\begin{bmatrix} 3 & 1 \\ 2 & 0 \\ 5 & 7 \end{bmatrix} = \begin{bmatrix} 24 & 8 \\ 16 & 0 \\ 40 & 56 \end{bmatrix} , \quad 5\begin{bmatrix} 1 & 0 & 0 \\ 0 & 2 & 0 \\ 0 & 0 & -3 \end{bmatrix} = \begin{bmatrix} 5 & 0 & 0 \\ 0 & 10 & 0 \\ 0 & 0 & -15 \end{bmatrix} .$$

The *multiplication of two matrices* is somewhat complicated, but it is an operation which plays a major role in matrix formulations of many mathematical relations. Matrix multiplication is defined only when two matrices are of certain dimensions. If the number of columns of **A** is equal to the number of rows of **B**, **A** is said to be *conformable* to **B** for multiplication **AB**. If **A** is $m{\times}r$ and **B** is $r{\times}n$, then the product **AB** is $m{\times}n$ that is, the number of rows of **A** by the number of columns of **B**. Multiplication **AB** is defined by

$$\underset{m\times r}{\mathbf{A}}\ \underset{r\times n}{\mathbf{B}} = \underset{m\times n}{\mathbf{C}} = (c_{ij}) = (a_{ik})(b_{kj}) = (\sum_{k=1}^{r} a_{ik}b_{kj}) .$$

The typical element c_{ij} of the product **AB** is the sum of the elementwise product of the ith row of **A** and the jth column of **B**. For instance,

$$\begin{bmatrix} a_{11} & a_{12} \\ a_{21} & a_{22} \end{bmatrix} \begin{bmatrix} b_{11} & b_{12} \\ b_{12} & b_{22} \end{bmatrix} = \begin{bmatrix} (a_{11}b_{11}+a_{12}b_{21}) & (a_{11}b_{12}+a_{12}b_{22}) \\ (a_{21}b_{11}+a_{22}b_{21}) & (a_{21}b_{12}+a_{22}b_{22} \end{bmatrix}$$

$$\begin{bmatrix} 1 & 2 \\ 3 & 4 \end{bmatrix} \begin{bmatrix} 8 & 6 \\ 7 & 5 \end{bmatrix} = \begin{bmatrix} (1\times8+2\times7) & (1\times6+2\times5) \\ (3\times8+4\times7) & (3\times6+4\times5) \end{bmatrix} = \begin{bmatrix} 22 & 16 \\ 52 & 38 \end{bmatrix}$$

$$\begin{bmatrix} 2 \\ 3 \end{bmatrix} \begin{bmatrix} 4 & 5 \end{bmatrix} = \begin{bmatrix} 8 & 10 \\ 12 & 15 \end{bmatrix} \qquad \begin{bmatrix} 2 & 3 \end{bmatrix} \begin{bmatrix} 4 \\ 5 \end{bmatrix} = 23 .$$

Let us look at some more examples of matrix and vector multiplications. Consider two vectors,

$$\mathbf{1}_n = \begin{bmatrix} 1 \\ 1 \\ 1 \\ \vdots \\ 1 \end{bmatrix}, \qquad \mathbf{x} = \begin{bmatrix} x_1 \\ x_2 \\ x_3 \\ \vdots \\ x_n \end{bmatrix}.$$

Then,

$$\mathbf{1}_n'\mathbf{1}_n = n, \quad \mathbf{1}_n'\mathbf{x} = \mathbf{x}'\mathbf{1}_n = \Sigma x_i \quad \mathbf{x}'\mathbf{x} = \Sigma x_i^2$$

$$\mathbf{1}_n\mathbf{1}_n' = \begin{bmatrix} 1 & 1 & 1 & \cdots & 1 \\ 1 & 1 & 1 & \cdots & 1 \\ 1 & 1 & 1 & \cdots & 1 \\ \vdots & \vdots & \vdots & \ddots & 1 \\ 1 & 1 & 1 & \cdots & 1 \end{bmatrix}.$$

Elements of a_{jj} of an nxn square matrix **A** are called *diagonal elements*, and the others, that is a_{ij}, $i \neq j$, are called *off-diagonal* elements. A square matrix, of which off-diagonal elements are all zero, is called a *diagonal matrix* and is often denoted by **D**. The following are examples of diagonal matrices.

$$\mathbf{D} = \begin{bmatrix} d_{11} & 0 & 0 \\ 0 & d_{22} & 0 \\ 0 & 0 & d_{33} \end{bmatrix}, \qquad \mathbf{D} = \begin{bmatrix} 3 & 0 & 0 \\ 0 & 1 & 0 \\ 0 & 0 & 5 \end{bmatrix}.$$

A diagonal matrix is often written in terms of the diagonal elements as **D** = (d_{jj}), or, more popularly, **D** = diag(d_{jj}). If all the diagonal elements are identical, the matrix is called the *scalar matrix*. For example, the following are scalar matrices:

$$\mathbf{D} = \begin{bmatrix} 3 & 0 \\ 0 & 3 \end{bmatrix}, \qquad \mathbf{D} = \begin{bmatrix} 7 & 0 & 0 \\ 0 & 7 & 0 \\ 0 & 0 & 7 \end{bmatrix}.$$

If the diagonal elements of a scalar matrix are equal to 1, the matrix is called the *identity matrix*, and indicated by **I**:

$$\mathbf{I} = \begin{bmatrix} 1 & 0 \\ 0 & 1 \end{bmatrix}, \quad \mathbf{I} = \begin{bmatrix} 1 & 0 & 0 \\ 0 & 1 & 0 \\ 0 & 0 & 1 \end{bmatrix}.$$

The identity matrix has the following properties. If the two identity matrices are of the same dimension, then $\mathbf{II} = \mathbf{I}^2 = \mathbf{I}$. Furthermore, if the matrices **I** and **A** are conformable for multiplications **IA** and **AI**, then **IA**=**A**, and **AI**=**A**. Thus, the identity matrix is very much like 1.

Now that matrix multiplication is introduced, let us look at special matrices called *elementary matrices*. Suppose **A** is $m \times n$ and **I** is the $m \times m$ identity matrix. Then,

1. Let $\mathbf{E}_{ij}$ be the matrix obtained from **I** by interchanging its ith and jth rows. Then $\mathbf{E}_{ij}\mathbf{A}$ is the $m \times n$ matrix obtained from **A** by interchanging its ith and jth rows. For example,

$$\text{if } \mathbf{A} = \begin{bmatrix} 1 & 2 & 3 & 4 \\ 4 & 5 & 6 & 7 \\ 8 & 9 & 0 & -1 \end{bmatrix}, \text{ and } \mathbf{E}_{23} = \begin{bmatrix} 1 & 0 & 0 \\ 0 & 0 & 1 \\ 0 & 1 & 0 \end{bmatrix},$$

$$\text{then } \mathbf{E}_{23}\mathbf{A} = \begin{bmatrix} 1 & 2 & 3 & 4 \\ 8 & 9 & 0 & -1 \\ 4 & 5 & 6 & 7 \end{bmatrix}.$$

2. Let $\mathbf{K}_i$ be the matrix obtained from **I** by replacing the 1 in the ith diagonal element by k ($k \neq 0$). Then, $\mathbf{K}_i\mathbf{A}$ is the matrix obtained from **A** by multiplying each element of the ith row by k. For example, if **A** is given as above and

$$\mathbf{K}_2 = \begin{bmatrix} 1 & 0 & 0 \\ 0 & 3 & 0 \\ 0 & 0 & 1 \end{bmatrix}, \text{ then } \mathbf{K}_2\mathbf{A} = \begin{bmatrix} 1 & 2 & 3 & 4 \\ 12 & 15 & 18 & 21 \\ 8 & 9 & 0 & -1 \end{bmatrix}.$$

3. Let $\mathbf{S}_{ij(k)}$ be the matrix obtained from **I** by replacing the 0 in the (i,j) position by k. Then, $\mathbf{S}_{ij(k)}\mathbf{A}$ is the matrix obtained from **A** by adding to the ith row k times the jth row. For example, if **A** is given as above and

$$\mathbf{S}_{12(4)} = \begin{bmatrix} 1 & 4 & 0 \\ 0 & 1 & 0 \\ 0 & 0 & 1 \end{bmatrix}, \text{ then } \mathbf{S}_{12(4)}\mathbf{A} = \begin{bmatrix} 17 & 22 & 27 & 32 \\ 4 & 5 & 6 & 7 \\ 8 & 9 & 0 & -1 \end{bmatrix}.$$

These matrices $\mathbf{E}_{ij}$, $\mathbf{K}_i$, and $\mathbf{S}_{ij(k)}$ are elementary transformation matrices. Premultiplication of **A** by one of these matrices provides elementary row transformation. Similarly, by postmultiplying **A** by these types of matrices, we obtain elementary column transformation. For example,

$$\begin{bmatrix} 1 & 4 & 8 \\ 2 & 5 & 9 \\ 3 & 6 & 10 \\ 4 & 7 & 11 \end{bmatrix} \begin{bmatrix} 1 & 0 & 0 \\ 0 & 0 & 1 \\ 0 & 1 & 0 \end{bmatrix} = \begin{bmatrix} 1 & 8 & 4 \\ 2 & 9 & 5 \\ 3 & 10 & 6 \\ 4 & 11 & 7 \end{bmatrix}$$

$$\begin{bmatrix} 1 & 4 & 8 \\ 2 & 5 & 9 \\ 3 & 6 & 10 \\ 4 & 7 & 11 \end{bmatrix} \begin{bmatrix} 1 & 0 & 0 \\ 0 & 3 & 0 \\ 0 & 0 & 1 \end{bmatrix} = \begin{bmatrix} 1 & 12 & 8 \\ 2 & 15 & 9 \\ 3 & 18 & 10 \\ 4 & 21 & 11 \end{bmatrix}$$

$$\begin{bmatrix} 1 & 4 & 8 \\ 2 & 5 & 9 \\ 3 & 6 & 10 \\ 4 & 7 & 11 \end{bmatrix} \begin{bmatrix} 1 & k & 0 \\ 0 & 1 & 0 \\ 0 & 0 & 1 \end{bmatrix} = \begin{bmatrix} 1 & 1k+4 & 8 \\ 2 & 2k+5 & 9 \\ 3 & 3k+6 & 10 \\ 4 & 4k+7 & 11 \end{bmatrix}.$$

These elementary transformations are useful in discussing such concepts as inverse, canonical form, and rank, topics to be discussed shortly. In the meantime, let us introduce a few other concepts.

Associated with any square matrix **A**, there exists a quantity, called the *determinant*, which is a function of the elements of **A**, and closely related to the so-called volume function and covariance. The determinant of **A** is indicated by $|\mathbf{A}|$. The determinant of a 2×2 matrix **A** is given by

$$|\mathbf{A}| = \begin{vmatrix} a_{11} & a_{12} \\ a_{21} & a_{22} \end{vmatrix} = a_{11}a_{22} - a_{12}a_{21} .$$

Notice that the determinant can be negative, an aspect which is different from the volume function. If the element of the ith row and the

jth column are deleted from an $n \times n$ square matrix **A**, the determinant of the remaining $(n-1)\times(n-1)$ matrix is called the *minor* of the element a_{ij}, and indicated by $|\mathbf{M}_{ij}|$. For instance, consider a 3x3 matrix **A**, where

$$\mathbf{A} = \begin{bmatrix} a_{11} & a_{12} & a_{13} \\ a_{21} & a_{22} & a_{23} \\ a_{31} & a_{32} & a_{33} \end{bmatrix}.$$

Then, the following minors are defined.

$$|\mathbf{M}_{11}| = \begin{vmatrix} a_{22} & a_{23} \\ a_{32} & a_{33} \end{vmatrix} = a_{22}a_{33} - a_{23}a_{32}$$

$$|\mathbf{M}_{12}| = \begin{vmatrix} a_{21} & a_{23} \\ a_{31} & a_{33} \end{vmatrix} = a_{21}a_{33} - a_{23}a_{31}$$

$$|\mathbf{M}_{13}| = \begin{vmatrix} a_{21} & a_{22} \\ a_{31} & a_{32} \end{vmatrix} = a_{21}a_{32} - a_{22}a_{31}$$

Similarly, $|\mathbf{M}_{21}|$, $|\mathbf{M}_{22}|$, $|\mathbf{M}_{23}|$, $|\mathbf{M}_{31}|$, $|\mathbf{M}_{32}|$, and $|\mathbf{M}_{33}|$ can be calculated.

The minor with sign $(-1)^{i+j}$ is called the *cofactor* of a_{ij} and indicated by the Greek letter α_{ij}, that is,

$$\alpha_{ij} = (-1)^{i+j}\,|\mathbf{M}_{ij}|\,.$$

For example, the cofactors of the above example are

$$\begin{aligned} \alpha_{11} &= (-1)^{1+1}\,|\mathbf{M}_{11}| = a_{22}a_{33} - a_{23}a_{32} \\ \alpha_{12} &= (-1)^{1+2}\,|\mathbf{M}_{12}| = -a_{21}a_{33} + a_{23}a_{31} \\ \alpha_{13} &= (-1)^{1+3}\,|\mathbf{M}_{13}| = a_{21}a_{32} - a_{22}a_{31} \end{aligned}\;.$$

The determinant of **A** can be expressed as a sum of the product of each element of a row (column) of **A** times its cofactor,

$$\mathbf{A} = \sum_{i=1}^{n} a_{ij}\alpha_{ij} = \sum_{j=1}^{n} a_{ij}\alpha_{ij}\,.$$

The following is an example of *column expansion* with $j = 3$:

$$\begin{vmatrix} 2 & 1 & 0 \\ 3 & 5 & 4 \\ 6 & 7 & 8 \end{vmatrix} \quad \begin{aligned} &= 0(-1)^{1+3}[3\times7-5\times6]+4(-1)^{2+3}[2\times7-1\times6] \ . \\ &\quad +8(-1)^{3+3}[2\times5-1\times3] = 0-32+56 = 24 \end{aligned}$$

Similarly, the *row expansion* with $i = 1$ is given by

$$\begin{vmatrix} 2 & 1 & 0 \\ 3 & 5 & 4 \\ 6 & 7 & 8 \end{vmatrix} \quad \begin{aligned} &= 2(-1)^{1+1}[5\times8-4\times7]+1(-1)^{1+2}[3\times8-4\times6] \ . \\ &\quad +0(-1)^{1+3}[3\times7-5\times6] = 24-0+0 = 24 \end{aligned}$$

When the matrix is of higher order than this, one would not usually calculate the determinant in this manner. There are other techniques more appropriate for larger matrices. A matrix **A** is said to be *singular* if $|\mathbf{A}|=0$ and *nonsingular* if $|\mathbf{A}| \neq 0$.

The *inverse* of a square matrix **A**, denoted by $\mathbf{A}^{-1}$, is defined for a non-singular matrix by the relation

$$\mathbf{A}^{-1}\mathbf{A} = \mathbf{A}\mathbf{A}^{-1} = \mathbf{I} \ .$$

Thus, the inverse is like the reciprocal in scalar arithmetic. The inverse of a diagonal matrix, $\mathbf{D} = \text{diag}(d_{ij})$, is given by $\mathbf{D}^{-1} = \text{diag}(1/d_{ij})$. For instance,

$$\mathbf{D} = \begin{bmatrix} 2 & 0 \\ 0 & 4 \end{bmatrix}, \text{ then } \mathbf{D}^{-1} = \begin{bmatrix} 0.50 & 0 \\ 0 & 0.25 \end{bmatrix}.$$

If you already have the matrix of cofactors of **A**, then the inverse of **A** is given by

$$\mathbf{A}^{-1} = \frac{(\alpha_{ij})'}{|\mathbf{A}|} \ .$$

There is a practical way to calculate the inverse of a matrix, using elementary matrices. Given **A**, premultiply (or postmultiply) it successively by elementary matrices until the product becomes the identity matrix. At that moment, the product of those elementary matrices is the inverse of **A**. To put this idea into practice, start with the augmented matrix [**A**, **I**], and transform this matrix by successive applications of elementary matrices until you obtain the matrix [I, B]. Then, the matrix B is the inverse of the matrix A. Study the following example of this procedure:

Given [A, I], change it to [I, A^{-1}], where

$$\mathbf{A} = \begin{bmatrix} 2 & 1 & 0 \\ 2 & 5 & 1 \\ 1 & 0 & 6 \end{bmatrix}.$$

The computation may be carried out as follows:

$$\begin{bmatrix} 2 & 1 & 0 & 1 & 0 & 0 \\ 2 & 5 & 1 & 0 & 1 & 0 \\ 1 & 0 & 6 & 0 & 0 & 1 \end{bmatrix} \rightarrow \begin{bmatrix} 2 & 1 & 0 & 1 & 0 & 0 \\ 2 & 5 & 1 & 0 & 1 & 0 \\ 2 & 0 & 12 & 0 & 0 & 2 \end{bmatrix} \rightarrow$$

$$\begin{bmatrix} 2 & 1 & 0 & 1 & 0 & 0 \\ 0 & 4 & 1 & -1 & 1 & 0 \\ 0 & -1 & 12 & -1 & 0 & 2 \end{bmatrix} \rightarrow \begin{bmatrix} 8 & 4 & 0 & 4 & 0 & 0 \\ 0 & 4 & 1 & -1 & 1 & 0 \\ 0 & -4 & 48 & -4 & 0 & 8 \end{bmatrix} \rightarrow$$

$$\begin{bmatrix} 8 & 0 & -1 & 5 & -1 & 0 \\ 0 & 4 & 1 & -1 & 1 & 0 \\ 0 & 0 & 48 & 0 & 0 & 8 \end{bmatrix} \rightarrow \begin{bmatrix} 8 & 0 & -1 & 5 & -1 & 0 \\ 0 & 4 & 1 & -1 & 1 & 0 \\ 0 & 0 & 6 & 0 & 0 & 1 \end{bmatrix} \rightarrow$$

$$\begin{bmatrix} 48 & 0 & -6 & 30 & -6 & 0 \\ 0 & 24 & 6 & -6 & 6 & 0 \\ 0 & 0 & 6 & 0 & 0 & 1 \end{bmatrix} \rightarrow \begin{bmatrix} 48 & 0 & 0 & 30 & -6 & 1 \\ 0 & 24 & 0 & -6 & 6 & -1 \\ 0 & 0 & 6 & 0 & 0 & 1 \end{bmatrix} \rightarrow$$

$$\begin{bmatrix} 1 & 0 & 0 & \frac{30}{48} & -\frac{6}{48} & \frac{1}{48} \\ 0 & 1 & 0 & -\frac{12}{48} & \frac{12}{48} & -\frac{2}{48} \\ 0 & 0 & 1 & \frac{0}{48} & \frac{0}{48} & \frac{8}{48} \end{bmatrix}.$$

Thus, the inverse of **A** is given by

$$\mathbf{A}^{-1} = \left(\frac{1}{48}\right) \begin{bmatrix} 30 & -6 & 1 \\ -12 & 12 & -2 \\ 0 & 0 & 8 \end{bmatrix}.$$

There are square matrices that have specific patterns. In particular, if a square matrix **A** can be expressed as

$$\mathbf{A} = \mathbf{T} + \alpha\mathbf{u}\mathbf{w}', \text{ then, } \mathbf{A}^{-1} = \mathbf{T}^{-1} - \frac{\alpha\mathbf{T}^{-1}\mathbf{u}\mathbf{w}'(\mathbf{T}^{-1})'}{1 + \alpha\mathbf{u}'\mathbf{T}^{-1}\mathbf{w}}.$$

For instance, consider the following matrix that can be decomposed as indicated:

$$\mathbf{A} = \begin{bmatrix} 3 & 2 & 2 \\ 2 & 3 & 2 \\ 2 & 2 & 3 \end{bmatrix} = \begin{bmatrix} 1 & 0 & 0 \\ 0 & 1 & 0 \\ 0 & 0 & 1 \end{bmatrix} + 2\begin{bmatrix} 1 \\ 1 \\ 1 \end{bmatrix}\begin{bmatrix} 1 & 1 & 1 \end{bmatrix}.$$

Then, using the above formula,

$$\mathbf{A}^{-1} = \mathbf{I}^{-1} - \frac{2\mathbf{I}^{-1}\mathbf{1}\mathbf{1}'\mathbf{I}^{-1}}{1 + 2\mathbf{1}'\mathbf{1}} = \mathbf{I} - \frac{2}{7}\mathbf{1}\mathbf{1}' = \frac{1}{7}\begin{bmatrix} 5 & -2 & -2 \\ -2 & 5 & -2 \\ -2 & -2 & 5 \end{bmatrix}.$$

There are many other topics that are important and useful. However, we will not inundate ourselves with too much technical information at this stage. Rather, let us now look at more practical topics.

5.2 CANONICAL REDUCTION AND EIGENEQUATION

A square matrix $\mathbf{T}$ is said to be *orthogonal* if its inverse is the same as its transpose. Thus, if $\mathbf{T}$ is orthogonal, then $\mathbf{T}\mathbf{T}' = \mathbf{I}$ since $\mathbf{T}' = \mathbf{T}^{-1}$. It is interesting to note that the determinant of an orthogonal matrix is either 1 or -1, which has an implication for the transformation problem to be discussed in this section. Let us look at Figure 5.1, in which the point P_1 with coordinates x_1 and x_2 is indicated. If we rotate axes X_1 and X_2 by the angle of θ to X_1^* and X_2^*, respectively, the new coordinates of P_1, that is, x_1^* and x_2^*, can be calculated by

(5.1)
$$\begin{bmatrix} x_1^* \\ x_2^* \end{bmatrix} = \begin{bmatrix} \cos\theta & \sin\theta \\ -\sin\theta & \cos\theta \end{bmatrix}\begin{bmatrix} x_1 \\ x_2 \end{bmatrix}$$

or,

(5.2)
$$\mathbf{u}^* = \mathbf{T}\mathbf{u},$$

where **T** is the 2×2 matrix, $\mathbf{u}^*$ and $\mathbf{u}$ are vectors consisting of elements x_i^* and x_i, respectively. The matrix **T** is orthogonal, for

$$\mathbf{TT}' = \begin{bmatrix} \cos\theta & \sin\theta \\ -\sin\theta & \cos\theta \end{bmatrix} \begin{bmatrix} \cos\theta & -\sin\theta \\ \sin\theta & \cos\theta \end{bmatrix} = \begin{bmatrix} 1 & 0 \\ 0 & 1 \end{bmatrix}, \tag{5.3}$$

since $\cos^2\theta + \sin^2\theta = 1$. Notice also that the determinant of **T**, $|\mathbf{T}|$, is $\cos^2\theta + \sin^2\theta = 1$.

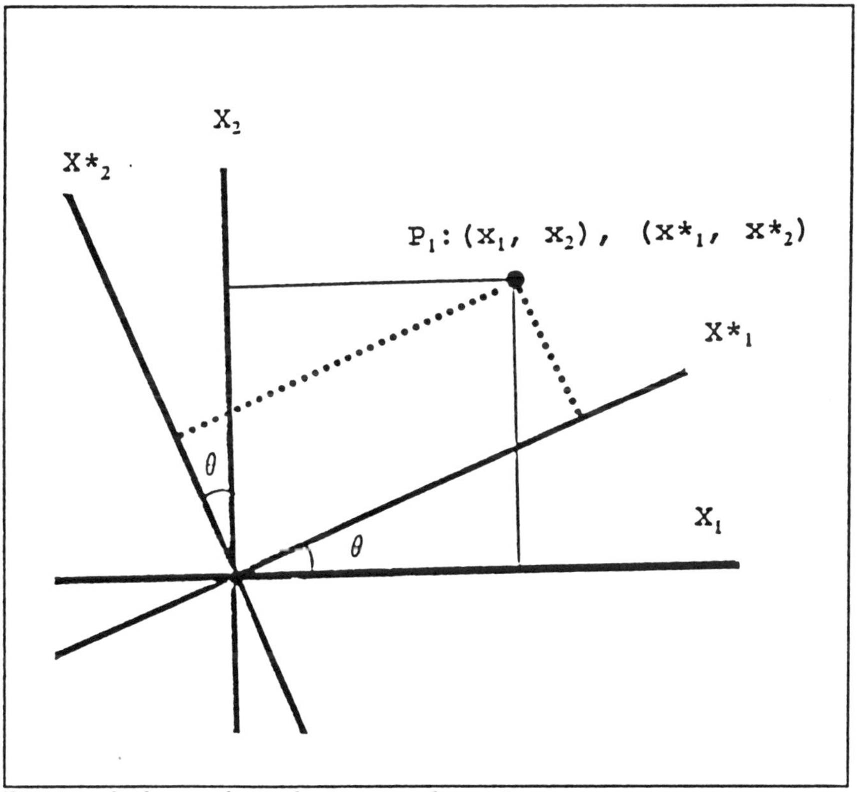

Fig. 5.1. Orthogonal coordinate transformation

This transformation of **u** to **u*** is called the *orthogonal coordinate transformation*. When the determinant of the transformation matrix **T** is 1, as is the case with the present example, we say that the orientation of axes is preserved (the two axes are rotated in the same direction by the same degree). If the determinant is -1, the orientation of axes is altered, meaning that one is rotated by θ and the other axis by θ plus 180 degrees.

Let **A** be an $n \times n$ symmetric matrix of real numbers and **x** be an $n \times 1$ vector of unknowns. Then, the scalar $\mathbf{x}'\mathbf{A}\mathbf{x}$ is called a *quadratic form* in **x**. For example, consider the following:

(5.4) $$ax_1^2 + 2bx_1x_2 + cx_2^2 = \begin{bmatrix} x_1 & x_2 \end{bmatrix} \begin{bmatrix} a & b \\ b & c \end{bmatrix} \begin{bmatrix} x_1 \\ x_2 \end{bmatrix} = \mathbf{x}'\mathbf{A}\mathbf{x}$$

Given a quadratic form, it is *always* possible through orthogonal coordinate transformation to express it in terms of only quadratic terms, namely,

(5.5) $$\lambda_1 x_1^{*2} + \lambda_2 x_2^{*2} = \begin{bmatrix} x_1^* & x_2^* \end{bmatrix} \begin{bmatrix} \lambda_1 & 0 \\ 0 & \lambda_2 \end{bmatrix} \begin{bmatrix} x_1^* \\ x_2^* \end{bmatrix} = \mathbf{x}^{*\prime}\Lambda\mathbf{x}^*.$$

where λ_1 and λ_2 are constants, and $\Lambda = \text{diag}(\lambda_j)$. It is known that this problem of transformation can be handled algebraically by solving the equation

(5.6) $$(\mathbf{A} - \lambda_j\mathbf{I})\mathbf{t}_j = \mathbf{0},$$

where $\mathbf{t}_j$ is the jth column of the desired orthogonal transformation matrix **T**. The above equation is called *eigenequation* of the quadratic form $\mathbf{x}'\mathbf{A}\mathbf{x}$; λ_j are called *eigenvalues* and $\mathbf{t}_j$ *eigenvectors*.[1]

The eigenequation is a set of *homogeneous equations* which are always satisfied when $\mathbf{t}_j = \mathbf{0}$, irrespective of **A**. Thus, this is called a *trivial solution*. To avoid this trivial solution, it is well known that the determinant of the coefficient matrix must vanish, that is,

(5.7) $$|\mathbf{A} - \lambda\mathbf{I}| = 0.$$

This is called the *characteristic equation*. In our example, the characteristic equation is

(5.8) $$|\mathbf{A} - \lambda_j\mathbf{I}| = \begin{vmatrix} (a-\lambda_j) & b \\ b & (c-\lambda_j) \end{vmatrix} = (a-\lambda_j)(c-\lambda_j) - b^2$$
$$= \lambda_j^2 - (a+c)\lambda_j + (ac-b^2) = 0.$$

Once eigenvalues λ_j are obtained, the corresponding eigenvectors $\mathbf{t}_j$ can be obtained by solving the eigenequation. The form without product terms,

---

[1] Eigenvalues are also called *latent roots*, *characteristic roots*, and *proper values*, and the corresponding eigenvectors *latent vectors*, *characteristic vectors*, and *proper vectors*.

$\mathbf{x}'\Lambda\mathbf{x}$, is called *canonical form*, and the axes related to the canonical form, X_1^* and X_2^*, are called *principal axes*.

The meaning of this *canonical reduction* becomes clear once we apply it to a numerical example. Consider the canonical reduction of $5x_1^2 + 8x_1x_2 + 5x_2^2 = 9$, that is,

(5.9)
$$\begin{bmatrix} x_1 & x_2 \end{bmatrix} \begin{bmatrix} 5 & 4 \\ 4 & 5 \end{bmatrix} \begin{bmatrix} x_1 \\ x_2 \end{bmatrix} = 9.$$

The characteristic equation is

(5.10)
$$\left| \begin{bmatrix} 5 & 4 \\ 4 & 5 \end{bmatrix} - \lambda \begin{bmatrix} 1 & 0 \\ 0 & 1 \end{bmatrix} \right| = \begin{vmatrix} (5-\lambda) & 4 \\ 4 & (5-\lambda) \end{vmatrix} = -(\lambda-9)(\lambda-1) = 0.$$

Thus, the two eigenvalues are 9 and 1. Hence, the canonical form is given by

$$9x_1^{*2} + x_2^{*2} = 9.$$

Examine the relation between the original form and its canonical form as shown in Figure 5.2. The principal axis X_2^* cuts through the points where the distance of the ellipse from the origin is the largest, and the principal axis X_1^* cuts through those where the distance is the smallest. These largest and smallest distances are respectively 3 and 1 that correspond to the squareroots of the two eigenvalues. Furthermore, the ellipse is symmetric with respect to the principal axes. The plane represented by the principal axes is called the *principal plane*. In previous chapters, principal component analysis (PCA) was mentioned. We can now define it in a precise manner: PCA is a technique to describe data in terms of principal axes or in the principal plane.

To find the corresponding eigenvectors, we must solve the following two equations, one for each eigenvalue:

$$\begin{bmatrix} (5-9) & 4 \\ 4 & (5-9) \end{bmatrix} \begin{bmatrix} t_{11} \\ t_{21} \end{bmatrix} = 0,$$

$$\begin{bmatrix} (5-1) & 4 \\ 4 & (5-1) \end{bmatrix} \begin{bmatrix} t_{12} \\ t_{22} \end{bmatrix} = 0.$$

We obtain $t_{11} = t_{21}$, and $t_{12} = -t_{22}$. Thus, we arbitrarily set $t_{11} = t_{21} = 1$, $t_{12} = 1$ and $t_{22} = -1$, and *normalize* them by using the formula

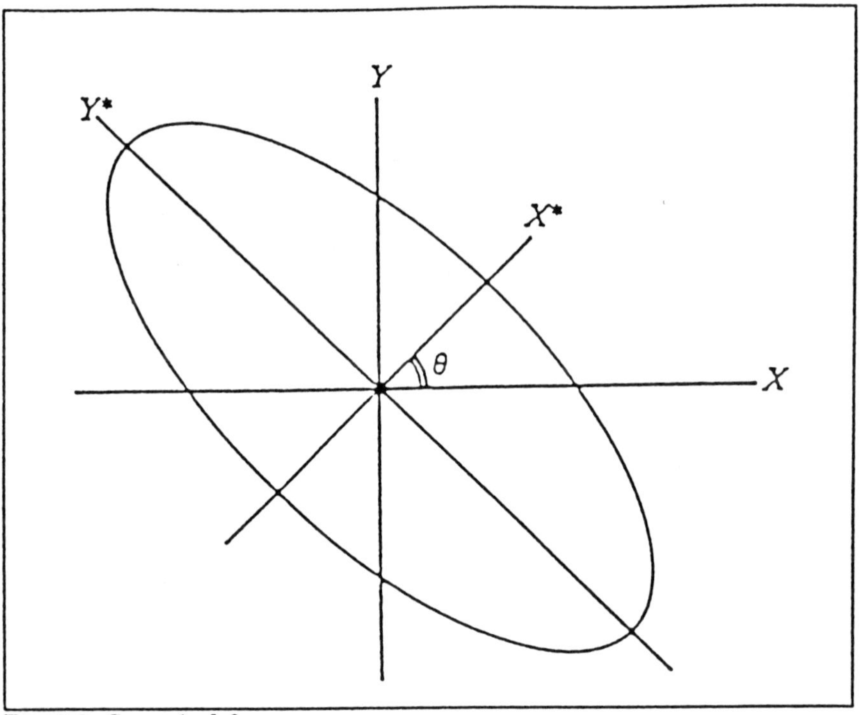

Fig. 5.2. Canonical form

$$t_{ij}^{*} = \frac{t_{ij}}{\sqrt{\sum_{i=1} t_{ij}^{2}}} . \tag{5.11}$$

Or, in the matrix notation, the normalized vector is given by $\mathbf{t}^*$, where $\mathbf{t}^* = \mathbf{t}/(\mathbf{t}'\mathbf{t})^{1/2}$. The matrix $\mathbf{T}$ with elements t_{ij}^* is orthogonal,

$$\mathbf{T} = \frac{1}{\sqrt{2}} \begin{bmatrix} 1 & 1 \\ 1 & -1 \end{bmatrix}. \tag{5.12}$$

The columns of $\mathbf{T}$ are eigenvectors, corresponding to the eigenvalues of 9 and 1. The orthogonal coordinate transformation for this canonical reduction is therefore given by

$$\begin{bmatrix} X_1^* \\ X_2^* \end{bmatrix} = \frac{1}{\sqrt{2}} \begin{bmatrix} 1 & 1 \\ 1 & -1 \end{bmatrix} \begin{bmatrix} X_1 \\ X_2 \end{bmatrix}. \tag{5.13}$$

Since cosθ and sinθ are equal to 1 over the square root of 2, the angle of this rotation θ is 45 degrees.

In data analysis, eigenvalues often play an important role as we will see in subsequent chapters. Let us introduce a useful concept that is related to eigenvalues. Suppose that the matrix **S** is symmetric. If all the eigenvalues are nonnegative, **S** is said to be *nonnegative definite* (n.n.d.). Furthermore, if all the eigenvalues are greater than zero, **S** is called *positive definite*, and if some of the eigenvalues of a nonnegative definite matrix is zero, it is called *positive semidefinite*. Alternatively, we can define a nonnegative definite matrix as a symmetric matrix **S** such that $\mathbf{x}'\mathbf{S}\mathbf{x}$ is greater than, or equal to, zero: If $\mathbf{x}'\mathbf{S}\,\mathbf{x} = 0$ only when $\mathbf{x} = \mathbf{0}$, then **S** is positive definite; and if $\mathbf{x}'\mathbf{S}\mathbf{x} = 0$ for the null vector $\mathbf{x}$ and some nonnull vectors $\mathbf{x}$, then **S** is positive semidefinite. We can further state that if **S** is positive definite, then its diagonal elements are all positive and the determinant is also positive, and that if **S** is positive semidefinite, all the diagonal elements are greater than, or equal to, zero, and the determinant is zero. In data analysis, including dual scaling, we often see positive semidefinite, or positive definite matrices: variance-covariance matrices, product-moment correlation matrices, and the matrix involved in the eigenequation of dual scaling are all such examples.

Let us look at one interesting application of the concept of positive semidefiniteness of a matrix to a problem of data analysis. Consider a 3×3 correlation matrix, and imagine that one of the elements is illegible. Can we set the bound on this missing value of the correlation, that is, can we find the maximal and the minimal values that this missing correlation can take? Let us indicate the missing value by x in the 3×3 correlation matrix **R**, where

$$\mathbf{R} = \begin{bmatrix} 1 & r_{12} & x \\ r_{21} & 1 & r_{23} \\ x & r_{32} & 1 \end{bmatrix}.$$

Note that the matrix is symmetric. Since **R** is positive semidefinite, it follows that its determinant is nonnegative. Thus, expanding the determinant of **R** and setting it to be greater than or equal to zero, we will be able to find the range for x, that is,

$$\begin{vmatrix} 1 & r_{12} & x \\ r_{21} & 1 & r_{23} \\ x & r_{32} & 1 \end{vmatrix} = (1)\begin{vmatrix} 1 & r_{23} \\ r_{32} & 1 \end{vmatrix} - (r_{12})\begin{vmatrix} r_{21} & r_{23} \\ x & 1 \end{vmatrix} + (x)\begin{vmatrix} r_{21} & 1 \\ x & r_{32} \end{vmatrix} \geq 0$$

After some algebraic manipulations, we arrive at the following expression:

$$r_{12}r_{23} - \sqrt{(r_{12}^2 - 1)(r_{23}^2 - 1)} \leq x \leq r_{12}r_{23} + \sqrt{(r_{12}^2 - 1)1(r_{23}^2 - 1)}$$

Thus, if $r_{12} = 0.6$ and $r_{23} = 0.3$, the missing correlation r_{13} can be anywhere between -0.58 and 0.94.

It is also useful to know that the eigenvalues of a symmetric matrix **A** are related to the trace and the determinant of **A** by the following relations:

$$tr(\mathbf{A}) = \sum_{i=1}^{n} \lambda_j , \quad |\mathbf{A}| = \prod_{i=1}^{n} \lambda_i$$

5.3 DERIVATION OF EIGENEQUATION

The eigenequation is very important in statistics because such analyses as factor analysis, principal component analysis (PCA), multiple discriminant analysis, and canonical analysis all involve solving eigenequations. It would be instructive, therefore, to look at an example where an eigenequation is derived in the course of statistical analysis. To do so, let us consider PCA.

Recall our discussion on the sum of weighted responses in chapter 4, and a brief introduction to PCA in Section 4.1. We also recognized that where the sum of squares of weights is 1, it has the effect of preserving the unit of measurement. We have observed above that the canonical reduction can be viewed in the context of orthogonal coordinate transformation, and that the sum of squares of each column of **T** is indeed equal to 1. Can you see therefore that PCA and canonical reduction are mathematically very similar? PCA seeks a set of such weights for variables that maximize the variance of the sum of the weighted responses, subject to the condition on the weights. This task is typically carried out by using *Lagrange's method of unknown multipliers*, which involves *partial differentiation* of a quadratic form. To make our discussion complete, let us digress a little and introduce *differentiation* of a function.

Let $y = f(x)$ be a continuous function in a given interval. Choose two points x_0 and x_1. The values of the function at these two points are respectively $y_0 = f(x_0)$ and $y_1 = f(x_1)$. Then the quotient

(5.14)
$$\frac{y_1 - y_0}{x_1 - x_0} = \frac{\Delta y}{\Delta x}$$

is known as the *difference quotient*, and indicates the slope of the straight line passing through the points (x_0,y_0) and (x_1,y_1). If the quotient has a limit as x_1 approaches x_0 (i.e., $x_1 \to x_0$), that limit is called the *derivative* of $f(x)$ at x_0, and is indicated by $f'(x_0)$. That is,

(5.15)
$$f'(x_0) = \lim_{x_1 \to x_0} \left(\frac{y_1 - y_0}{x_1 - x_0} \right) = \lim_{\Delta x \to 0} \left(\frac{\Delta y}{\Delta x} \right) = \lim_{\Delta x \to 0} \left(\frac{f(x_0 + \Delta x) - f(x_0)}{\Delta x} \right).$$

The derivative is the *slope* of $f(x)$ at the point (x_0,y_0). Therefore, if $f(x)$ is a quadratic function and if $f'(x) = 0$, then the point (x_0,y_0) is the maximum (highest point) or the minimum (lowest point), depending whether the function is concave or convex.

Let $u = f(x,y)$ be a continuous function of x and y containing a point (x_0,y_0). Suppose that we consider u as a function of only x in the neighborhood of (x_0,y_0), say $u = f(x_0+h, y_0)$. Then, if the limit

(5.16)
$$\lim_{h \to 0} \left(\frac{f(x_0 + h, y_0) - f(x_0, y_0)}{h} \right)$$

exists, it is called the *partial derivative* of $f(x,y)$ with respect to x at the point (x_0,y_0). The partial derivative is indicated by

$$\frac{\partial f}{\partial x}, \quad \frac{\partial f}{\partial y} \; .$$

The partial derivative in this case tells us how the function changes as x varies while y is kept constant or fixed. In the search for a maximum or a minimum of $f(x,y)$, we set the partial derivative of f with respect to x equal to zero and the partial derivative of f with respect to y zero. The solutions for these two equations, x^* and y^*, say, indeed provide the extremum of the function, which is given by $f(x^*, y^*)$.

There is a well-known basic rule for the derivative of a polynomial function of x with various degrees, say,

(5.17)
$$f(x) = ax^n + bx^{n-1} + \ldots + qx^3 + rx^2 + sx + t \; .$$

The derivative of the above function with respect to x is given by

$$f'(x) = nax^{n-1} + (n-1)bx^{n-2} + \ldots + 3qx^2 + 2rx + s. \tag{5.18}$$

Therefore, the derivative of $5x^3+2x^2+9x+20$ with respect to x can easily be obtained by the rule given in Equation 5.18 as $15x^2+4x+9$. Let us now apply this rule to the function of more than one variable and to the problem of partial differentiation. Consider the function of three variables given by

$$f(x_1,x_2,x_3) = 2x_1^2 + 3x_1x_2 + 5x_3^2 .$$

The partial derivatives of f with respect to x_1, x_2, and x_3 are

$$\frac{\partial f}{\partial x_1} = 4x_1 + 3x_2, \quad \frac{\partial f}{\partial x_2} = 3x_1, \quad \frac{\partial f}{\partial x_3} = 10x_3 .$$

In this case, regard f as a function of the vector $\mathbf{x}$ and introduce the expression

$$\frac{\partial f}{\partial \mathbf{x}} = \begin{bmatrix} \frac{\partial f}{\partial x_1} \\ \frac{\partial f}{\partial x_2} \\ \frac{\partial f}{\partial x_3} \end{bmatrix} = \begin{bmatrix} 4x_1+3x_2 \\ 3x_1 \\ 10x_3 \end{bmatrix}.$$

To generalize, let $\mathbf{x}$ be an $n \times 1$ vector of variables and f be a function of the elements of $\mathbf{x}$. Then we define the following:

$$\frac{\partial f}{\partial \mathbf{x}} = \begin{bmatrix} \frac{\partial f}{\partial x_1} \\ \frac{\partial f}{\partial x_2} \\ \cdot \\ \cdot \\ \cdot \\ \frac{\partial f}{\partial x_n} \end{bmatrix}, \quad \frac{\partial f}{\partial \mathbf{x}'} = \left[\frac{\partial f}{\partial x_1}, \frac{\partial f}{\partial x_2}, \ldots, \frac{\partial f}{\partial x_n}\right]. \tag{5.19}$$

Consider a general quadratic function

$$\mathbf{y} = \mathbf{x}'\mathbf{A}\mathbf{x} + 2\mathbf{b}'\mathbf{x} + c\ ,$$

where $\mathbf{x}$ is an $n\times1$ vector of variables, $\mathbf{A}$ an $n\times n$ *symmetric* matrix of constants, $\mathbf{b}$ an $n\times1$ vector of constants, and c a scalar. Then

$$\frac{\partial y}{\partial \mathbf{x}} = \frac{\partial}{\partial \mathbf{x}}(\mathbf{x}'\mathbf{A}\mathbf{x} + 2\mathbf{b}'\mathbf{x} + c) = \frac{\partial(\mathbf{x}'\mathbf{A}\mathbf{x})}{\partial \mathbf{x}} + \frac{\partial(2\mathbf{b}'\mathbf{x})}{\partial \mathbf{x}} + \frac{\partial c}{\partial \mathbf{x}} = 2\mathbf{A}\mathbf{x} + 2\mathbf{b}\ .$$

Similarly,

$$\frac{\partial y}{\partial \mathbf{x}'} = 2\mathbf{x}'\mathbf{A} + 2\mathbf{b}'\ .$$

With this much preparation, let us get back to the problem of PCA. Suppose that N students wrote n science tests, and that the $N\times n$ matrix of scores, $\mathbf{X}$, was obtained. We now wish to generate a linear combination of n test scores in such a way that the variance of N composite scores be a maximum, under the condition that the sum of squares of the weights is 1. Let $\mathbf{y}$ be an $N\times1$ vector of such composite scores and $\mathbf{w}$ an $n\times1$ vector of weights for the n tests.

$$\mathbf{y} = \mathbf{X}\mathbf{w}, \quad \mathbf{w}'\mathbf{w} = 1\ .$$

The variance of $\mathbf{y}$ is defined as

$$\sum_{i=1}^{N}\frac{(y_i - m_y)^2}{N-1} = s_y^2\ , \tag{5.20}$$

where m_y is the mean (average) of y_i. Thus the variance is the sum of squares of *deviation scores* from the mean divided by (N-1). Although we should go on, let us spend a little more time on some basics. The vector $\mathbf{y}$ can be decomposed into the vector of deviation scores and the vector of the mean

$$\mathbf{y} = (\mathbf{y} - m_y\mathbf{1}) + m_y\mathbf{1}\ .$$

But,

$$m_y = \sum_{i=1}^{N}\frac{y_i}{N} = \frac{\mathbf{1}'\mathbf{y}}{\mathbf{1}'\mathbf{1}} = (\mathbf{1}'\mathbf{1})^{-1}\mathbf{1}'\mathbf{y}\ ,$$

so that

$$\mathbf{y} = (\mathbf{y} - \mathbf{1}m_y) + \mathbf{1}m_y = (\mathbf{y} - \mathbf{1}(\mathbf{1}'\mathbf{1})^{-1}\mathbf{1}'\mathbf{y}) + \mathbf{1}(\mathbf{1}'\mathbf{1})^{-1}\mathbf{1}'\mathbf{y} .$$

Let us use the following notation:

$$\mathbf{P_m} = \mathbf{1}(\mathbf{1}'\mathbf{1})^{-1}\mathbf{1}' , \quad \mathbf{P_d} = (\mathbf{I} - \mathbf{P_m}) .$$

Then,

(5.21) $$\mathbf{y} = (\mathbf{I} - \mathbf{P_m})\mathbf{y} + \mathbf{P_m}\mathbf{y} = \mathbf{P_d}\mathbf{y} + \mathbf{P_m}\mathbf{y} .$$

$\mathbf{P}_\mathrm{m}$ and $\mathbf{P}_\mathrm{d}$ are called *projection operators* for the mean subspace and the deviation subspace respectively. More formally defined, an *orthogonal projection operator* is a symmetric (i.e., $\mathbf{A}'$=$\mathbf{A}$) *idempotent* (i.e., $\mathbf{AA} = \mathbf{A}$, that is, $\mathbf{A}^2 = \mathbf{A}$) matrix. If $\mathbf{P}$ is a projection operator, so is $(\mathbf{I}$-$\mathbf{P})$. Furthermore, notice in the decomposition that $\mathbf{P}_\mathrm{d}\mathbf{P}_\mathrm{m} = (\mathbf{I} - \mathbf{P}_\mathrm{m})\mathbf{P}_\mathrm{m} = \mathbf{P}_\mathrm{m} - \mathbf{P}_\mathrm{m}^2 = \mathbf{O}$. Because of this property, these two projection operators are said to be *disjoint* and this expression of $\mathbf{y}$ is referrred to as an orthogonal decomposition of $\mathbf{y}$.

Let us get back to the variance of $\mathbf{y}$. Since $\mathbf{y} = \mathbf{P}_\mathrm{d}\mathbf{y} + \mathbf{P}_\mathrm{m}\mathbf{y}$, the variance can be expressed as

(5.22) $$s_y^2 = \frac{\mathbf{y}'\mathbf{P}_\mathrm{d}'\mathbf{P}_\mathrm{d}\mathbf{y}}{N-1} = \frac{\mathbf{y}'\mathbf{P}_\mathrm{d}\mathbf{y}}{N-1} = \frac{\mathbf{w}'\mathbf{X}'\mathbf{P}_\mathrm{d}\mathbf{X}\mathbf{w}}{N-1} = \mathbf{w}'\mathbf{V}\mathbf{w} ,$$

where $\mathbf{V}$ is $\mathbf{X}'\mathbf{P}_\mathrm{d}\mathbf{X}/(N - 1)$, and is called the *variance-covariance matrix* of n variables.

Our task is to determine $\mathbf{w}$ so as to maximize $\mathbf{w}'\mathbf{V}\mathbf{w}$, subject to the constraint that $\mathbf{w}'\mathbf{w} = 1$. This problem is typically handled by *Lagrange's method of unknown multipliers*. It defines *Lagrange's function*, which is in the present case given by

(5.23) $$Q(\mathbf{w},\lambda) = \mathbf{w}'\mathbf{V}\mathbf{w} - \lambda(\mathbf{w}'\mathbf{w} - 1) ,$$

where λ is called *Lagrange's multiplier*.[2] The function Q has two unknowns, $\mathbf{w}$ and λ. To find $\mathbf{w}$ that maximizes s_y^2, subject to $\mathbf{w}'\mathbf{w}$ being 1, we differentiate Q with respect to the two unknowns, set the two partial derivatives equal to zeros, and solve the equations for the two unknowns.

---

[2] Notice that as long as the constraint (i.e., $\mathbf{w}'\mathbf{w} = 1$) is satisfied, the function Q is the variance that is to be maximized. This is true, irrespective of the sign in front of Lagrange's multiplier. To be consistent with mathematical conditions for maxima and minima of a function, however, the minus sign in front of the multiplier is preferred.

This is Lagrange's method of unknown multiplier(s). Following the method, we obtain

(5.24) $$\frac{\partial Q}{\partial \mathbf{w}} = 2\mathbf{Vw} - 2\lambda\mathbf{w} = \mathbf{0}, \quad \frac{\partial Q}{\partial \lambda} = (\mathbf{w}'\mathbf{w}-1) = 0.$$

The first equation of (5.24) can be written as $\mathbf{Vw} = \lambda\mathbf{w}$, and the second equation is nothing but the constraint that $\mathbf{w}'\mathbf{w} = 1$. Thus, if we premultiply both sides by $\mathbf{w}'$, we obtain

(5.25) $$\lambda = \frac{\mathbf{w}'\mathbf{Vw}}{\mathbf{w}'\mathbf{w}} = \mathbf{w}'\mathbf{Vw} = s_y^2 .$$

In other words, the Lagrange multiplier is nothing but the variance that we wish to maximize. The first equation of (5.24) can further be rewritten as

(5.26) $$(\mathbf{V}-\lambda\mathbf{I})\mathbf{w} = \mathbf{0}.$$

This is an eigenequation! The characteristic equation must be solved first. Since the matrix $\mathbf{V}$ is $n \times n$, there are n eigenvalues, of which the largest value, indicated by λ_1 is what we want. Once λ_1 is obtained, the eigenvector $\mathbf{w}_1$, associated with it and satisfying the constraint, is calculated by substituting λ_1 in the eigenequation. The first *principal component score vector* $\mathbf{y}_1$ can be obtained by $\mathbf{y}_1 = \mathbf{Xw}_1$. There are a number of numerical methods to solve eigenequations. If you want to extract only the first few largest eigenvalues and the corresponding eigenvectors, the *power method* may be preferred to other methods that extract all the solutions—the method of reciprocal averages is an example of a power method. See the next section for the method.

Suppose that n eigenvalues λ_i and corresponding *normalized* eigenvectors $\mathbf{w}_i$ (i.e., $\mathbf{w}_i'\mathbf{w}_i = 1$, $i = 1,2,\ldots,n$ are obtained. Let $\mathbf{W} = [\mathbf{w}_1, \mathbf{w}_2, \ldots, \mathbf{w}_n]$ and $\Lambda = \text{diag}(\lambda_1, \lambda_2, \ldots, \lambda_n)$. Then $\mathbf{VW} = \mathbf{W}\Lambda$. Since $\mathbf{W}'\mathbf{W} = \mathbf{I}$, we obtain

(5.27) $$\mathbf{V} = \mathbf{W}\Lambda\mathbf{W}', \quad \text{or} \quad \mathbf{W}'\mathbf{VW} = \Lambda .$$

The above is called a *similarity transformation.*

A symmetric and positive (semi)definite matrix is called a Gramian matrix. If the $n \times n$ matrix $\mathbf{V}$ is Gramian, it can always be expressed as a product of a matrix and its transpose, that is, $\mathbf{V} = \mathbf{A}\,\mathbf{A}'$, of which the Young-Householder theorem (see Torgerson, 1958) states that

1. The total number of nonzero eigenvalues, t, indicates the number of dimensions required to accommodate the configuration of n variables, and

2. The matrix **A** is $n \times t$, and is the matrix of coordinates of n variables in t-dimensional Euclidean space.

The decomposition of **V** into **A A′** can easily be obtained from similarity transformation as follows:

$$\mathbf{V} = \mathbf{W}\mathbf{V}\mathbf{W}' = \mathbf{W}\Lambda^{1/2}\Lambda^{1/2}\mathbf{W}' = \mathbf{A}\mathbf{A}'$$

In other words, **A** is the matrix **W** rescaled by the square roots of the eigenvalues.

Another form of matrix decomposition is the expression referred to as *singular value decomposition* (Stewart, 1973) or the *Eckart-Young decomposition* (Eckart & Young, 1936). This concept can be viewed as a generalized version of similarity transformation, for it works with a square or a rectangular matrix. The singular value demposition states that given an $n \times m$ rectangular matrix **F**, one can always decompose it into the following expression

$$\mathbf{F} = \mathbf{Y}\Lambda\mathbf{X}', \tag{5.28}$$

where $\mathbf{Y}'\mathbf{Y} = \mathbf{I}_n$, $\mathbf{X}'\mathbf{X} = \mathbf{I}_m$, and $\Lambda = \text{diag}(\lambda_1, \Lambda_2, \ldots, \lambda_t)$, with t being the rank of **F** and $\mathbf{I}_k$ being the $k \times k$ identity matrix. Furthermore, it should be noted that **Y** and **X** are respectively an $n \times t$ matrix of eigenvectors of $\mathbf{F}\mathbf{F}'$ and an $m \times t$ matrix of eigenvectors of $\mathbf{F}'\mathbf{F}$, and that Λ is the $t \times t$ diagonal matrix with the diagonal elements being equal to the square roots of eigenvalues of $\mathbf{F}\mathbf{F}'$.[3] Dual scaling is an application of singular value decomposition to various types of categorical data. Since we are typically interested only in a few major solutions, it is useful to know a practical way to extract the solution associated with the largest eigenvalue. Let us discuss such a method.

5.4 FINDING THE MAXIMAL EIGENVALUE

When we are interested only in the first few largest eigenvalues and associated eigenvectors, the power method can be very practical because it extracts solutions one by one, starting with the largest eigenvalue, then the second largest eigenvalue, and so on in the order of the magnitude until we decide to stop the extraction. The method of reciprocal averages as dis-

[3] Eigenvalues of $\mathbf{F}'\mathbf{F}$ are exactly the same as those of $\mathbf{F}\mathbf{F}'$.

cussed in chapter 4 is an example of a power method. In this section, we will look at the rationale of the power method and its procedure as applied to a symmetric matrix.[4]

Let us first introduce a few more concepts. A set of n ($m\times1$) vectors ($\mathbf{a}_1$, $\mathbf{a}_2$,. . . , $\mathbf{a}_n$) is said to be *linearly dependent* if and only if there exists a set of numbers (c_1,c_2,. . . , c_n), not all of which are zero, such that

$$c_1\mathbf{a}_1 + c_2\mathbf{a}_2 + \ldots + c_n\mathbf{a}_n = \mathbf{0}. \tag{5.29}$$

For example, vectors

$$\begin{bmatrix}2\\3\\1\end{bmatrix}, \begin{bmatrix}1\\4\\2\end{bmatrix}, \begin{bmatrix}3\\2\\0\end{bmatrix}$$

are linearly dependent, for the set of numbers (2, -1, -1) yields

$$2\begin{bmatrix}2\\3\\1\end{bmatrix} - \begin{bmatrix}1\\4\\2\end{bmatrix} - \begin{bmatrix}3\\2\\0\end{bmatrix} = \begin{bmatrix}0\\0\\0\end{bmatrix}.$$

If a set of vectors is not linearly dependent, it is said to be *linearly independent*. In other words, a set of vectors is linearly independent if and only if Equation 5.29 implies that $c_1 = c_2 = \ldots = c_n = 0$. For example, vectors

$$\begin{bmatrix}1\\0\\0\end{bmatrix}, \begin{bmatrix}0\\1\\0\end{bmatrix}, \begin{bmatrix}0\\0\\1\end{bmatrix}$$

are linearly independent, for the equation that

$$c_1\begin{bmatrix}1\\0\\0\end{bmatrix} + c_2\begin{bmatrix}0\\1\\0\end{bmatrix} + c_3\begin{bmatrix}0\\0\\1\end{bmatrix} = \begin{bmatrix}c_1\\c_2\\c_3\end{bmatrix} = \begin{bmatrix}0\\0\\0\end{bmatrix}$$

indicates that $c_1 = c_2 = c_3 = 0$. In terms of linear independence of vectors, the rank of the matrix $\mathbf{A}$ can be defined as the maximum number of linearly independent vectors that one can construct from the set ($\mathbf{a}_1$,$\mathbf{a}_2$,. . .

[4] The method of reciprocal averages handles a rectangular (nonsymmetric) matrix, as well as a symmetric matrix.

$, \mathbf{a}_n$). From the given example of linearly independent vectors, it is possible to generalize the result and to state that any $n \times 1$ vector can be expressed as a linear combination of n linearly independent $n \times 1$ vectors.

In matrix algebra, there is a theorem that the eigenvalue of the pth power of a matrix is the pth power of the eigenvalue of the matrix, that is,

$$\mathbf{V}^p\mathbf{u} = \lambda^p\mathbf{u} \,. \tag{5.30}$$

Suppose $\mathbf{V}$ is $n \times n$ and all eigenvalues are distinct and different from zero. Then, because the n eigenvectors are linearly independent, any arbitrary $n \times 1$ vector $\mathbf{b}_0$ can be expressed as a linear combination of eigenvectors $\mathbf{u}_i$,

$$\mathbf{b}_0 = \sum_{i=1}^{n} c_i \mathbf{u}_i \,, \tag{5.31}$$

where c_i are constants. Since $\mathbf{Vu} = \lambda\mathbf{u}$, we obtain the relation

$$\mathbf{V}\mathbf{b}_0 = \sum_{i=1}^{n} \lambda_i c_i \mathbf{u}_i \,. \tag{5.32}$$

Let us consider the sequence of vectors, $\mathbf{b}_0$, $\mathbf{V}\mathbf{b}_0 = \mathbf{b}_1$, $\mathbf{V}\mathbf{b}_1 = \mathbf{V}^2\mathbf{b}_0 = \mathbf{b}_2$, $\mathbf{V}\mathbf{b}_2 = \mathbf{V}^3\mathbf{b}_0 = \mathbf{b}_3, \ldots, \mathbf{V}\mathbf{b}_{p-1} = \mathbf{V}^p\mathbf{b}_0 = \mathbf{b}_p$. Thus, from (5.31) and (5.32), we obtain the following expression:

$$\mathbf{b}_p = \mathbf{V}^p\mathbf{b}_0 = \sum_{i=1}^{n} \lambda_1^p c_i \mathbf{u}_i \,. \tag{5.33}$$

Let us assume that the eigenvalues are ordered as $\lambda_1 > \lambda_2 > \ldots > \lambda_n$ and rewrite (5.33) as follows:

$$\mathbf{b}_p = \lambda_1^p \left[c_1\mathbf{u}_1 + \sum_{j=2}^{n} \left(\frac{\lambda_j}{\lambda_1} \right)^p c_j \mathbf{u}_j \right] . \tag{5.34}$$

Therefore, as the value of p increases, the ratios of the two eigenvalues, raised to the power p, become smaller and smaller, and in the limit we obtain the following:

$$\lim_{p \to \infty} \mathbf{b}_p = \lambda_1^p c_1 \mathbf{u}_1 \,. \tag{5.35}$$

Thus, if p is sufficiently large, the largest eigenvalue can be approximated by

(5.36)
$$\lambda_1 = \frac{\mathbf{b}_p'\mathbf{b}_p}{\mathbf{b}_p'\mathbf{b}_{p-1}}.$$

Once λ_1 is obtained, the corresponding eigenvector can be obtained by solving the eigenequation. However, there exists a numerical method that yields the maximal eigenvalue and the corresponding eigenvector at the same time. This method proceeds as follows:

Step 1. Multiply the symmetric matrix $\mathbf{V}$ by an arbitrary vector (but not the vector $\mathbf{0}$) $\mathbf{b}_0$, and indicate the product by $\mathbf{b}_1$, that is, $\mathbf{Vb}_0 = \mathbf{b}_1$.

Step 2. Divide the elements of $\mathbf{b}_1$ by the greatest absolute value of the elements in $\mathbf{b}_1$, say k_1, and indicate the resulting vector by $\mathbf{a}_1$, that is, $\mathbf{b}_1/k_1 = \mathbf{a}_1$.

Step 3. Compute $\mathbf{Va}_1 = \mathbf{b}_2$, and standardize $\mathbf{b}_2$ in the same way as in **Step 1**, that is, $\mathbf{b}_2/k_2 = \mathbf{a}_2$, where k_2 is the greatest absolute value of the elements in $\mathbf{b}_2$.

Step 4. Repeat the calculation of $\mathbf{Va}_{j-1} = \mathbf{b}_j$, $\mathbf{b}_j/k_j = \mathbf{a}_j$, until $\mathbf{a}_{j-1}$ becomes identical or almost identical with $\mathbf{a}_j$.

Step 5. When the difference between $\mathbf{a}_{j-1}$ and $\mathbf{a}_j$ becomes negligible, $\mathbf{a}_j$ is an eigenvector, associated with the largest eigenvalue of $\mathbf{V}$, which is given by k_j .

Why by k_j? Because that $\mathbf{b}_j/k_j = \mathbf{a}_j$ means $\mathbf{b}_j = k_j\mathbf{a}_j$. Thus, if $\mathbf{a}_{j-1} = \mathbf{a}_j$, $\mathbf{Va}_{j-1} = \mathbf{Va}_j = \mathbf{b}_j = k_j\mathbf{a}_j$. In other words, $\mathbf{Va}_j = k_j\mathbf{a}_j$, indicating that k_j and $\mathbf{a}_j$ are respectively an eigenvalue and a corresponding eigenvector. Is this clear now?

At this stage, any vector proportional to $\mathbf{a}_j$ can be adopted as a solution vector. However, in most problems, the unit of weights is specified and known. For instance, suppose that the sum of squares of the weights is t. Then, one can transform $\mathbf{a}_j$ to $\mathbf{u}$ by the formula

(5.37) $$\mathbf{u} = \sqrt{\frac{t}{\mathbf{a}_j'\mathbf{a}_j}}\,\mathbf{a}_j .$$

Notice that $\mathbf{u}'\mathbf{u} = t$. Therefore, the largest eigenvalue and the corresponding eigenvector such that the sum of squares of the elements is equal to t are given by k_j and $\mathbf{u}$, respectively.

This method is very similar to the method of reciprocal averages (MRA), which can now be presented in matrix notation:

1. $\mathbf{x}$ = an arbitrary initial trial vector for the columns of the data matrix $\mathbf{F}$.
2. $\mathbf{y} = \mathbf{F}\,\mathbf{x}$.
3. m_y = mean = $\mathbf{f}_r'\mathbf{y}/f_t$.
4. $\mathbf{y}_1 = \mathbf{y} - m\mathbf{1}$.
5. $\mathbf{y}_2 = \mathbf{y}_1/k$, where k is the largest absolute value of the elements of $\mathbf{y}_1$.
6. $\mathbf{x} = \mathbf{F}'\mathbf{y}$.
7. m_x = mean = $\mathbf{f}_c'\mathbf{x}/f_t$.
8. $\mathbf{x}_1 = \mathbf{x} - m_x\mathbf{1}$.
9. $\mathbf{x}_2 = \mathbf{x}_1/k$, where k is the largest absolute value of elements of $\mathbf{x}_1$.

*** Reciprocate the above averaging process until the two vectors $\mathbf{y}_2$ and $\mathbf{x}_2$ stabilize.

10. Optimal vectors are given by $\mathbf{y} = c_r\mathbf{y}_2$, and $\mathbf{x} = c_c\mathbf{x}_2$, where

$$c_r = \sqrt{\frac{f_r}{\mathbf{y}_2'\mathbf{D}_r\mathbf{y}_2}}, \quad c_c = \sqrt{\frac{f_t}{\mathbf{x}_2'\mathbf{D}_c\mathbf{x}_2}}, \quad \mathbf{D}_r = \mathrm{diag}(f_{i.}) \,, \; \mathbf{D}_c = \mathrm{diag}(f_{.j})$$

Express the final vectors $\mathbf{y}$ and $\mathbf{x}$ by $\mathbf{y}_1$ and $\mathbf{x}_1$, respectively, to indicate that they are optimal weight vectors of solution 1. The squared correlation ratio and the product-moment correlation can be calculated by

$$\eta_1^2 = \frac{\mathbf{x}_1'\mathbf{F}'\mathbf{D}_r^{-1}\mathbf{F}\mathbf{x}_1}{\mathbf{x}_1'\mathbf{D}_c\mathbf{x}_1} = \frac{\mathbf{y}_1'\mathbf{F}\mathbf{D}_c^{-1}\mathbf{F}'\mathbf{y}_1}{\mathbf{y}_1'\mathbf{D}_r\mathbf{y}_1},$$

$$\rho_1 = \frac{\mathbf{y}_1'\mathbf{F}\mathbf{x}_1}{f_t}$$

The solution set is given by $[\rho_1 , \mathbf{y}_1, \mathbf{x}_1]$. The residual matrix is indicated by $\mathbf{F}_1$, where $\mathbf{F}_1$ is

$$\mathbf{F}_1 = \mathbf{F} - \frac{\mathbf{D}_r(\rho_1 \mathbf{y}_1 \mathbf{x}_1')\mathbf{D}_c}{f_t}.$$

Subject $\mathbf{F}_1$ to the same iterative process, and obtain solution 2, that is, [ρ_2, $\mathbf{y}_2$, $\mathbf{x}_2$], and the residual matrix $\mathbf{F}_2$, that is,

$$\mathbf{F}_2 = \mathbf{F} - \frac{\mathbf{D}_r(\sum_{k=1}^{2} \rho_k \mathbf{y}_k \mathbf{x}_k')\mathbf{D}_c}{f_t}.$$

This process continues to extract more solutions.

Many of the topics discussed in this chapter will be used in chapter 6 to derive a procedure of dual scaling. For this reason, the current chapter will prove useful in understanding the details of the mathematical operations involved in dual scaling.

CHAPTER 6
Mathematics of Dual Scaling

On the basis of the preparations so far, we are now ready to look at a formal derivation of dual scaling. As you may have guessed from the many names for this quantification method and from the earlier discussion of the many ideas behind dual scaling, there are a number of ways to formulate it. This chapter contains a detailed description of one of them. Some readers may find this chapter too technical. If this is the case, feel free to skip this chapter, for the basic idea of the method has already been provided in chapters 2 and 4.

For the benefit of those who want to know the details of the method, however, chapters 5 and 6 are included in this part. Many formulas are presented in this chapter, occasionally without sufficient background information. However, once the basic idea is understood, technical matters can be mastered through self-directed endeavor, and this chapter should serve as such a guide by offering a bird's-eye view of a mathematical treatment of dual scaling.

6.1 OPTIMAL SOLUTION

In chapter 4, we looked at different approaches to dual scaling: the *approach by principal component analysis*, in which the object is to determine weights for categories so as to maximize the variance of the composite scores; the *analysis-of-variance approach*, in which category weights are determined so as to maximize the between-row (column) sum of squares relative to the total sum of squares, that is, the squared correlation ratio; the *bivariate correlation approach*, in which the product-moment correlation between responses weighted by row weights and those by column weights is maximized; and the *simultaneous linear regression approach*, in which row and column weights are determined so as to make the regression of rows on columns and the regression of columns on rows simultaneously linear. All of these approaches and others[1] lead to the identical quantification. In chapter 6, we look at only one of them, namely, the *analysis of variance approach*.

In chapter 4, we looked at the decomposition of the total sum of squares into the between and the within-sums of squares, that is, $SS_t = SS_b + SS_w$. This decomposition is a basis for the one-way analysis of variance, and when data are continuous, the decomposition is uniquely determined. In contrast, when data are categorical, relative magnitudes of those sums of squares vary, depending on weights given to the categories. In other words, one can manipulate weights to change the relative magnitude of each sum of squares. Dual scaling chooses the weights in a particular way: the weights are determined in such a way that the ratio SS_b/SS_t, that is, the squared correlation ratio η^2, be a maximum. In chapter 4, data in Table 4.3 were expressed in terms of column (option) weights (Table 4.6), and then expressions of SS_t and SS_b in terms of such weights were obtained. Our task is to determine weights x_j so as to maximize η^2.

Let us follow the entire process of dual scaling, using the data in Table 4.10, that is, the data used to illustrate the method of reciprocal averages. Let us assign unknown weights x_1, x_2, and x_3 to three categories, good, average, and poor, respectively. Since the origin and the unit of quantified data are arbitrary, it is customary to set the sum of weighted responses and the sum of squares of weighted responses equal to 0 and f_t (the total number of responses), respectively:

(6.1) $$\sum_{i=1}^{n}\sum_{j=1}^{m} f_{ij}x_j = 0,$$

(6.2) $$\sum_{i=1}^{n}\sum_{j=1}^{m} f_{ij}x_j^2 = f_t.$$

The choice of the origin by (6.1) makes the average of the weighted responses zero, eliminating the correction term, C (see chapter 4), from the expressions of SS_t and SS_b. These sums of squares for the data in Table 4.10 are thus simplified and are given by:

(6.3) $$SS_t = 10\times1^2 + 11\times2^2 + 8\times3^2,$$

(6.4) $$SS_b = \frac{(x_1+3x_2+6x_3)^2}{10} + \frac{(3x_1+5x_{21}+2x_3)^2}{10} + \frac{(6x_1+3x_2)^2}{10}.$$

[1] See Gifi (1990), Greenacre (1984), and Nishisato (1980a).

Let us introduce the following matrix notation:

$\mathbf{F} = (f_{ij})$ = the data matrix.
$\mathbf{f}_r$ = the vector of row totals of $\mathbf{F}$.
$\mathbf{f}_c$ = the vector of column totals of $\mathbf{F}$.
$\mathbf{D}_r$ = the diagonal matrix with row totals in the main diagonal.
$\mathbf{D}_c$ = the diagonal matrix with column totals in the main diagonal.
$\mathbf{y}$ = a vector of weights for the rows of $\mathbf{F}$.
$\mathbf{x}$ = a vector of weights for the columns of $\mathbf{F}$.
f_t = the sum of the elements of $\mathbf{F}$, that is, $\Sigma\Sigma f_{ij}$.

In the current example, we have

$$\mathbf{F} = \begin{bmatrix} 1 & 3 & 6 \\ 3 & 5 & 2 \\ 6 & 3 & 0 \end{bmatrix}, \quad f_r = \begin{bmatrix} 10 \\ 10 \\ 9 \end{bmatrix}, \quad \mathbf{f}_c = \begin{bmatrix} 10 \\ 11 \\ 8 \end{bmatrix},$$

$$\mathbf{D}_r = \begin{bmatrix} 10 & 0 & 0 \\ 0 & 10 & 9 \\ 0 & 0 & 9 \end{bmatrix}, \quad \mathbf{D}_c = \begin{bmatrix} 10 & 0 & 0 \\ 0 & 11 & 0 \\ 0 & 0 & 8 \end{bmatrix}, \quad f_t = 29 ,$$

$$\mathbf{y} = \begin{bmatrix} y_1 \\ y_2 \\ y_3 \end{bmatrix}, \quad \mathbf{x} = \begin{bmatrix} x_1 \\ x_2 \\ x_3 \end{bmatrix}.$$

In terms of matrix notation, those Equations 6.1, 6.2, 6.3, and 6.4 can be expressed respectively as:

(6.5) $$\mathbf{f}_c'\mathbf{x} = 0 ;$$

(6.6) $$\mathbf{x}'\mathbf{D}_c\mathbf{x} = f_t ;$$

(6.7) $$SS_t = \mathbf{x}'\mathbf{D}_c\mathbf{x} ;$$

(6.8) $$SS_b = \mathbf{x}'\mathbf{F}'\mathbf{D}_r^{-1}\mathbf{F}\mathbf{x} .$$

Notice how simple these expressions are, and that they can be used for any data matrix. Because of this simplicity, the rest of this chapter will use mainly the matrix notation. If you find this too difficult, review chapter 5, or try a small numerical example to translate the scalar notation to the matrix notation. For instance, you can verify Equations 6.7 and 6.8 in the following way:

$$\mathbf{x}'\mathbf{D}_c\mathbf{x} = \begin{bmatrix} x_1 & x_2 & x_3 \end{bmatrix} \begin{bmatrix} 10 & 0 & 0 \\ 0 & 11 & 0 \\ 0 & 0 & 8 \end{bmatrix} \begin{bmatrix} x_1 \\ x_2 \\ x_3 \end{bmatrix}$$

$$= \begin{bmatrix} 10x_1 & 11x_2 & 8x_3 \end{bmatrix} \begin{bmatrix} x_1 \\ x_2 \\ x_3 \end{bmatrix} = 10x_1^2 + 11x_2^2 + 8x_3^2 = (6.3) \ .$$

$$\mathbf{x}'\mathbf{F}'\mathbf{D}_r^{-1}\mathbf{F}\mathbf{x} = \begin{bmatrix} x_1 & x_2 & x_3 \end{bmatrix} \begin{bmatrix} 1 & 3 & 6 \\ 3 & 5 & 3 \\ 6 & 2 & 0 \end{bmatrix} \begin{bmatrix} \frac{1}{10} & 0 & 0 \\ 0 & \frac{1}{10} & 0 \\ 0 & 0 & \frac{1}{9} \end{bmatrix} \begin{bmatrix} 1 & 3 & 6 \\ 3 & 5 & 2 \\ 6 & 3 & 0 \end{bmatrix} \begin{bmatrix} x_1 \\ x_2 \\ x_3 \end{bmatrix} .$$

Carry out the remainder of this numerical verification step by step. You will find this expression to be identical with Equation 6.4.

The object of dual scaling in this approach is to determine $\mathbf{x}$ so as to maximize the squared correlation ratio η^2, which is given by SS_b/SS_t. This problem can be handled in two ways: one is to maximize SS_b/SS_t in terms of $\mathbf{x}$, and the other is to maximize SS_b, subject to the condition that $SS_t = f_t$ as indicated by Equation 6.2. Both lead to the same outcome. A widely used procedure is the latter one, which is handled by the well-known Lagrange method of unknown multipliers. See its application to formulating principal component analysis (PCA) as discussed in chapter 5. You will note throughout the ensuing derivation of dual scaling that all the mathematical tools to be used here (e.g., quadratic forms, Lagrangian function, partial differentiation, eigenequation) are the same as those for PCA that we looked at in chapter 5. Now, in the Lagrange method, the objective function for optimization is called the *Lagrangian function*, which is in the present example given by:

$$(6.9) \qquad Q(\mathbf{x},\lambda) = SS_b - \lambda(SS_t - f_t) = \mathbf{x}'\mathbf{F}'\mathbf{D}_r^{-1}\mathbf{F}\mathbf{x} - \lambda(\mathbf{x}'\mathbf{D}_c\mathbf{x} - f_t) \ .$$

The function Q has two unknowns, $\mathbf{x}$ and λ, and we wish to determine $\mathbf{x}$ and λ so as to maximize Q. Following the standard procedure for maximizing a quadratic function (see chapter 5), we set the partial derivatives of Q with respect to $\mathbf{x}$ and λ equal to zeros:

(6.10) $$\frac{\partial Q}{\partial \mathbf{x}} = 2\mathbf{F}'\mathbf{D}_r^{-1}\mathbf{F}\mathbf{x} - 2\lambda\mathbf{D}_c\mathbf{x} = \mathbf{0}\ ,$$

(6.11) $$\frac{\partial Q}{\partial \lambda} = \mathbf{x}'\mathbf{D}_c\mathbf{x} - f_t = 0.$$

The first formula can be rewritten as $\mathbf{F}'\mathbf{D}_r^{-1}\mathbf{F}\mathbf{x} = \lambda\mathbf{D}_c\mathbf{x}$. If we premultiply both sides of this expression by $\mathbf{x}'$, we obtain $\mathbf{x}'\mathbf{F}'\mathbf{D}_r^{-1}\mathbf{F}\mathbf{x} = \lambda\mathbf{x}'\mathbf{D}_c\mathbf{x}$, that is,

(6.12) $$\lambda = \frac{\mathbf{x}'\mathbf{F}'\mathbf{D}_r^{-1}\mathbf{F}\mathbf{x}}{\mathbf{x}'\mathbf{D}_c\mathbf{x}} = \eta^2\ .$$

Thus, the unknown multiplier λ is nothing but the squared correlation ratio that we want to maximize. Therefore, Equation 6.10 can be rewritten as

(6.13) $$(\mathbf{F}'\mathbf{D}_r^{-1}\mathbf{F} - \eta^2\mathbf{D}_c)\mathbf{x} = \mathbf{0}\ .$$

Moreover, Formula 6.11 is nothing but the constraint 6.2 itself. Equation 6.13 is called the *generalized eigenequation*. Notice the slight difference between Equation 6.13 and the standard form of eigenequation such as Equation 5.26 for PCA. Any expression of the form $(\mathbf{A}\text{-}\lambda\mathbf{B})\mathbf{x}\text{=}\mathbf{0}$, in which $\mathbf{B}$ is different from the identity matrix, is referred to as a generalized eigenequation.

To arrive at the standard form, the simplest transformation is to premultiply Equation 6.13 by the inverse of $\mathbf{D}_c$ to yield

(6.14) $$(\mathbf{D}_c^{-1}\mathbf{F}'\mathbf{D}_r^{-1}\mathbf{F} - \eta^2\mathbf{I})\mathbf{x} = \mathbf{0}\ .$$

This expression, however, is not a preferred one when you want to extract more than one solution (dimension, component) because it involves solving an eigenequation of a nonsymmetric matrix that requires an extra amount of computations. In our example,

$$\mathbf{F}'\mathbf{D}_r^{-1}\mathbf{F} = \begin{bmatrix} 1 & 3 & 6 \\ 3 & 5 & 3 \\ 6 & 2 & 0 \end{bmatrix} \begin{bmatrix} \frac{1}{10} & 0 & 0 \\ 0 & \frac{1}{10} & 0 \\ 0 & 0 & \frac{1}{9} \end{bmatrix} \begin{bmatrix} 1 & 3 & 6 \\ 3 & 5 & 2 \\ 6 & 3 & 0 \end{bmatrix} = \begin{bmatrix} 5.0 & 3.8 & 1.2 \\ 3.8 & 4.4 & 2.8 \\ 1.2 & 2.8 & 4.0 \end{bmatrix}$$

so that

$$\mathbf{D}_c^{-1}\mathbf{F}'\mathbf{D}_r^{-1}\mathbf{F} = \begin{bmatrix} 0.50 & 0.38 & 0.12 \\ 0.35 & 0.40 & 0.25 \\ 0.15 & 0.35 & 0.50 \end{bmatrix}.$$

It is obvious that this is not symmetric. To arrive at the standard form, therefore, we will use the following transformation. Let

(6.15)
$$\mathbf{B} = \mathbf{D}_r^{-1/2}\mathbf{F}\,\mathbf{D}_c^{-1/2} .$$

Then, note that if $\mathbf{D}_c^{1/2}$ is taken outside the bracket of (6.13) we obtain $(\mathbf{F}'\mathbf{D}_r^{-1}\mathbf{F}\mathbf{D}_c^{-1/2} - \eta^2\mathbf{D}_c^{1/2})\mathbf{D}^{1/2}\mathbf{x} = \mathbf{0}$. By premultiplying this by $\mathbf{D}_c^{-1/2}$, we obtain $(\mathbf{D}_c^{-1/2}\mathbf{F}'\mathbf{D}_r^{-1}\mathbf{F}\mathbf{D}_c^{-1/2} - \eta^2\mathbf{I})\mathbf{D}_c^{1/2}\mathbf{x} = \mathbf{0}$. Noting that $\mathbf{D}_c^{-1/2}\mathbf{F}'\mathbf{D}_r^{-1}\mathbf{F}\mathbf{D}_c^{-1/2} = (\mathbf{D}_c^{-1/2}\mathbf{F}'\mathbf{D}_r^{-1/2})(\mathbf{D}_r^{-1/2}\mathbf{F}\mathbf{D}_c^{-1/2}) = \mathbf{B}'\mathbf{B}$, and setting $\mathbf{D}_c^{1/2}\mathbf{x} = \mathbf{w}$, we obtain the following eigenequation in the standard form:

(6.16)
$$(\mathbf{B}'\mathbf{B} - \eta^2\mathbf{I})\mathbf{w} = \mathbf{0} .$$

In our example,

$$\mathbf{B}'\mathbf{B} = \begin{bmatrix} 0.5000 & 0.3623 & 0.1342 \\ 0.3623 & 0.4000 & 0.2985 \\ 0.1342 & 0.2985 & 0.5000 \end{bmatrix}.$$

This is indeed symmetric. Equation 6.16 can be written as

$$\left(\begin{bmatrix} 0.5000 & 0.3623 & 0.1342 \\ 0.3623 & .4000 & 0.2985 \\ 0.1342 & 0.2985 & 0.5000 \end{bmatrix} - \eta^2 \begin{bmatrix} 1 & 0 & 0 \\ 0 & 1 & 0 \\ 0 & 0 & 1 \end{bmatrix} \right) \begin{bmatrix} w_1 \\ w_2 \\ w_3 \end{bmatrix} = \begin{bmatrix} 0 \\ 0 \\ 0 \end{bmatrix},$$

$$\begin{aligned} (0.5000-\eta^2)w_1 + 0.3623w_2 + 0.1342w_3 &= 0 \\ 0.3623w_1 + (0.4000-\eta^2)w_2 + 0.2985w_3 &= 0 \\ 0.1324w_1 + 0.2985w_2 + (0.5000-\eta^2)w_3 &= 0 . \end{aligned}$$

Eigenequation 6.16 has two *trivial solutions*:

(6.17)
$$\text{Trivial solution I:} \quad \mathbf{w} = \mathbf{0} .$$

It is obvious that 6.17 is a solution. But, as we have already discussed in chapter 5, this solution is avoided by solving the characteristic equation, that is, $|\mathbf{B}'\mathbf{B} - \eta^2\mathbf{I}| = 0$.

(6.18)
$$\text{Trivial solution II:} \quad \mathbf{w} = \mathbf{D}_c^{1/2}\mathbf{1}, \text{ that is, } \mathbf{x} = \mathbf{1} .$$

In this case, the corresponding eigenvalue η^2 is 1. This can easily be shown as follows. First, rewrite Equation 6.16 as

(6.19) $$\mathbf{B}'\mathbf{B}\mathbf{w} = \eta^2\mathbf{w} .$$

Then, if we substitute $\mathbf{w} = \mathbf{D}_c^{1/2}\mathbf{1}$ for the left-hand side of Equation 6.19, we obtain

$$\mathbf{B}'\mathbf{B}\mathbf{w} = \mathbf{D}_c^{-1/2}\mathbf{F}'\mathbf{D}_r^{-1}\mathbf{F}\mathbf{D}_c^{-1/2}\mathbf{D}_c^{1/2}\mathbf{1} = \mathbf{D}_c^{-1/2}\mathbf{F}'\mathbf{D}_r^{-1}\mathbf{F}\mathbf{1} .$$

But $\mathbf{F1} = \mathbf{f}_r$ so that $\mathbf{D}_r^{-1}\mathbf{f}_r = \mathbf{1}$. Then,

(6.20) $$\mathbf{D}_c^{-1/2}\mathbf{F}'\mathbf{1} = \mathbf{D}_c^{-1/2}\mathbf{f}_c = \mathbf{D}_c^{1/2}\mathbf{1} .$$

The substitution of $\mathbf{w}=\mathbf{D}_c^{1/2}\mathbf{1}$ for the right-hand side of Equation 6.19 gives us $\eta^2\mathbf{D}_c^{1/2}\mathbf{1}$. Comparing this with Equation 6.20, we conclude that $\eta^2 = 1$ when $\mathbf{w}=\mathbf{D}_c\mathbf{1}$. That $\mathbf{x} = \mathbf{1}$ amounts to putting all the columns (categories) into one, making both SS_w and SS_b zeros, hence no information to analyze. Furthermore, this "solution" vector does not satisfy the condition that $\mathbf{f}_t'\mathbf{x} = 0$. Thus, we must eliminate the second trivial solution from our computation. To carry out this task, the following theorem is relevant:

Lagrange's Theorem (Rao, 1965): Let **G** be any square matrix of order *n* and rank *m*>0, and **p** and **q** be column vectors such that **p'Gq** is not zero. Then, the residual matrix $\mathbf{G}_1$ is exactly of rank *m*-1, where

(6.21) $$\mathbf{G}_1 = \mathbf{G} - \mathbf{G}\mathbf{q}(\mathbf{p}'\mathbf{G}\mathbf{q})^{-1}\mathbf{p}'\mathbf{G} .$$

We can use this formula to eliminate the second trivial solution from **B'B** of Equation 6.16. Recall that our matrix **B'B** is symmetric and that the trivial solution is an eigenvector. Therefore, $\mathbf{p} = \mathbf{q} = \mathbf{a}$, say, and $\mathbf{G}\mathbf{a} = \lambda\mathbf{a}$. An application of Equation 6.21 to an eigenequation yields the following formula for eliminating the effect of a solution from an eigenequation:

(6.22) $$\mathbf{G}_1 = \mathbf{G} - \lambda\mathbf{a}(\mathbf{a}'\mathbf{a})^{-1}\mathbf{a}' .$$

If we *normalize* **a** to **a*** by

(6.23) $$\mathbf{a}^* = \frac{\mathbf{a}}{\sqrt{\mathbf{a}'\mathbf{a}}} ,$$

that is, $\mathbf{a}'^*\mathbf{a}^* = 1$, then

(6.24) $$\mathbf{G}_1 = \mathbf{G} - \lambda\mathbf{a}^*\mathbf{a}^{*\prime} .$$

This formula is used to extract the contribution of each successive solution in dual scaling, and as such Formula 6.24 is important to remember.

Applying Formula 6.22 to our problem of eliminating the second trivial solution from the equation, we calculate the residual matrix, indicated by $\mathbf{C}_1$, noting $\mathbf{a}$ in (6.22) is $\mathbf{D}_c^{1/2}\mathbf{1}$, as

$$\mathbf{C}_1 = \mathbf{B}'\mathbf{B} - \frac{\mathbf{D}_c^{1/2}\mathbf{1}\mathbf{1}'\mathbf{D}_c^{1/2}}{f_t}. \qquad (6.25)$$

Note that in deriving Formula 6.25 we also used the relations $\eta^2 = 1$ and $\mathbf{w}'\mathbf{w} = \mathbf{1}'\mathbf{D}_c^{1/2}\mathbf{D}_c^{1/2}\mathbf{1} = f_t$. The last term of the right-hand side of Formula 6.25 in the current example is given as

$$\frac{\mathbf{D}_c^{1/2}\mathbf{1}\mathbf{1}'\mathbf{D}_c^{1/2}}{f_t} = \begin{bmatrix} 0.3448 & 0.3617 & 0.3084 \\ 0.3617 & 0.3793 & 0.3235 \\ 0.3084 & 0.3235 & 0.2759 \end{bmatrix}.$$

Thus, the residual matrix $\mathbf{C}_1$ that is free from the effect of the second trivial solution is given as

$$\mathbf{C}_1 = \begin{bmatrix} 0.1552 & 0.0006 & -0.1742 \\ 0.0006 & 0.0207 & -0.0250 \\ -0.1742 & -0.0250 & 0.2241 \end{bmatrix}.$$

Following the procedure of the power method discussed in Section 5.4, we may find the maximal value of η^2 and the corresponding eigenvector of $\mathbf{C}_1$. Let us use [1 0 -1] as the initial vector $\mathbf{b}_0$. Then,

$$\mathbf{C}_1\mathbf{b}_0 = \begin{bmatrix} 0.1552 & 0.0006 & -0.1742 \\ 0.0006 & 0.0207 & -0.0250 \\ -0.1742 & -0.0250 & 0.2241 \end{bmatrix} \begin{bmatrix} 1 \\ 0 \\ -1 \end{bmatrix} = \begin{bmatrix} 0.3294 \\ 0.0256 \\ -0.3983 \end{bmatrix}.$$

The largest absolute value in $\mathbf{b}_1$ is 0.3983. Thus $k_1 = 0.3983$, and

$$\mathbf{b}_1^* = \frac{\mathbf{b}_1}{0.3983} = \begin{bmatrix} 0.8270 \\ 0.0643 \\ -1.0000 \end{bmatrix}.$$

Therefore,

$$\mathbf{C}_1\mathbf{b}_1^* = \begin{bmatrix} 0.3026 \\ 0.0268 \\ -0.3698 \end{bmatrix} = \mathbf{b}_2, \quad k_2 = 0.3698 .$$

Similarly,

$$\mathbf{b}_2^* = \frac{\mathbf{b}_2}{0.3698} = \begin{bmatrix} 0.8181 \\ 0.0725 \\ -1.0000 \end{bmatrix}.$$

$$\mathbf{C}_1\mathbf{b}_2^* = \begin{bmatrix} 0.3012 \\ 0.0270 \\ -0.3685 \end{bmatrix} = \mathbf{b}_3, \quad k_3 = 0.3685$$

$$\mathbf{b}_3^* = \frac{\mathbf{b}_3}{0.3685} = \begin{bmatrix} 0.8174 \\ 0.0733 \\ -1.0000 \end{bmatrix}$$

$$\mathbf{C}_1\mathbf{b}_3^* = \begin{bmatrix} 0.3011 \\ 0.0270 \\ -0.3683 \end{bmatrix} = \mathbf{b}_4, \quad k_4 = 0.3683$$

$$\mathbf{b}_4^* = \frac{\mathbf{b}_4}{0.3683} = \begin{bmatrix} 0.8175 \\ 0.0733 \\ -1.0000 \end{bmatrix}$$

$$\mathbf{C}_1\mathbf{b}_4^* = \begin{bmatrix} 0.3011 \\ 0.0270 \\ -0.3683 \end{bmatrix} = \mathbf{b}_5, \quad k_5 = 0.3683$$

$$\mathbf{b}_5^* = \frac{\mathbf{b}_5}{0.3683} = \begin{bmatrix} 0.8175 \\ 0.0733 \\ -1.0000 \end{bmatrix}.$$

Since $\mathbf{b}_4 = \mathbf{b}_5$, the iterative process has converged to the solution, and $k_5 = \eta_1^2 = 0.3683$. That k_5 is the largest eigenvalue is explained in chapter 5 [see Equation 5.34 and discussion of Step 5]. To obtain the corresponding weight vector that satisfies Equation 6.6, the vector $\mathbf{b}_5$* must be rescaled.

Equation 6.6 states that $\mathbf{x}'\mathbf{D}_c\mathbf{x} = f_t$. But $\mathbf{x}'\mathbf{D}_c\mathbf{x} = \mathbf{w}'\mathbf{w}$, and $f_t = 29$. Therefore,

$$\mathbf{w} = \sqrt{\frac{29}{\mathbf{b}_s^{*\prime}\mathbf{b}_s^{*}}}\ \mathbf{b}_s^{*} = \begin{bmatrix} 3.4029 \\ 0.3051 \\ -4.1626 \end{bmatrix}.$$

The optimal weight vector $\mathbf{x}_1$ is given by

$$\mathbf{x}_1 = \mathbf{D}_c^{-1/2}\mathbf{w} = \begin{bmatrix} \frac{1}{\sqrt{10}} & 0 & 0 \\ 0 & \frac{1}{\sqrt{11}} & 0 \\ 0 & 0 & \frac{1}{\sqrt{8}} \end{bmatrix} \begin{bmatrix} 3.4029 \\ 0.3051 \\ -4.1626 \end{bmatrix} = \begin{bmatrix} 1.0761 \\ 0.0920 \\ -1.4717 \end{bmatrix}.$$

Therefore, optimal weights for categories good, average, and poor are 1.0761, 0.0920, and -1.4717. What about the weight vector $\mathbf{y}$ for the teachers? We can start in the same way as we have shown, except that we now look at $\mathbf{y}$ for $\mathbf{x}$ and the between-column sum of squares for the between-row sum of squares. The squared correlation ratio η^2 is now defined as the ratio of the between-column sum of squares to the total sum of squares, and this ratio is maximized with respect to $\mathbf{y}$ under the constraints $\mathbf{f}_r'\mathbf{y} = 0$ and $\mathbf{y}'\mathbf{D}_r\mathbf{y} = f_t$. Following the same procedure as before, we obtain

(6.26) $$(\mathbf{BB}' - \eta^2\mathbf{I})\mathbf{D}_r^{1/2}\mathbf{y} = \mathbf{0}.$$

There are two trivial solutions as before, $\mathbf{D}_r^{1/2}\mathbf{y} = \mathbf{0}$ and $\mathbf{y} = \mathbf{1}$. The second trivial vector always yields $\eta^2 = 1$. So every process is parallel to the previous problem. Indeed, there exists complete symmetry or *duality* between the two problems. Thus, rather than repeating the same process, let us present an important relation between $\mathbf{x}$ and $\mathbf{y}$. As mentioned briefly in chapter 4, the following are called *dual relations* (Nishisato, 1980a) or *substitution formulas* (Benzécri et al., 1973) between the two weight vectors of any solutions. Let η be the positive square root of η^2. Then,

(6.27) $$\mathbf{y} = \left(\frac{1}{\eta}\right)\mathbf{D}_r^{-1}\mathbf{F}\mathbf{x},$$

(6.28) $$\mathbf{x} = \left(\frac{1}{\eta}\right)\mathbf{D}_c^{-1}\mathbf{F}'\mathbf{y},$$

or,

$$\eta \mathbf{y} = \mathbf{D}_r^{-1}\mathbf{F}\mathbf{x} \,, \tag{6.29}$$

$$\eta \mathbf{x} = \mathbf{D}_c^{-1}\mathbf{F}'\mathbf{y} \,, \tag{6.30}$$

Using Equation 6.27, we obtain the weight vector for the rows as

$$\begin{bmatrix} y_1 \\ y_2 \\ y_3 \end{bmatrix} = \frac{1}{\sqrt{0.3683}} \begin{bmatrix} \frac{1}{10} & 0 & 0 \\ 0 & \frac{1}{10} & 0 \\ 0 & 0 & \frac{1}{9} \end{bmatrix} \begin{bmatrix} 1 & 3 & 6 \\ 3 & 5 & 2 \\ 6 & 3 & 0 \end{bmatrix} \begin{bmatrix} 1.0761 \\ 0.0920 \\ -1.4717 \end{bmatrix} = \begin{bmatrix} -1.2322 \\ 0.1227 \\ 1.2326 \end{bmatrix} .$$

Thus, Teachers 1, 2, and 3 receive respectively scores -1.2322, 0.1227, and 1.2326. Teacher 3 is the best and Teacher 1 the worst of the three.

The weight vectors and the squared correlation ratio we have obtained are indicated respectively by $\mathbf{x}_1$, $\mathbf{y}_1$, and η_1^2. The subscript refers to solution 1, or dimension 1, or component 1, whichever you may wish to use. In this book, we will use "solution." The solution obtained above is optimal in the sense that it maximizes the objective function η^2, subject to the conditions Equations 6.5 and 6.6. The optimal solution, however, does not generally explain the data in an exhaustive manner. The possibility exists that it leaves a substantial amount of information in the data unaccounted for. This calls for the extraction of more solutions, that is, *multidimensional decomposition* of the data.

6.2 MULTIDIMENSIONAL DECOMPOSITION

The question that should be asked at this stage of analysis is whether Solution 1 is sufficient to explain the data or if it exhaustively analyzes the information in the data. Mathematically it is usually the case that the number of possible nontrivial solutions from an $n \times m$ contingency table such as the current example is min(n-1,m-1), that is, the smaller of the two quantities in the parentheses. In the present example, the data matrix is 3×3 so that the number of possible solutions is 2. When the data matrix is large (e.g., 75×50), however, the number of possible solutions may not be of much interest because there are too many solutions to examine. Instead, most researchers would be interested in the question of how many solutions are needed to explain the major portion of the total information. There are two matters relevant to this question: (a) the percentage of total information explained by a subset of solutions, and (b) the approximation of the data matrix by a subset of solutions.

(a) Percentage of Total Information Explained: The sum of the diagonal elements of the matrix $\mathbf{C}_1$, that is, the *trace* of $\mathbf{C}_1$ written as $tr(\mathbf{C}_1)$, is referred to as the *total information*. It is known that $tr(\mathbf{C}_1) = \Sigma\eta_j^2$. Thus, we can define the percentge of the total information explained by Solution j, δ_j, as follows:

$$\delta_j = \frac{100\eta_j^2}{tr(\mathbf{C}_1)} . \tag{6.31}$$

In the present example, $tr(\mathbf{C}_1) = 0.1552+0.0207+0.2241 = 0.4000$, and $\eta_1^2 = 0.3683$ so that $\delta_1 = 100\times0.3683/0.4000 = 92.075\%$. This is a very large value, indicating that the first solution captures the major portion of information, contained in two solutions. In using this statistic for different types of categorical data, however, one should be aware of some precautional remarks on difficulty in interpreting this statistic, which are presented in chapter 17.

(b) Order-k Approximation: When there is no information available about the distribution of responses in the two-way table, the best we can do to predict the response distribution in the table is to use the case in which the rows and the columns of the table are statistically independent. Let us call this the *order-0 approximation*, $\mathbf{F}_{(0)}$. This approximation is nothing but the (second) trivial solution. Suppose that we proceed with scaling and extract the first solution. Then the data approximated by both the order-0 approximation (the trivial solution) and the first solution is called the *order-1 approximation*, $\mathbf{F}_{(1)}$. When the second solution is included, it is referred to as the *order-2 approximation*, $\mathbf{F}_{(2)}$. Thus, the *order-k approximation*, $\mathbf{F}_{(k)}$, is the one based on the trivial solution and the first k solutions, and it is given by

$$\mathbf{F}_{(k)} = \left(\frac{1}{f_t}\right)\mathbf{D}_r\left(\mathbf{1}\mathbf{1}' + \sum_{j=1}^{k} \rho_j \mathbf{y}_j \mathbf{x}_j'\right)\mathbf{D}_c . \tag{6.32}$$

The corresponding residual matrix, that is, the data matrix from which the trivial solution and the first k solutions are removed, is given by

$$\mathbf{F}_{res(k)} = \mathbf{F} - \mathbf{F}_{(k)} . \tag{6.33}$$

In the current example, $\mathbf{F}$, $\mathbf{F}_{(0)}$, $\mathbf{F}_{(1)}$, and $\mathbf{F}_{res(1)}$ are given as follows:

$$\mathbf{F} = \begin{bmatrix} 1 & 3 & 6 \\ 3 & 5 & 2 \\ 6 & 3 & 0 \end{bmatrix},$$

$$\mathbf{F}_{(0)} = \begin{bmatrix} 3.45 & 3.79 & 2.76 \\ 3.45 & 3.79 & 2.76 \\ 3.10 & 3.41 & 2.48 \end{bmatrix},$$

$$\mathbf{F}_{(1)} = \begin{bmatrix} 0.67 & 3.53 & 5.79 \\ 3.72 & 3.82 & 2.46 \\ 5.60 & 3.65 & -0.25 \end{bmatrix},$$

$$\mathbf{F}_{res(1)} = \begin{bmatrix} 0.33 & -0.53 & 0.21 \\ -0.72 & 1.18 & -0.46 \\ 0.40 & -0.65 & 0.25 \end{bmatrix}.$$

Although the first solution accounts for the substantial amount of information as seen from the statistic δ_1 and two matrices, $\mathbf{F}_{(1)}$ and $\mathbf{F}_{res(1)}$, let us extract one more solution.

In Section 6.1, we discussed how the contribution of a solution can be eliminated from a matrix. Using the procedure, we can eliminate the contribution of the first solution from the matrix $\mathbf{C}_1$. Let us indicate by $\mathbf{C}_2$ the residual matrix after eliminating the first solution (η^2, $\mathbf{w}_1$). In the current example,

$$\mathbf{C}_2 = \mathbf{C}_1 - \eta^2 \frac{\mathbf{w}_1\mathbf{w}_1'}{\mathbf{w}_1'\mathbf{w}_1}$$

$$(6.34) \quad = \begin{bmatrix} 0.1552 & 0.0006 & -0.1742 \\ 0.0006 & 0.0207 & -0.0250 \\ -0.1742 & -0.0250 & 0.2241 \end{bmatrix} - \begin{bmatrix} 0.1471 & 0.0132 & -0.1700 \\ 0.0132 & 0.0012 & -0.0161 \\ -0.1799 & -0.0161 & 0.2201 \end{bmatrix}$$

$$= \begin{bmatrix} 0.0081 & -0.0126 & 0.0057 \\ -0.0126 & 0.0195 & -0.0089 \\ 0.0057 & -0.0089 & 0.0040 \end{bmatrix}.$$

This residual matrix is now subjected to the same iterative method to extract the second solution. Although the initial trial vector can be arbitrary, we can save computation time by using the available information. In the present example, the first solution shows that $w_1 > w_2 > w_3$. It is therefore unlikely that the second solution shows the same pattern as the first one. Thus, we may reverse the pattern and use the following trial vector for the second solution, namely, $\mathbf{b}'_0 = [-1 \;\; 0 \;\; 1]$. The iterative procedure starts with multiplication of $\mathbf{C}_2$ by $\mathbf{b}_0$,

$$\mathbf{C}_2\mathbf{b}_0 = \begin{bmatrix} 0.0081 & -0.0126 & 0.0057 \\ -0.0126 & 0.0195 & -0.0089 \\ 0.0057 & -0.0089 & 0.0040 \end{bmatrix} \begin{bmatrix} -1 \\ 0 \\ 1 \end{bmatrix} = \begin{bmatrix} -0.0024 \\ 0.0037 \\ -0.0017 \end{bmatrix} = \mathbf{b}_1 .$$

The largest absolute value in $\mathbf{b}_1$ is 0.0037, which is k_1. Thus, we obtain the next vector and then continue the process as before until convergence is obtained,

$$\frac{\mathbf{b}_1}{k_1} = \begin{bmatrix} -0.6486 \\ 1.0000 \\ -0.4595 \end{bmatrix} = \mathbf{b}_1^*$$

$$\mathbf{C}_2\mathbf{b}_1^* = \begin{bmatrix} -0.0205 \\ 0.0318 \\ -0.0144 \end{bmatrix} = \mathbf{b}_2, \quad k_2 = 0.0318$$

$$\frac{\mathbf{b}_2}{k_2} = \begin{bmatrix} -0.6447 \\ 1.0000 \\ -0.4528 \end{bmatrix} = \mathbf{b}_2^*$$

$$\mathbf{C}_2\mathbf{b}_2^* = \begin{bmatrix} -0.0204 \\ 0.0317 \\ -0.0144 \end{bmatrix} = \mathbf{b}_3, \quad k_3 = 0.0317$$

$$\mathbf{C}_2\mathbf{b}_3^* = \begin{bmatrix} -0.0204 \\ 0.0317 \\ -0.0144 \end{bmatrix} = \mathbf{b}_4 = \mathbf{b}_3 .$$

Thus, $\mathbf{b}_3^* = \mathbf{b}_4^*$. The eigenvector that satisfies the unit constraint is therefore obtained by

$$\mathbf{w}_2 = \sqrt{\frac{f_t}{\mathbf{b}_4^{*\prime}\mathbf{b}_4^*}}\, b_4^* = \begin{bmatrix} -2.7222 \\ 4.2304 \\ -1.9219 \end{bmatrix}.$$

Thus, the second optimal weight vector $\mathbf{x}_2$ is given by

$$\mathbf{x}_2 = \mathbf{D}_c^{-1/2}\mathbf{w}_2 = \begin{bmatrix} -0.8608 \\ 1.2755 \\ -0.6795 \end{bmatrix}.$$

Then $\mathbf{y}_2$ can be calculated by Equation 6.27, namely

$$\mathbf{y}_2 = \frac{1}{\sqrt{0.0317}}\mathbf{D}_r^{-1}\mathbf{F}\mathbf{x}_2 = \begin{bmatrix} -0.6242 \\ 1.3683 \\ -0.8352 \end{bmatrix}.$$

Note that $\eta_1^2 + \eta_2^2 = tr(\mathbf{C}_1)$. In other words, two solutions exhaustively account for the entire information in the data. Therefore, the order-2 approximation reproduces $\mathbf{F}$ exactly, resulting in the residual matrix $\mathbf{F}_{res(2)} = \mathbf{0}$, the null matrix. In general, the above process will continue to extract more solutions. Once η_2^2, $\mathbf{y}_2$, and $\mathbf{x}_2$ are obtained, calculate

(6.35) $$\mathbf{C}_3 = \mathbf{C}_2 - \eta_2^2\mathbf{w}_2\mathbf{w}_2'/\mathbf{w}_2'\mathbf{w}_2 ,$$

and subject $\mathbf{C}_3$ to the iterative process to determine η_3^2, $\mathbf{y}_3$, and $\mathbf{x}_3$. Thus, the general scheme involves successive applications of the iterative method to

(6.36) $$\mathbf{C}_{r+1} = \mathbf{C}_r - \eta_r^2\mathbf{w}_r\mathbf{w}_r'/\mathbf{w}_r'\mathbf{w}_r ,$$

where $\mathbf{w}_r = \mathbf{D}_c^{1/2}\mathbf{x}_r$. When all the possible solutions, say t solutions, are obtained, we can put them in matrix forms as

(6.37) $$\Lambda^2 = \mathrm{diag}[1 \;\; \eta_1^2 \;\; \eta_2^2 \;\; . \;\; . \;\; . \;\; . \;\; \eta_t^2] ,$$

(6.38) $$\mathbf{Y} = [\mathbf{1} \;\; \mathbf{y}_1 \;\; \mathbf{y}_2 \;\; . \;\; . \;\; . \;\; \mathbf{y}_t] ,$$

(6.39) $$\mathbf{X} = [\mathbf{1} \;\; \mathbf{x}_1 \;\; \mathbf{x}_2 \;\; . \;\; . \;\; . \;\; \mathbf{x}_t] .$$

Then, in terms of these matrices, the input data can be expressed as

$$\mathbf{F} = f_t \mathbf{Y} \Lambda \mathbf{X}' \ . \tag{6.40}$$

Notice that Equation 6.40 is the singular value decomposition of the data matrix $\mathbf{F}$, except that our choice of the unit resulted in the multiplier f_t on the right-hand side of (6.40).

6.3 SOME INTERESTING RELATIONS

Dual scaling can be formulated by optimizing several different criteria, sometimes called *objective functions*. All of them, however, result in the canonical form, singular value decomposition, or dual relations. These different, but mathematically equivalent, formulations are typically carried out using the constraint that the sum of weighted responses is zero, and that the sum of squares of weighted responses is constant. In the previous section, we used these constraints. We can consider other constraints as well. For instance, some researchers used order constraints on column weights of a contingency table, and some set the minimal weight to be zero and the maximal one. Interested readers are referred to some earlier studies.[2] Within the context of the current formulation, we can identify a number of interesting mathematical properties of dual scaling. Let us look at some of them.

(a) Dual Relations. We have already looked at dual relations between row weights and column weights as the single defining characteristic of the method. The same property holds for principal component analysis.[3]

(b) Structural Equation. Dual relations define expressions for ρ, $\mathbf{y}$, and $\mathbf{x}$ for each optimal solution. We can express the data matrix in terms of the entire solutions, that is, the structural equation for $\mathbf{F}$,

$$\mathbf{F} = \frac{1}{f_t}\mathbf{D}_r(\mathbf{1}\mathbf{1}' + \sum_{k=1}^{t} \rho_k \mathbf{y}_k \mathbf{x}_k')\mathbf{D}_c \ , \tag{6.41}$$

---

[2] For instance, Bradley, Katti and Coons (1962), Fisher (1948), Healy & Goldstein (1976), Goldstein (1987), Nishisato (1980a), Nishisato and Arri (1975), Tanaka and Asano (1978), Tanaka and Kodake (1980), and Tenenhaus (1988).

[3] See, for example, Gower (1966).

where t is the rank of **F** minus 1, that is, $t = r(\mathbf{C}_1)$. As we have already seen, this equation can be used to approximate the input data by a subset of solutions. Bénzecri et al. (1973) calls it the formula for reconstitution.

(c) Trivial Solution II for the Contingency Table. As shown in chapter 7, the second trivial solution is equivalent to the solution expected when the rows and the columns of a contingency table are statistically independent. Under this statistical independence, the frequency for the ith row and the jth column is given by $f_{i.}f_{.j}/f_t$, or the matrix of **F** expected under this condition is expressed as $\mathbf{f}_r\mathbf{f}_c'/f_t$. Try to show that the following equation is true.

$$\mathbf{D}_c^{-1/2}(\mathbf{F} - \frac{\mathbf{f}_r\mathbf{f}_c'}{f_t})'\mathbf{D}_r^{-1}(\mathbf{F} - \frac{\mathbf{f}_r\mathbf{f}_c'}{f_t})\mathbf{D}_c^{-1/2} = \mathbf{D}_c^{-1/2}\mathbf{F}'\mathbf{D}_r^{-1}\mathbf{F}\,\mathbf{D}_c^{-1/2} - \frac{\mathbf{D}_c^{1/2}\,\mathbf{1}\mathbf{1}'\mathbf{D}_c^{1/2}}{f_t} \, . \tag{6.42}$$

(d) Trivial Solution II for Multiple-Choice Data. When you have a subject-by-option matrix of (1, 0) response patterns as data, we can show that the second trivial solution is equivalent to column means only if rows and columns of **F** have constant marginals. The vector of column means of **F** can be expressed as $\mathbf{m}_c = \mathbf{F}'\mathbf{1}(\mathbf{1}'\mathbf{1})^{-1}$. Thus, the matrix of deviations from the column means is given by

$$\mathbf{F} - \mathbf{1}m_c = \mathbf{F} - \mathbf{1}(\mathbf{1}'\mathbf{1})^{-1}\mathbf{1}'\mathbf{F} = \mathbf{F} - \mathbf{P}_m\mathbf{F} = (\mathbf{I} - \mathbf{P}_m)\mathbf{F} = \mathbf{F}_d \, . \tag{6.43}$$

Try to show that the following equation is true for any matrix **F**, provided that $\mathbf{D}_r = n\mathrm{I}$ and $\mathbf{D}_c = (nN/m)\mathrm{I}$:

$$\mathbf{D}_c^{-1/2}\mathbf{F}_d'\mathbf{D}_r^{-1}\mathbf{F}_d\mathbf{D}_c^{-1/2} = \mathbf{D}_c^{-1/2}\mathbf{F}'\mathbf{D}_r^{-1}\mathbf{F}\mathbf{D}_c^{-1/2} - \mathbf{D}_c^{1/2}\mathbf{1}\mathbf{1}'\mathbf{D}_c^{1/2} \, . \tag{6.44}$$

(e) Trivial Solution for Paired Comparison, Rank-Order, and Successive Categories Data. In the case of these so-called dominance data, original data are first transformed to dominance numbers, that is, the number of times a stimulus is judged preferred to (greater, or better, or more serious than) the remaining stimuli minus the number of times the other stimuli are preferred to the stimulus. Dominance numbers of stimuli are calculated within each subject, and in consequence the sum of dominance numbers of each subject is zero. Because of this condition of zero row marginals, paired comparison, rank-order, and successive categories data are all free from the Trivial Solution II. In other words, in these cases, the second trivial solution does not exist.

(f) Principle of Equivalent Partitioning. One can partition any row into proportional rows, or any column into proportional columns, without changing the distribution of the squared correlation ratio. Furthermore, the set of distinct weights from the original data matrix is preserved in the partitioned matrix. In other words, if you have a 3x3 contingency table and decide to partition each row into equal halves, thus creating a 6x3 table, you will still obtain two squared correlation ratios which are identical with those from the original matrix, and three distinct weights for the six rows which are the same as the three weights for the rows of the original matrix. This principle is used for the forced classification procedure in chapter 13. Another interesting application of this principle is to partition a contingency table in such a way that all row marginals and column marginals are 1. For instance, compare the two equivalent matrices $\mathbf{F}_1$ and $\mathbf{F}_2$ given below; one is 2x2 (original), and the other is 6x6, partitioned in such a way that all the marginals are 1.

$$\mathbf{F}_1 = \begin{bmatrix} 1 & 3 \\ 2 & 0 \end{bmatrix},$$

$$\mathbf{F}_2 = \frac{1}{12}\begin{bmatrix} 1 & 1 & 1 & 3 & 3 & 3 \\ 1 & 1 & 1 & 3 & 3 & 3 \\ 1 & 1 & 1 & 3 & 3 & 3 \\ 1 & 1 & 1 & 3 & 3 & 3 \\ 4 & 4 & 4 & 0 & 0 & 0 \\ 4 & 4 & 4 & 0 & 0 & 0 \end{bmatrix}.$$

The maximized squared correlations from $\mathbf{F}_1$ and $\mathbf{F}_2$ are the same and equal to 0.50. Weights for rows ($\mathbf{y}$) and columns ($\mathbf{x}$) of $\mathbf{F}_1$ are $(y_i) = (-0.71, 1.41)$ and $(x_j) = (1.00, -1.00)$, and those of $\mathbf{F}_2$ are $(y_i) = (-0.71, -0.71, -0.71, -0.71, 1.41, 1.41)$ and $(x_j) = (1.00, 1.00, 1.00, -1.00, -1.00, -1.00)$. This example shows that $\mathbf{F}_1$ and $\mathbf{F}_2$ can be said to be equivalent. Examine $\mathbf{F}_2$ closely. Can you see the relations of derived weights to the distribution of responses in $\mathbf{F}_2$?

(g) Principle of Constant Proportionality. Consider the multiple-choice data matrix partitioned into n items,

$$\mathbf{F} = [\mathbf{F}_1, \mathbf{F}_2, \ \ldots \ , \mathbf{F}_j, \ \ldots \ , \mathbf{F}_n]$$

where $\mathbf{F}_j$ is the Nxm_j incidence matrix for item j, and $\mathbf{F}_j\mathbf{1} = \mathbf{1}$. Let us indicate by $\mathbf{D}_{cj}$ and $\mathbf{x}_j$ respectively the diagonal matrix of column totals of $\mathbf{F}_j$ and the vector of weights for m_j options of item j. In this chapter, we defined the total sum of squares, SS_t and the between-row (between-column) sum of squares, SS_b. These quantities are defined for the entire data set. We can partition SS_t into the contributions of individual items as

$$SS_t = SS_{t(1)} + SS_{t(2)} + \ . \ . \ . \ + SS_{t(j)} + \ . \ . \ . \ + SS_{t(n)} ,$$

where

$$SS_{t(j)} = \mathbf{x}_j'\mathbf{D}_{cj}\mathbf{x}_j \ .$$

Similarly, SS_b can be partitioned as

$$SS_b = SS_{b(1)} + SS_{b(2)} + \ . \ . \ . \ + SS_{b(j)} + \ . \ . \ . \ + SS_{b(n)} ,$$

where

$$SS_{b(j)} = \mathbf{x}_j'\mathbf{F}_j'\mathbf{D}_r^{-1}\mathbf{F}\ \mathbf{x} = \frac{1}{n}\mathbf{x}_j'\mathbf{F}_j'\mathbf{F}\ \mathbf{x} \ .$$

We can also partition SS_w into the sum of the contributions of individual items. The principle of constant proportionality states that for each dual scaling solution the following stationality holds:

(6.45)
$$\frac{SS_{b(1)}}{SS_{t(1)}} = \frac{SS_{b(2)}}{SS_{t(2)}} = \ . \ . \ . \ = \frac{SS_{b(j)}}{SS_{t(j)}} = \ . \ . \ . \ = \frac{SS_{b(n)}}{SS_{t(n)}} = \frac{SS_b}{SS_t} = \eta^2 \ .$$

Note also that

(6.46)
$$\frac{SS_{b(j)}}{SS_{t(j)}} = 1 - \frac{SS_{w(j)}}{SS_{t(j)}} = 1 - \frac{SS_w}{SS_t} \ .$$

Thus, if you want to know how good your subjective option weights are, you can check it by examining constant proportionality of the weighted responses.

(h) Squared Correlation Ratio and Item Ststistics. It is important to know the relations between the maximized squared correlation ratio and other item statistics. The square of the product-moment correlation between scores of item j, $\mathbf{F}_j\mathbf{x}_j$, and total scores $\mathbf{Fx}$ is given by

(6.47)
$$r_{jt}^2 = \frac{(\mathbf{x}_j'\mathbf{F}_j'\mathbf{F}\ x)^2}{\mathbf{x}_j'\mathbf{D}_{cj}\mathbf{x}_j\ \mathbf{x}'\mathbf{F}'\mathbf{x}} = \frac{n\eta^2\mathbf{x}_j'\mathbf{F}_j'\mathbf{F}\mathbf{x}}{\mathbf{x}'\mathbf{F}'\mathbf{F}\mathbf{x}} \ .$$

But,

$$\mathbf{x}_j'\mathbf{F}_j'\mathbf{F}\mathbf{x} = n\eta^2\mathbf{x}_j'\mathbf{D}_{cj}\mathbf{x} ,$$

$$\mathbf{x}'\mathbf{F}'\mathbf{F}\mathbf{x} = n\eta^2\mathbf{x}'\mathbf{D}\mathbf{x} .$$

Therefore,

$$r_{jt}^2 = \mathbf{x}_j'\mathbf{D}_{cj}\mathbf{x}_j\left[\frac{n\eta^2}{\mathbf{x}'\mathbf{D}\mathbf{x}}\right]. \tag{6.48}$$

Since the quantity in the bracket is constant for the entire data set, we conclude that r_{jt}^2 is proportional to $\mathbf{x}_j'\mathbf{D}_{cj}\mathbf{x}_j$, that is,

$$r_{jt}^2 \propto \mathbf{x}_j'\mathbf{D}_{cj}\mathbf{x}_j . \tag{6.49}$$

Further, we obtain that

$$\sum_{j=1}^{n} r_{jt}^2 = \frac{\mathbf{x}'\mathbf{F}'\mathbf{F}\mathbf{x}}{\mathbf{x}'\mathbf{D}\mathbf{x}} = n\eta^2 ,$$

hence

$$\eta^2 = \frac{\sum_{j=1}^{n} r_{jt}^2}{n} . \tag{6.50}$$

The maximized squared correlation ratio is equal to the average of the squared item total correlations. This relation suggests that by discarding those items which are poorly correlated with the total score one can increase η^2.

The internal consistency reliability, also referred to as Kuder-Richardson reliability, Cronbach's coefficient of generalizability, or Cronbach's α, is defined by

$$\alpha = \left(\frac{n}{n-1}\right)\left(1 - \frac{SS_t - SS_n}{nSS_b}\right), \tag{6.51}$$

where SS_n, SS_b, and SS_t are respectively the between-item sum of squares, the between-subject sum of squares, and the total sum of squares. In dual scaling, SS_n is set to zero; therefore, we obtain that

$$\eta^2 = \frac{1}{1+(n-1)(1-\alpha)}, \tag{6.52}$$

or

$$\alpha = 1-\frac{1-\eta^2}{(n-1)\eta^2}. \tag{6.53}$$

Lord (1958) concluded that maximizing η^2 is equivalent to maximizing α. Notice the effects of n on η^2 and α. When α is fixed, η^2 becomes smaller as the value of n increases; when η^2 is fixed, α becomes larger as n increases. Thus, by discarding many poor items from data, we may not be able to increase α because the homogeneity of items is attained at the expense of the number of items.

There are many other interesting statistics we can look at. However, let us leave them for later in other chapters. There is one outstanding problem that we should keep in mind. It is graphical display. Although the topic is discussed in detail in chapter 14, it seems important to draw our attention to the problem now. The following is just an outline of the problem.

6.4 CAUTIONARY NOTES ON JOINT GRAPHICAL DISPLAY

Dual scaling of the contingency table, multiple-choice data, and sorting data determines weights for rows, y_i, and those for columns, x_j, in such a way that data weighted by y_i and those by x_j attain the maximal correlation ρ. This objective tempts us to plot y_i and x_j, which is called joint graphical display. Greenacre (1984) called it a between-set plot. The problem is that y_i and x_j span the same space only when $\rho = 1$. Otherwise, the space for y_i and that for x_j are different, thus the computation of the distance between y_i and x_j is not justifiable. If $\rho = 1$, we can calculate the between-set distance, but it would not be of much use because there will be a perfect match between points y_i and x_j. In other words, if y_1, y_2, y_3 and x_1, x_2, x_3, x_4 are obtained from a 3x4 contingency table and $\rho = 1$, then each one of y_i must be equal to one of x_j, and each one of x_j must be equal to one of y_i. In this case, two points y_i and x_j are either identical or different, and joint graphical display would not be needed for the interpretation of the results. Thus, our concern is with the case when $\rho \neq 1$.

The general practice has been to obtain joint display and carry out between-set comparisons of points in a subjective manner, without distance calculations. However, as we see in chapter 14, this practice can mislead the researcher, for two points from a row and a column (i.e., two points from different sets) which appear close to each other in joint display may in fact be two widely separated points.

One logical approach to this tricky problem is to project one set of points (e.g., y_i) onto the space of the other set (e.g., x_j), and plot the projected points and the other set of points in the joint graph. Recall dual relations between y_i and x_j. The projection of y_i onto the space of x_j is given by ρy_i, and the projection of x_j onto the space of y_i is given by ρx_j, that is,

(6.54)
$$\rho y_i = \frac{\sum_{j=1}^{m} f_{ij} x_j}{f_{i.}} \quad \text{and} \quad \rho x_j = \frac{\sum_{i=1}^{n} f_{ij} y_i}{f_{.j}} ,$$

or in matrix notation,

(6.55)
$$\rho \mathbf{y} = \mathbf{D}_r^{-1}\mathbf{F}\mathbf{x} \quad \text{and} \quad \rho \mathbf{x} = \mathbf{D}_c^{-1}\mathbf{F}'\mathbf{y} .$$

Thus, the plot of $\mathbf{x}$ and $\rho\mathbf{y}$ uses the same space and is justifiable. Similarly, the plot of $\rho\mathbf{x}$ and $\mathbf{y}$ is correct. From the practical point of view, however, these correct and justifiable plots have a problem and are not as useful as one may expect. The problem stems from the well-known theorem that the norm of a vector $\mathbf{a}$ is greater than or equal to the norm of its projection $\mathbf{Pa}$, that is $\sqrt{\mathbf{a}'\mathbf{a}} \geq \sqrt{\mathbf{a}'\mathbf{Pa}}$. In our case, $\sqrt{\mathbf{x}'\mathbf{x}} \geq \rho\sqrt{\mathbf{y}'\mathbf{y}}$ and $\sqrt{\mathbf{y}'\mathbf{y}} \geq \rho\sqrt{\mathbf{x}'\mathbf{x}}$. Thus, if we plot ($\mathbf{x}$, $\rho\mathbf{y}$), ρy_i tends to be closer to the origin of the graph than x_j. The same applies to the plot of ($\mathbf{y}$, $\rho\mathbf{x}$): $\rho\mathbf{x}$ is closer to the origin than $\mathbf{y}$. Therefore the idea of projections of one set onto the other is not very satisfactory. An alternative approach to joint graphical display is needed, and is presented and discussed in chapter 14.

Lastly, we should note that we must deal with similar problems when dominance data (e.g., paired comparison, rank-order, successive categories data) are analyzed. Please refer to chapter 14 for further details on graphical display.

PART II
Incidence Data

Part 2 contains three chapters, dealing with contingency/frequency data, multiple-choice data, and sorting data. These three types of categorical data are closely related in the way in which they are anlayzed by dual scaling, that is, a straightforward application of the basic principle of the quantification method. These data are absolute measurements, given in the form of presence or absence, or frequencies, and are grouped together under the name *incidence data*. Out of the three data types, contingency tables and multiple-choice data are the two main data types that have been considered and thoroughly investigated for quantification by the other variants of this general technique, especially the French *analyse des correspondances*.

The reason for distinguishing these types of data from others will become clear when you look at and examine data types that are grouped under the name *dominance data*, to be discussed in Part 3. Dominance data are ordinal measurements that indicate only equality or inequality (greater than or smaller than) of variables, and present fundamentally distinct problems from those associated with incidence data. We have chosen incidence data to start with, so that you may become familiar with the standard use of dual scaling before you look at some of its variations.

CHAPTER 7
Contingency/Frequency Tables

7.1 SPECIAL PROBLEMS

Contingency/frequency tables are the most straightforward type of data to handle by dual scaling discussed in this book. Let us first recapitulate what dual scaling will do with this data type. In summary form, dual scaling will determine weights for the rows, y_i, and weights for the columns, x_j, of the contingency or frequency table in such a way that:

1. The row-column association in the data will be maximally explained;
2. Analysis will continue, as necessary, to extract more dimensions to explain other distinct patterns of association;
3. The correlation between the rows and the columns will be a maximum;
4. The regression of y_i on x_j and the regression of x_j on y_i will be simultaneously linear; and
5. Both between-row and between-column discriminations (variations) will be maximal.

These characteristics have been discussed in previous chapters. In the case of Item 1, however, we need to discuss the trivial solution. Although the trivial solution has already been mentioned briefly in chapter 6, it is nevertheless useful to look at it from a different point of view, that is, in relation to the analysis of the contingency table. Suppose that you asked two groups of 100 people each how often they read *Business Week* (often, sometimes, never) and the *Financial Times* (often, sometimes, never), and that you obtained the results shown in Table 7.1.

In both groups, 50% of people read the *Financial Times* often, 30% sometimes, and 20% never. Similarly, both groups showed the same percentages of 40%, 30%, and 30% for the three response categories for the question about *Business Week*. These percentages are called *marginal frequencies*, and this example shows that the *Financial Times* and *Business Week* have identical marginal frequencies for the two groups. When we consider combinations of categories (e.g., often for the *Financial Times* and

TABLE 7.1
Contingency Tables from Two Groups

		(Group 1) FT					(Group 2) FT			
		a	b	c	Total		a	b	c	Total
	a	20	12	8	40	a	35	5	0	40
BW	b	15	9	6	30	b	9	20	1	30
	c	15	9	6	30	c	6	5	19	30
	Total	50	30	20	100		50	30	20	100

Note: a = often; b = sometimes; c = never: BW = *Business Week*; FT = *Financial Times*

sometimes for *Business Week*), the corresponding frequencies are called *joint frequencies*. When we tabulate the joint frequency tables (contingency tables) for the two groups, they are quite different. Group 1 shows an example in which rows and columns of the table are said to be *statistically independent*, meaning that for this group of people as a whole the *Financial Times* and *Business Week* do not have a common appeal. In other words, your knowledge of someone in the group who reads *Business Week* often would not help you to figure out whether the same person reads the *Financial Times* often. Simply put, you cannot make any inference from one to the other. This is the condition of *statistical independence*.

Checking data for statistical independence is relatively simple. For instance, look at Group 1 and the distribution of responses over the three categories of *Business Week* with respect to each category of the *Financial Times*: there are 50 in "often" for the *Financial Times*, who are distributed as 20, 15, and 15 (i.e., 40%, 30%, 30%) over the three categories of *Business Week*; there are 30 in "sometimes" for the *Financial Times*, who are again distributed over the three *Business Week* categories with percentages of 40%, 30%, and 30%, respectively; there are 20 in "never" for the *Financial Times*, who are again equally distributed over the three *Business Week* categories, that is, 40%, 30%, and 30%. Thus, knowledge of someone's response to *Financial Times* would not help you guess their response to *Business Week*, and vice versa.[1]

We often describe this condition as *joint frequencies being proportional to the marginal frequencies*. You can also find the following relation: when rows and columns are statistically independent, the joint frequency f_{ij} in row i and column j is equal to the product of the marginal row frequency f_i

[1] If you reverse the argument, you note that for each category of *Business Week*, you see 50%, 30% and 20% of responses for often, sometimes, and never of the *Financial Times*.

and the marginal column frequency f_j, divided by the total frequency f_t. For instance, 20 in cell (1,1) = (50×40)/100, 12 in cell (1,2) = (30×40)/100, and so on. You can see that all the nine elements of the table for Group 1 can be expressed by this formula of the product of the corresponding marginals divided by the total, that is, $f_i f_j / f_t$, or, in matrix notation, $\mathbf{f}_r \mathbf{f}_c' / f_t$. To simplify the notation, let us indicate these two terms by $f_{ij(0)}$ and $\mathbf{F}_0$, respectively. That is,

$$f_{ij(0)} = \frac{f_{i.}\, f_{.j}}{f_t}, \quad \text{and} \quad \mathbf{F}_0 = \frac{\mathbf{f}_r\, \mathbf{f}_c'}{f_t}.$$

Look at the data for Group 2. It is easy to see that the concept of statistical independence does not apply to this table. The three columns are not proportional. This necessarily means that the same applies to the three rows. When rows and columns are not statistically independent, you can say that, in the present example, the *Financial Times* and *Business Week* have common appeal to these 100 subjects. For instance, it may mean that if someone reads the *Financial Times* often, the person is likely to read *Business Week* often. This kind of common appeal is called the *association* between rows and columns of the contingency/frequency table.

From the preceding discussion, you know that the portion of the data that can be attributed to the row-column association can be expressed as the observed table minus the table of frequencies expected when the rows and the columns are statistically independent. In our example, it is clear that if the data are obtained from Group 1, the table of frequencies that results from the association consists of all zeros. In contrast, if the observed data are from Group 2, the table of frequencies that results from the association is given by Table 7.2.

TABLE 7.2
Table of Association

		FT			
		a	b	c	Total
	a	15	-7	-8	0
BW	b	-6	11	-5	0
	c	-9	-4	13	0
	Total	0	0	0	0

Did you follow how this table was obtained? It is simple: Group 2 minus Group 1. Consider another example in which the survey investigated distributions of male and female readers of three magazines (Table 7.3).

TABLE 7.3
Contingency Table: Genders and Magazines

		Magazine			
		A	B	C	Total
Sex	F	80	40	0	120
	M	20	20	40	80
	Total	100	60	40	200

The expected table of frequencies, when the rows and the columns are statistically independent, can be calculated as in Table 7.4.

TABLE 7.4
Frequencies When Gender and Magazine are Statistically Independent

	A	B	C	Total
F	120×100/200 = 60	120×60/200 = 36	120×40/200 = 24	120
M	80×100/200 = 40	80×60/200 = 24	80×40/200 = 16	80
	100	60	40	200

Therefore, the table of frequencies resulting from the row-column association is obtained by subtracting the elements of Table 7.4 from the corresponding elements in Table 7.3, resulting in Table 7.5.

TABLE 7.5
Observed Minus Expected Frequencies

	A	B	C	Total
F	20	4	-24	0
M	-20	-4	24	0
Total	0	0	0	0

Dual scaling analyzes this table of association. You may wonder how this is related to the examples of dual scaling you saw in the previous chapters. To fill the gap between dual scaling of this table of association and dual scaling of the original table of observed frequencies, a convincing explanation is in order. Do you recall the dual relations of Formula 4.8? It so happens that the information expected when there exists statistical independence between rows and columns of the table does indeed satisfy the dual relations. More specifically, you should be aware of the following well-known fact: statistical independence corresponds to the case in which $y_i = x_j = 1$ for all rows and columns, and in which $\eta = 1$. This applies to any contingency/frequency table, independent of a data matrix, that is, it has nothing to do with the distribution of responses in the data. This is therefore called a *trivial solution*, and is eliminated from reporting of the results. You should also know that assigning the same weight to all the rows (or columns) neither maximizes the between-column (row) sum of squares, nor satisfies the condition that the sum of the weighted responses is zero. So, you must discard such a solution that does not reflect the effect of quantification.[2]

Another question you may have is why the method of reciprocal averages, discussed in the previous chapter, was applied not to the table of association, that is, $\mathbf{F} - \mathbf{F}_0$, but to the table of observed response frequencies, $\mathbf{F}$. It was because we carefully avoided obtaining this trivial solution by choosing an initial set of weights and entering the intermediate weights by subtracting the mean from them. To have a general, and preferred, method, dual scaling should be applied to the table of association. You can therefore say that dual scaling is a technique to decompose row-column association in the data into many components or patterns of relationship.

Now that this problem of decomposition has been brought up, it would be appropriate to present a formula for the structure of f_{ij}. There is a well-known formula in statistics which presents one way of expressing how f_{ij} can be described.[3] Within the framework of dual scaling, it is called the structural equation (see chapter 6), or reconstitution formula (Benzécri et al., 1973). Using our notation, the observed frequency f_{ij} in cell (i,j) can be expressed in terms of the trivial solution as discussed above and dual scaling solutions (ρ_k, y_{ki}, x_{kj}) as follows:

$$f_{ij} = f_{ij(0)}[1 + \rho_1 y_{1i} x_{1j} + \rho_2 y_{2i} x_{2j} + \cdots + \rho_k y_{ki} x_{kj} + \cdots + \rho_t y_{ti} x_{tj}],$$

where the subscripts 1, 2, . . . , k, . . . , t indicate solutions 1, 2, . . . , k, . . . , t, respectively; t is usually equal to the smaller of $(n\text{-}1)$ and $(m\text{-}1)$ of the $n \times m$ table; and ρ_k is the same as η_k, as discussed in chapter 4, that is, the

---

[2] The trivial solution discussed is exactly the same as what was debated in chapter 6. If you replace the data matrix $\mathbf{F}$ with the residual matrix $\mathbf{F} - \mathbf{F}_0$ in forming SS_b, you will get the matrix that is free from the trivial solution:

$$\mathbf{D}_c^{-1/2}(\mathbf{F}\text{-}\mathbf{F}_0)'\mathbf{D}_r^{-1}(\mathbf{F}\text{-}\mathbf{F}_0)\mathbf{D}_c^{-1/2} = \mathbf{D}_c^{-1/2}\mathbf{F}'\mathbf{D}_r^{-1}\mathbf{F}\mathbf{D}_c^{-1/2} - \mathbf{D}_c^{1/2}\mathbf{1}\mathbf{1}'\mathbf{D}_c^{1/2}.$$

[3] See, for example, Lancaster (1958), Dempster (1969), and Benzécri et al. (1973).

positive squareroot of η_k^2. Note that the first term in the square bracket, 1, indicates the contribution of the trivial solution as we discussed earlier. The trivial solution is given by ($\rho = 1, y_i = 1, x_j = 1$), so that $\rho y_i x_j = 1$. When there exists statistical independence between rows and columns, we have $\rho_1 = \rho_2 = \ldots = \rho_k = \ldots = \rho_t = 0$: so that $f_{ij} = f_i f_j / f_t$. Notice, however, that the trivial solution is "trivial" not because the same expression can be obtained under statistical independence, but because it provides ρ to be equal to one no matter what data you may have.

As mentioned above, dual scaling decomposes association in the data, hence using the following formula:

$$f_{ij}^* = f_{ij} - f_{ij(0)} = f_{ij(0)}[\rho_1 y_{1i} x_{1j} + \rho_2 y_{2i} x_{2j} + \ldots + \rho_k y_{ki} x_{kj} + \ldots + \rho_t y_{ti} x_{tj}].$$

This is an important formula to remember.[4] As Bock (1960) stated, we determine y_{ki} and x_{kj} using the *least-squares principle*. Although other approaches to dual scaling were presented in chapter 4, the least-squares principle defines the problem in the following way. Determine ρ_1, y_1, and x_1 in such a way that the "loss function" Ψ_1 should be a minimum, under appropriate constraints on ρ_1, y_{1i}, and x_{1j} (*e.g.*, $0 < \rho_1 < 1$, $\Sigma f_i y_{1i}^2 = \Sigma f_j x_{1j}^2 = f_t$), where

$$\Psi_1 = \sum\sum (f_{ij} - f_{ij(0)} - \rho_1 y_{1i} x_{1j})^2$$

In other words, the first solution (ρ_1, y_{1i}, x_{1j}) that minimizes Ψ_1 explains the maximum amount of association in f_{ij}^*. For this reason, Nishisato (1980a) uses the term *order-k approximation to* f_{ij} as presented in chapter 6.

1. **Order-0 approximation** $= f_{ij(0)}$
2. **Order-1 approximation** $= f_{ij(1)} = f_{ij(0)}(1 + \rho_1 y_{1i} x_{1j})$
3. **Order-2 approximation** $= f_{ij(2)} = f_{ij(1)} + f_{ij(0)} \rho_2 y_{2i} x_{2j}$

The order-2 approximation is the solution that, given the first solution, minimizes Ψ_2, where

---

[4] In matrix notation,

$$\mathbf{F}_{res} = \mathbf{F} - \mathbf{F}_0 = \frac{1}{f_t}\mathbf{D}_r[\rho_1 \mathbf{y}_1 \mathbf{x}_1' + \rho_2 \mathbf{y}_2 \mathbf{x}_2' + \ldots + \rho_t \mathbf{y}_t \mathbf{x}_t']\mathbf{D}_c = \frac{1}{f_t}\mathbf{D}_r \mathbf{Y} \Lambda \mathbf{X}' \mathbf{D}_c,$$

where $\mathbf{D}_r$ and $\mathbf{D}_c$ are diagonal matrices of row totals and column totals of $\mathbf{F}$, respectively, $\mathbf{Y} = [\mathbf{y}_1, \mathbf{y}_2, \ldots, \mathbf{y}_t]$, $\mathbf{X} = [\mathbf{x}_1, \mathbf{x}_2, \ldots, \mathbf{x}_t]$, and $\Lambda = \text{diag}(\rho_1, \rho_2, \ldots, \rho_t)$.

$$\Psi_2 = \sum\sum(f_{ij} - f_{ij(1)} - f_{ij(0)}\rho_2 y_{2i} x_{2j})^2$$

These solutions can be obtained by applying the method of reciprocal averages, first to $f_{ij}*$, and then to each successive residual table, that is, to the difference between $f_{ij}*$ and the order-k approximation.

7.2 APPLICATIONS

In this section, we present two examples of dual scaling as applied to contingency/frequency tables.

7.2.1 Theory of Evolution?

Table 7.6 is a part of data reported in Stebbins (1950). Eleven varieties of barley were planted at 10 agricultural experimental stations in the United States. An equal number of seeds of each variety were planted. At harvesting time, all the seeds were mixed and 500 seeds were randomly sampled and sorted into the 11 varieties. These 500 seeds were then planted, and 500 randomly chosen seeds were then sorted into the varieties. This process was repeated a number of times as indicated in the table. Table 7.6 is a part of the data, and shows the final counting of six varieties of barley at six locations. In spite of the fact that the distribution of seeds in Year 1 was perfectly even, the above table shows that after a number of years different varieties of barley have more or less identified particular favored locations. For instance, Moro is dominated by White Smyrna with 189 seeds out of 500. In fact, each location is distinctly dominated by one variety, namely, Arlington by Coast & Trebi, Ithaca by Manchuria, St. Paul by Hanchen, Mocasin by White Smyrna, Moro by White Smyrna, and Davis by Coast & Trebi. There are, however, other combinations which are not insignificant. Thus, it would be interesting to see what dual scaling can do to identify relative positions of those varieties of barley and locations of agricultural experimental stations.

Let us look at the frequencies expected when rows and columns are statistically independent, referred to as the "order-0 approximation." As discussed earlier, the frequency of cell (i,j) under statistical independence is given by $f_i f_j / f_t$, or in matrix notation, $\mathbf{f}_r\mathbf{f}_c'/f_t$, as shown in Table 7.7.

The differences between the observed frequencies and those expected under statistical independence are joint frequencies due to association between rows and columns, quantities that dual scaling decomposes into orthogonal components. Table 7.8 shows this matrix of association. Since

TABLE 7.6
Barley Data (Stebbins, 1950)

Variety of Barley	Location: Arlington, VA 4 years	Ithaca, NY 12 years	St. Paul, MN 10 years	Mocasin, MT 12 years	Moro, OR 12 years	Davis, CA 4 years
CT	446	57	83	87	6	362
H	4	34	305	19	4	34
WS	4	0	4	241	489	65
Man	1	343	2	21	0	0
G	13	9	15	58	0	1
Mel	4	0	0	4	0	27

Notes: CT = Coast & Trebi; H = Hanchen; WS = White Smyrna;
Man = Manchuria; G = Gatami; Mel = Meloy

TABLE 7.7
Order-0 Approximation to the Observed Frequencies

Barley	Arlington	Ithaca	St. Paul	Mocasin	Moro	Davis
CT	179.19	168.18	163.25	155.28	189.45	185.65
H	68.85	64.62	62.73	59.66	72.79	71.33
WS	138.23	129.73	125.93	119.78	146.13	143.20
Man	63.17	59.29	57.55	54.74	66.79	65.45
G	16.53	15.51	15.05	14.32	17.47	17.12
Mel	6.02	5.65	5.49	5.22	6.37	6.24

TABLE 7.8
Matrix of Association (Observed Minus Expected)

Barley	Arlington	Ithaca	St. Paul	Mocasin	Moro	Davis
CT	266.81	-111.18	-72.28	-76.25	-183.45	176.35
H	-64.85	-30.62	245.34	-43.73	-68.79	-37.33
WS	-134.23	-129.73	-115.78	115.07	342.87	-78.20
Man	-62.17	283.71	-52.74	-36.55	-66.79	-65.45
G	-3.53	-6.51	0.68	42.95	-17.47	-16.12
Mel	-2.02	-5.65	-5.22	-1.49	-6.37	20.76

the data matrix is 6×6, the total number of possible solutions is in general five. To see what they are, let us extract all the solutions. Table 7.9 lists the squared correlation ratio η_k^2, the percentage of each contribution δ_k, and the cumulative percentage of each of the five solutions. If you look at the squared correlation ratios, you notice that the first three correlation ratios were comparatively large, and the fourth correlation ratio drops rather

TABLE 7.9
η_j^2, δ_j, and $\Sigma\delta_j$ of Five Solutions

Solution	η_j^2	δ_j	$\Sigma\delta_j$
1	0.7352	38.18%	38.18%
2	0.6325	32.85%	71.03%
3	0.4802	24.94%	95.97%
4	0.0586	3.04%	99.01%
5	0.0190	0.99%	100%

suddenly, suggesting that the first three solutions are main components for the data set. Indeed, the cumulative δ_k seems to support this conjecture.[5] Therefore, we examine only the first three solutions.

TABLE 7.10
Normed and Adjusted Row and Column Weights

Row Weights	Normed			Adjusted		
Solution	1	2	3	1	2	3
Row 1	0.13	0.86	-0.88	0.11	0.68	-0.61
Row 2	0.58	0.93	2.14	0.49	0.73	1.49
Row 3	-1.28	-0.82	0.20	-1.10	-0.65	0.14
Row 4	1.86	-1.71	-0.25	1.60	-1.36	-0.17
Row 5	-0.19	-0.09	0.25	-0.16	-0.07	0.17
Row 6	-0.12	0.69	-1.04	-0.05	0.69	-0.72
Column Weights	Normed			Adjusted		
Solution	1	2	3	1	2	3
Column 1	0.13	1.02	-1.17	0.11	0.81	-0.81
Column 2	1.75	-1.43	-0.20	1.50	-1.14	-0.14
Column 3	0.52	0.84	2.06	0.45	0.85	1.43
Column 4	-0.70	-0.33	0.06	-0.60	-0.33	0.04
Column 5	-1.45	-0.79	0.29	-1.25	-0.79	0.20
Column 6	-0.05	0.64	-0.77	-0.04	0.64	-0.53

In chapter 6, the distinction was made between *normed* and *adjusted* vectors of weights. Recall that normed weight vectors $\mathbf{y}_k$ and $\mathbf{x}_k$ are scaled to $\mathbf{y}_k'\mathbf{D}_r\mathbf{y}_k = \mathbf{x}_k'\mathbf{D}_c\mathbf{x}_k = f_t$, irrespective of the value of η_k^2, $k = 1, 2, \ldots, t$. We

[5] The interpretation of δ_k is tricky, especially when it is used for the contingency table and multiple-choice data. See Chapter 17 for thorough information on δ_k.

have just seen above that the relative importance of each solution is judged by the magnitude of η_k^2. Hence we have introduced adjusted weight vectors $\mathbf{y}_k^*$ and $\mathbf{x}_k^*$ such that $\mathbf{y}_k^* = \eta_k \mathbf{y}_k$ and $\mathbf{x}_k^* = \eta_k \mathbf{x}_k$, that is, $\mathbf{y}_k^{*\prime}\mathbf{D}_r\mathbf{y}_k^* = \mathbf{x}_k^{*\prime}\mathbf{D}_c\mathbf{x}_k^* = \eta_k^2 f_t$. Table 7.10 lists both normed and adjusted weights of the three solutions.

Considering that only the first three solutions seem to have substantial contributions to the total information, let us plot only the three solutions, using adjusted weights (Figure 7.1). By looking at the coordinates (weights) and these graphs, one can construct Table 7.11 that presents clusters of row and column variables. As was mentioned in chapter 6, the space for $\mathbf{y}_k$ and the space for $\mathbf{x}_k$ do not match perfectly, hence overlaying the two graphs is only a means to arrive at rough clustering of rows and columns. At this stage of analysis, we would need an expert in agriculture to explain, for example, what common conditions exist among those places in a cluster—what agricultural conditions do Davis, California, and Arlington, Virginia, share? What about the relation between Moro, Oregon, and Mocasin, Montana? Without expert knowledge about the substantive area, we cannot give an appropriate interpretation to the question of why two locations are suited to one variety. All we can do is present the full results for some future interpretion.

TABLE 7.11
Clusters of Rows and Columns

Variety of Barley	Location
Mancuria	Ithaca
Hanchen	St. Paul
White Smyrna, Gatami	Moro, Mocasin
Coast & Trebi, Meloy	Arlington, Davis

Some of the analytical results reported in this book were prepared by the DUAL3 Statistical Package (Appendix A). It may be of interest for the reader to see an output of the present example from the DUAL3 program. Only the first three solutions are presented below. In this output, the following terms are used:

The total variance accounted for = The sum of all the squared correlation ratios

Maximum product-moment correlation = ρ = The positive square root of η^2

Percentage homogeneity = $\delta \times$ total variance = $100\eta^2$

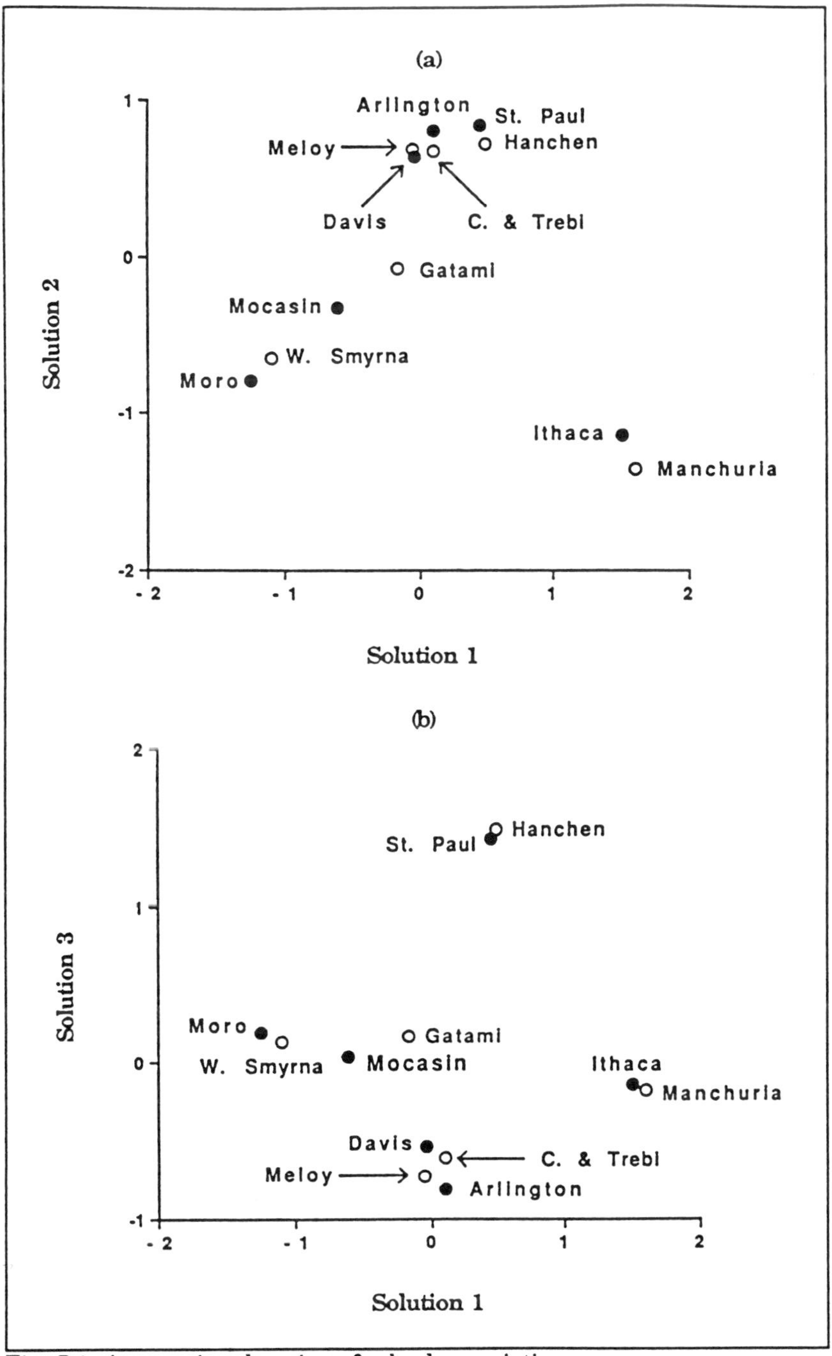

Fig. 7.1. Appropriate locations for barley varieties.

DUAL3CF Output of Barley Data
(The first three solutions only)

DUAL SCALING
CONTINGENCY/FREQUENCY TABLES

TITLE OF STUDY: Stebbins' Barley Data

*** INPUT DATA ***

446	57	83	87	6	362
4	34	305	19	4	34
4	0	4	241	489	65
1	343	2	21	0	0
13	9	15	58	0	1
4	0	0	4	0	27

*** ROW TOTALS ***

1041	400	803	367	96	35

*** COLUMN TOTALS ***

472	443	409	430	499	489

*** ORDER 0 APPROXIMATION ***

179.2	168.2	155.3	163.2	189.4	185.6
68.9	64.6	59.7	62.7	72.8	71.3
129.7	119.8	125.9	146.1	143.2	
63.2	59.3	54.7	57.6	66.8	65.4
16.5	15.5	14.3	15.1	17.5	17.1
6.0	5.7	5.2	5.5	6.4	6.2

*** RESIDUAL MATRIX ***

266.8	-111.2	-72.3	-76.2	-183.4	176.4
-64.9	-30.6	245.3	-43.7	-68.8	-37.3
-134.2	-129.7	-115.8	115.1	342.9	-78.2
-62.2	283.7	-52.7	-36.6	-66.8	-65.4
-3.5	-6.5	0.7	42.9	-17.5	-16.1
-2.0	-5.7	-5.2	-1.5	-6.4	20.8

CHI-SQUARE DUE TO ROW-COLUMN ASSOCIATION
= 5279.896 WITH 25 DEGREES OF FREEDOM
SIGNIFICANT AT THE 0.05 LEVEL
*** THE TOTAL VARIANCE TO BE ACCOUNTED FOR = 1.925564 ***

SOLUTION 1

SQUARED CORRELATION RATIO = 0.73524
MAXIMUM PRODUCT-MOMENT CORRELATION = 0.85746
PERCENTAGE HOMOGENITY = 73.52
DELTA (TOTAL VARIANCE ACCOUNTED FOR): PARTIAL = 38.18
CUMULATIVE = 38.18

*** OPTIMAL WEIGHT VECTORS ***

	*** ROWS ***			*** COLUMNS ***	
	NORMED	WEIGHTED		NORMED	WEIGHTED
1	0.1281	0.1098	1	0.1316	0.1128
2	0.5771	0.4948	2	1.7502	1.5007
3	-1.2781	-1.0960	3	0.5203	0.4461
4	1.8647	1.5989	4	-0.7000	-0.6002
5	-0.1868	-0.1602	5	-1.4535	-1.2464
6	-0.1198	-0.1027	6	-0.0489	-0.0420

*** ORDER 1 APPROXIMATION ***

181.8	200.5	164.1	150.7	159.2	184.7
73.3	120.6	75.0	41.0	20.4	69.6
118.3	-119.1	51.5	222.5	378.9	150.9
76.5	225.2	100.3	-6.9	-88.4	60.3
16.2	11.2	13.1	16.7	21.5	17.3
5.9	4.6	4.9	5.9	7.3	6.3

*** RESIDUAL MATRIX ***

264.2	-143.5	-81.1	-63.7	-153.2	177.3
-69.3	-86.6	230.0	-22.0	-16.4	-35.6
-114.3	119.1	-47.5	18.5	110.1	-85.9
-75.5	117.8	-98.3	27.9	88.4	-60.3
-3.2	-2.2	1.9	41.3	-21.5	-16.3
-1.9	-4.6	-4.9	-1.9	-7.3	20.7

CHI-SQUARE DUE TO ROW-COLUMN ASSOCIATION
= 3635.282 WITH 9 DEGREES OF FREEDOM
SIGNIFICANT AT THE 0.05 LEVEL

SOLUTION 2

SQUARED CORRELATION RATIO = 0.63246
MAXIMUM PRODUCT-MOMENT CORRELATION = 0.79528
PERCENTAGE HOMOGENITY = 63.25
DELTA (TOTAL VARIANCE ACCOUNTED FOR): PARTIAL = 32.85
CUMULATIVE = 71.03

*** OPTIMAL WEIGHT VECTORS ***

	*** ROWS ***			*** COLUMNS ***	
	NORMED	WEIGHTED		NORMED	WEIGHTED
1	0.8571	0.6816	1	1.0209	0.8119
2	0.9265	0.7368	2	-1.4349	-1.1412
3	-0.8195	-0.6517	3	1.0626	0.8450
4	-1.7057	-1.3565	4	-0.4186	-0.3329
5	-0.0941	-0.0748	5	-0.9875	-0.7853
6	0.8641	0.6872	6	0.8016	0.6375

*** ORDER 2 APPROXIMATION ***

306.5	36.0	276.6	104.1	31.7	286.1
125.1	52.3	121.7	21.7	-32.5	111.7
26.3	2.2	-31.5	256.9	473.0	76.1
-11.0	340.6	21.4	25.8	1.0	-10.8
14.9	12.8	12.0	17.2	22.8	16.2
10.2	-0.9	8.8	4.3	3.0	9.7

*** RESIDUAL MATRIX ***

139.5	21.0	-193.6	-17.1	-25.7	75.9
-121.1	-18.3	183.3	-2.7	36.5	-77.7
-22.3	-2.2	35.5	-15.9	16.0	-11.1
12.0	2.4	-19.4	-4.8	-1.0	10.8
-1.9	-3.8	3.0	40.8	-22.8	-15.2
-6.2	0.9	-8.8	-0.3	-3.0	17.3

CHI-SQUARE DUE TO ROW-COLUMN ASSOCIATION
= 2738.044 WITH 7 DEGREES OF FREEDOM
SIGNIFICANT AT THE 0.05 LEVEL

SOLUTION 3

SQUARED CORRELATION RATIO = 0.48024
MAXIMUM PRODUCT-MOMENT CORRELATION = 0.69299
PERCENTAGE HOMOGENITY = 48.02
DELTA (TOTAL VARIANCE ACCOUNTED FOR): PARTIAL = 24.94
CUMULATIVE = 95.97

*** OPTIMAL WEIGHT VECTORS ***

	*** ROWS ***			*** COLUMNS ***	
	NORMED	WEIGHTED		NORMED	WEIGHTED
1	-0.8792	-0.6093	1	-1.1737	-0.8134
2	2.1436	1.4855	2	-0.1973	-0.1367
3	0.2014	0.1396	3	2.0636	1.4301
4	-0.2495	-0.1729	4	0.0600	0.0416
5	0.2500	0.1733	5	0.2944	0.2040
6	-1.0380	-0.7194	6	-0.7675	-0.5319

*** ORDER 3 APPROXIMATION ***

434.6	56.2	81.4	98.2	-2.3	372.9
5.1	33.3	304.6	27.2	-0.7	30.4
3.7	-1.4	3.0	257.9	479.0	60.7
1.8	342.6	1.8	25.2	-2.4	-2.2
11.6	12.3	17.1	17.4	23.7	14.0
15.3	-0.1	1.0	4.1	1.6	13.2

*** RESIDUAL MATRIX ***

11.4	0.8	1.6	-11.2	8.3	-10.9
-1.1	0.7	0.4	-8.2	4.7	3.6
0.3	1.4	1.0	-16.9	10.0	4.3
-0.8	0.4	0.2	-4.2	2.4	2.2
1.4	-3.3	-2.1	40.6	-23.7	-13.0
-11.3	0.1	-1.0	-0.1	-1.6	13.8

CHI-SQUARE DUE TO ROW-COLUMN ASSOCIATION
= 1790.082 WITH 5 DEGREES OF FREEDOM
SIGNIFICANT AT THE 0.05 LEVEL
** NOTE: DELTA CUMULATIVE EXCEEDS 95% **
*** END OF ANALYSIS ***

7.2.2 Rorschach Responses and Induced Moods

Swiss psychiatrist Rorschach suggested in 1921 that the way a person perceived inkblots would indicate important aspects of his or her personality makeup, and proposed a projective technique for analyzing personality: the subject is shown a series of inkblots and is asked to tell what each inkblot looks like, or what the subject sees in the inkblot; the investigator then analyzes the responses in terms of what they are and how they are formed (e.g., global picture, detailed picture, dynamic or static picture). Garmize and Rycklak (1964) postulated some relations between Rorschach responses and moods of the subjects, and collected Rorschach responses under six types of experimentally induced moods. Table 7.12 shows joint frequencies of responses.

TABLE 7.12
Rorschach Responses and Induced Moods (Garmize and Rychlak, 1964)

Rorschach	Induced Moods					
Response	Fear	Anger	Depression	Ambition	Security	Love
Bat	33	10	18	1	2	6
Blood	10	5	2	1	0	0
Butterfly	0	2	1	26	5	18
Cave	7	0	13	1	4	2
Clouds	2	9	30	4	1	6
Fire	5	9	1	2	1	1
Fur	0	3	4	5	5	21
Mask	3	2	6	2	2	3
Mountains	2	1	4	1	18	2
Rocks	0	4	2	1	2	2
Smoke	1	6	1	0	1	0

This table reveals comparatively more confusion in the row-column correspondence, which may indicate that the association patterns may be multidimensional, that is, multifarious. To see this aspect, let us obtain the table of associations (Table 7.13), that is, the table of observed frequencies minus expected frequencies, $\mathbf{F}\text{-}\mathbf{F}_0$, or if you prefer, a residual matrix (see chapter 6). The best indicator of multidimensionality is perhaps the distribution of eigenvalues, that is, η_k^2 in dual scaling. Since the data matrix is 11×6, five solutions (i.e., min[n-1,m-1]) are obtained, and the distribution of η_k^2 is as shown in Table 7.14. Recall that in our first example there was a sudden drop in the value of η^2 between the third and the fourth solutions, suggesting that we should look at only three solutions. In this second example, there does not seem to be such a sudden drop.

Therefore, let us examine all the solutions and see if we can interpret each one of them.

TABLE 7.13
Residual Table: Association

Response	Induced Moods					
	Fear	Anger	Depression	Ambition	Security	Love
Bat	20.11	-0.44	1.22	-8.01	-6.39	-6.49
Blood	6.68	2.32	-2.32	-1.32	-2.16	-3.21
Butterfly	-9.58	-5.75	-11.47	19.31	-1.23	8.73
Cave	2.03	-4.03	6.53	-2.47	0.76	-2.82
Clouds	-7.58	1.25	17.53	-2.69	-5.23	-3.27
Fire	1.50	6.17	-3.56	-0.44	-1.28	-2.39
Fur	-7.00	-2.67	-5.11	0.11	0.44	14.22
Mask	-0.32	-0.68	1.68	-0.32	-0.16	-0.21
Mountains	-3.16	-3.18	-2.71	-2.60	14.64	-2.99
Rocks	-2.03	2.36	-0.64	-0.42	0.68	0.04
Smoke	-0.66	4.66	-1.16	-1.16	-0.08	-1.61

TABLE 7.14
Statistics for Five Solutions

Solution	η_j^2	δ_j	$\Sigma\delta_j$
1	0.4633	42.72%	42.72%
2	0.2505	23.10%	65.82%
3	0.1704	15.71%	81.53%
4	0.1281	11.81%	93.35%
5	0.0721	6.65%	100.00%

Table 7.15 lists adjusted weight vectors of five solutions. For purposes here, we are using adjusted weights because they reflect relative magnitudes of η. When the data matrix is relatively large, it is helpful to introduce an arbitrary cutting point, for instance, the absolute value of 0.3000, for each solution. In this case, any values between -0.2999 and 0.2999 are eliminated from the table as shown in Table 7.16. Another way of simplifying the task of interpretation is to plot variables in the graph and label only those which are above the cutting point, as shown in Figure 7.2.[6]

[6] For 5 solutions, there are 10 combinations of 2 solutions (i.e., solutions 1 & 2, 1 & 3, . . ., 4 & 5) for graphical display. However, for most practical purposes, plotting 1 vs. 2, 1 vs. 3, 1 vs 4, and 1 vs. 5, or 1 vs. 2, 2 vs. 3, 3

TABLE 7.15
Adjusted Weights

	Solution				
Response	1	2	3	4	5
Bat	0.70	0.16	0.15	-0.34	-0.08
Blood	0.87	0.35	0.60	-0.18	0.07
Butterfly	-1.17	0.44	0.17	-0.15	0.32
Cave	0.37	-0.30	-0.44	-0.34	0.11
Clouds	0.23	0.08	-0.75	0.30	0.13
Fire	0.42	0.30	0.63	0.59	0.07
Fur	-0.78	0.08	-0.08	0.01	-0.67
Mask	0.05	-0.04	-0.21	-0.04	0.02
Mountains	-0.34	-1.54	0.31	-0.04	0.11
Rocks	-0.11	-0.15	0.13	0.70	-0.08
Smoke	0.57	0.05	0.57	1.25	0.00
Mood					
Fear	0.87	0.17	0.39	-0.48	-0.03
Anger	0.44	0.23	0.35	0.75	-0.02
Depression	0.39	-0.09	-0.68	0.02	0.10
Ambition	-1.03	0.51	0.15	-0.12	0.48
Security	-0.42	-1.27	0.29	0.00	0.05
Love	-0.76	0.23	-0.09	-0.08	-0.47

From the plot of Solution 1 and Solution 2, we can see clearly that mood "security" and response "mountains" are closely located, that is, clustered. From the plot of Solutions 2 and 3, we can identify a cluster, consisting of mood "depression" and responses "cave" and "clouds." The plot of Solutions 3 and 4 shows a cluster of mood "anger" and responses "smoke," "fire," and "rocks," and a cluster of mood "fear" and responses "cave" and "bat." Finally, the second quadrant of the first graph (i.e., Solution 1 and Solution 2) is divided into two clusters in the last graph, one consisting of mood "ambition" and response "butterfly" and the other of mood "love" and response "fur." It would be interesting to summarize dual scaling results in comparison with the relations predicted by Garmize and Rycklak as shown in Table 7.17. In interpreting the results, it is important to remember that these relations were extracted not from the clear-cut distinction in weights but from the subjective judgment of relatively large weights. Thus,

---

vs. 4, and 4 vs. 5, would suffice to identify the main contributors to individual solutions.

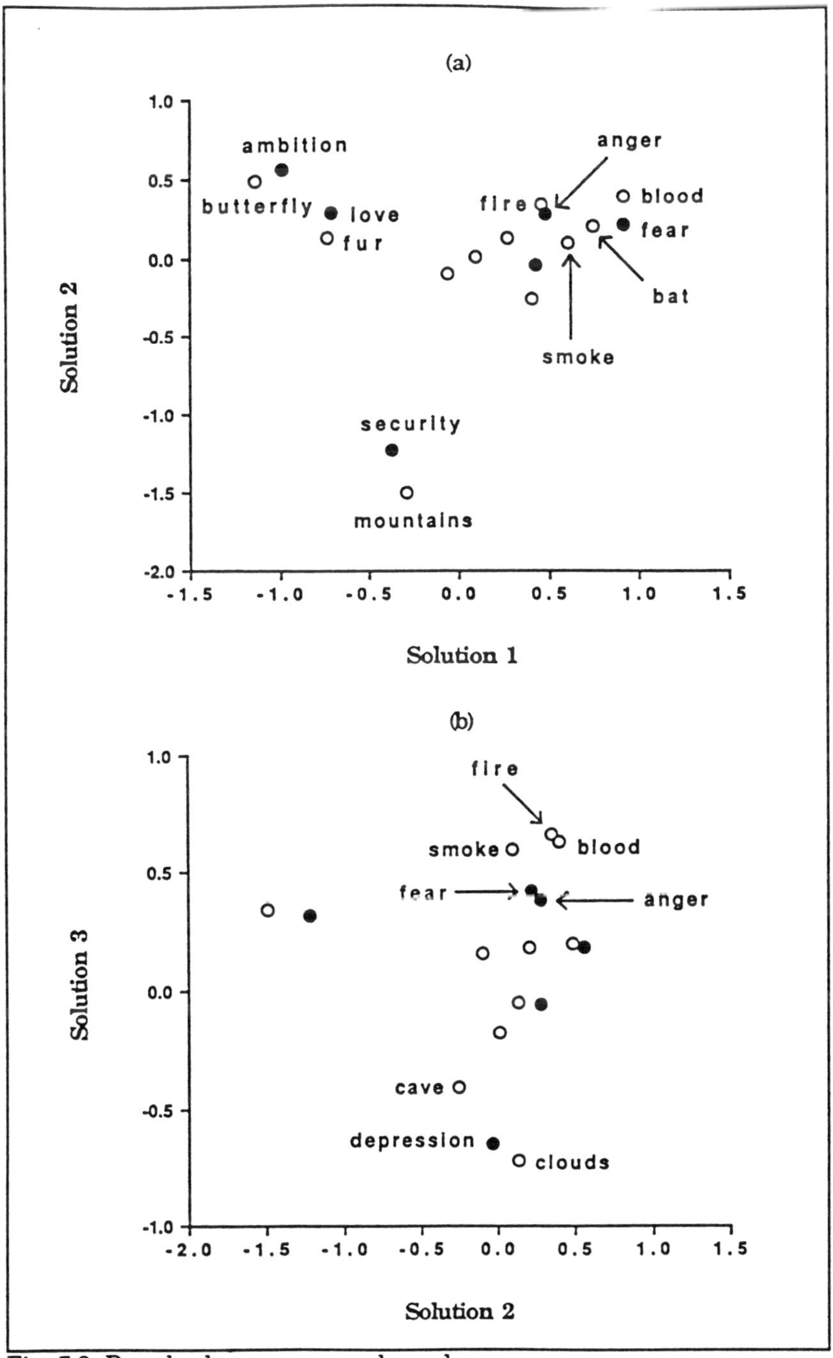

Fig. 7.2. Rorschach responses and moods.

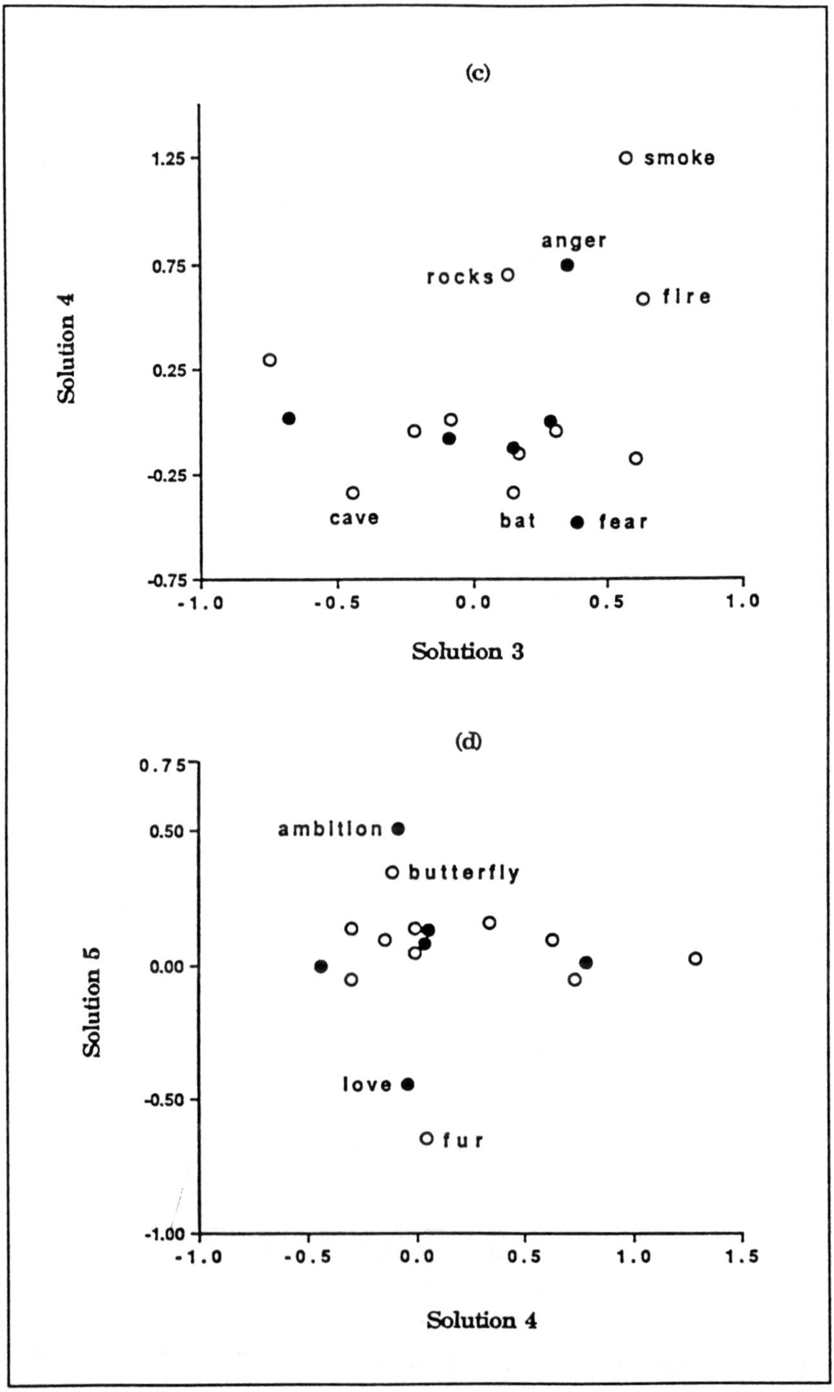
(c)
1.25
0.75
0.25
-0.25
-0.75
Solution 4
smoke
anger
rocks
fire
cave
bat
fear
-1.0
-0.5
0.0
0.5
1.0
Solution 3
(d)
0.75
0.50
0.00
-0.50
-1.00
Solution 5
ambition
butterfly
love
fur
-1.0
-0.5
0.0
0.5
1.0
1.5
Solution 4

the relations derived therein refer to major patterns or tendencies instead of to the exact one-to-one correspondence. Thus, responses associated with love, for example, are not restricted to fur only.

TABLE 7.16
Table of Main Contributors to Each Solution

	Solution				
Response	1	2	3	4	5
Bat	0.70			-0.34	
Blood	0.87	0.35	0.60		
Butterfly	-1.17	0.44			0.32
Cave	0.37		-0.44	-0.34	
Clouds		-0.75			
Fire	0.42		0.63	0.59	
Fur	-0.78				-0.67
Mask					
Mountains	-0.34	-1.54		0.31	
Rocks				0.70	
Smoke	0.57		0.57	1.25	
Mood					
Fear	0.87		0.39	-0.48	
Anger	0.44		0.35	0.75	
Depression	0.39		-0.68		
Ambition	-1.03	0.51			0.48
Security	-0.42	-1.27			
Love	-0.76				-0.47

TABLE 7.17
Summary of Relations
Hypothesized by Garmize and Rychlak, and Those by Dual Scaling

Relations Hypothesized by Garmize and Rychlak

Mood	**Rorschach Response**
Fear	Bat, Blood, Rocks
Depression	Clouds, Mask, Smoke
Ambition	Butterfly
Love	Fur, Cave, Rocks

Relations Obtained by Dual Scaling

Mood	**Rorschach Response**
Security	Mountains
Fear	Bat, Cave
Anger	Smoke, Rock, Fire
Ambition	Butterfly
Love	Fur
Depression	Clouds, Cave

In this chapter, we have looked at probably the most popular application of dual scaling, that is, to the contingency table. Even though dual scaling yields informative statistics, the interpretation of the results must still depend on the experience of the data analyst. It is also clear that the interpretation can be quite subjective, as we may infer from the two examples. In the first example, we have not presented any interpretation of why Arlington, VA, and Davis, CA, are conducive to barley varieties Meloy and Coast & Trebi. It should be obvious that dual scaling provides only statistics for you to interpret, and you must bear the burden of the final stage of interpreting the results.

CHAPTER 8
Multiple-choice Data

8.1 SPECIAL PROBLEMS

In French correspondence analysis (CA), there is a distinction between CA and multiple correspondence analysis (MCA). CA applies to the contingency table, and MCA to multiple-choice data. Although dual scaling can be applied to both data types without much attention to the differences between them, it is instructive to know their similarities and differences. This section will look at them first as an introduction to dual scaling of multiple-choice data.

8.1.1 Relation Between Contingency Tables and Multiple-Choice Data

Multiple-choice data are abundant in such areas as marketing research, personality testing, and job performance evaluation. The data can be regarded as an extension of contingency table data: while contingency tables are typically obtained by asking two multiple-choice questions (see chapters 3 and 7), multiple-choice data are obtained by asking subjects two or more multiple-choice questions. Contingency tables consist of joint frequencies tabulated in the form of the options of one item by the options of the other item, while multiple-choice data are typically represented in the subjects-by-options form of (1,0) response patterns. To clarify similarities and differences between contingency table and multiple-choice data, let us consider the following contingency table (Table 8.1), obtained by asking two questions:

Question 1: Were you born in this country?
Yes ( )   No ( )
Question 2: What is your position on the death penalty?
For ( )   Unsure ( )   Against ( )

Noting that Table 8.1 is a summary of responses to the two questions from 13 subjects, we can construct from Table 8.1, without loss of information, a 13x5 table of response patterns (incidence matrix) as seen in Table 8.2. This is a typical form of multiple-choice data. Can you see how Table 8.2 is constructed from Table 8.1? In Table 8.2, entries are 1s and 0s, the former indicating choices and the latter nonchoices. Thus, the response pattern (1, 0, 1, 0, 0) for the set (yes, no, for, unsure, against) means that the subject chose yes and against for the two questions. Subjects 1, 2, and 3 show this pattern, and the frequency of this pattern, 3, constitutes the element (1, 1) (i.e., combination of yes and for) of Table 8.1. Is this clear now? If not, try converting Table 8.2 back to Table 8.1 by counting frequencies of six response combinations, (yes, for), (yes, unsure), (yes, against), (no, for), (no, unsure), and (no, against). This exercise will probably show you the relation between the two tables. In matrix notation, the relation between the two can be expressed by a formula. Let **C** be the contingency table, and $\mathbf{F}_1$, $\mathbf{F}_2$ respectively the response-pattern matrices for questions 1 and 2. Then,

$$\mathbf{C} = \mathbf{F}_1'\mathbf{F}_2 \,. \tag{8.1}$$

TABLE 8.1
Contingency Table

		Q.2			
		For	Unsure	Against	Total
Q.1	Yes	3	2	1	6
	No	1	2	4	7
	Total	4	4	5	13

Notice that the column marginals of Table 8.2 correspond to the row marginals and the column marginals of Table 8.1, and that the row marginals of Table 8.2 indicate the number of multiple-choice questions. The grand total of Table 8.1 is 13, and that of Table 8.2 is twice that number, 26. Let us also note that Table 8.2 can be constructed from Table 8.1, and that Table 8.1 can be constructed from Table 8.2, indicating no loss of information in one format or the other.

In spite of the interchangeability between the two formats, the multiple-choice format (e.g., Table 8.2) always yields more solutions than the contingency-table format (e.g., Table 8.1). Recall that the maximum number of nontrivial solutions from a pxq contingency table is

TABLE 8.2
Multiple-Choice Data

Subject	Q.1 Yes	No	Q.2 For	Unsure	Against	Total
1	1	0	1	0	0	2
2	1	0	1	0	0	2
3	1	0	1	0	0	2
4	1	0	0	1	0	2
5	1	0	0	1	0	2
6	1	0	0	0	1	2
7	0	1	1	0	0	2
8	0	1	0	1	0	2
9	0	1	0	1	0	2
10	0	1	0	0	1	2
11	0	1	0	0	1	2
12	0	1	0	0	1	2
13	0	1	0	0	1	2
Total	6	7	4	4	5	26

(8.2) $$N_{max} = \min(p,q) - 1 ,$$

where min(*p*,*q*) means the smaller value of p and q. In contrast, N_{max} for multiple-choice data is different. Let us indicate by *n* the number of multiple-choice questions and by *m* the total number of options over *n* items. Then N_{max} for multiple-choice data is given by

(8.3) $$N_{max} = m - n .$$

Thus, the values of N_{max} for Table 8.1 and Table 8.2 are respectively 2 - 1 and 5 - 2, that is, 1 and 3.

As you may have guessed from the interchangeability of the two formats, [min(*p*,*q*)-1] solutions from the *p*×*q* contingency tables are contained in (*m* - *n*) solutions from the corresponding multiple-choice data. Notice, however, that the squared correlation ratios η_c^2 from the contingency table and η_m^2 from multiple-choice data are not identical, but are related by

(8.4) $$\eta_c^2 = (2\eta_m^2 - 1)^2 .$$

Let us look at the normed weights of Solution 1 obtained from the two formats (Table 8.3). Nishisato (1980a) shows how cell scores of the contingency table, say c_{ij}, can be calculated. If we indicate by y_i and x_j the weights of row i and column j, respectively, then the cell score c_{ij} is given by

$$c_{ij} = \frac{y_i + x_j}{t} \qquad (8.5)$$

where t is the square root of $2(\eta_c+1)$. In the present example, $t = 1.7082$. Therefore, the six cell scores are

$c_{11} = (1.0801 + 1.2605)/1.7082 = 1.3702$
$c_{12} = (1.0801 + 0.1681)/1.7082 = 0.7307$
$c_{13} = (1.0801 - 1.1429)/1.7082 = -0.0368$
$c_{21} = (-0.9258 + 1.2605)/1.7082 = 0.1959$
$c_{22} = (-0.9258 + 0.1681)/1.7082 = -0.4436$
$c_{23} = (-0.9258 - 1.1429)/1.7082 = -1.2110$

TABLE 8.3
Normed Weights from Contingency Table and Response-Pattern Table

Contingency		Response Pattern		
row	column	row	(Subject)	column
1.0801 (yes)	1.2605 (for)	1.3702	(1)	1.0801 (yes)
-0.9258 (no)	0.1681 (unsure)	1.3702	(2)	-0.9258 (no)
	-1.1429 (unsure)	1.3702	(3)	1.2605 (for)
		0.7307	(4)	0.1681 (against)
		0.7307	(5)	-1.1429 (against)
		-0.0368	(6)	
		0.1959	(7)	
		-0.4436	(8)	
	($\eta_c^2 = 0.2107$)	-0.4436	(9)	($\eta_m^2 = 0.7295$)
		-1.2110	(10)	
		-1.2110	(11)	
		-1.2110	(12)	
		-1.2110	(13)	

Compare these cell scores with subjects' scores from response patterns (multiple-choice data) in Table 8.3. You can see that these cell scores are identical to the scores for subjects.

Another way to relate the contingency table to the response-pattern representation is to create a hybrid of the two formats as shown in Table 8.4. Can you see how Table 8.4 is constructed? The simplest way to generate Table 8.4 is to use Table 8.2. For instance, pattern (1, 0, 1, 0, 0) appears three times in Table 8.2, which is represented as (3, 0, 3, 0, 0) in

TABLE 8.4
Hybrid of Contingency and Responde-Pattern Tables

	Q.1		Q.2		
	Yes	No	For	Unsure	Against
1*	3	0	3	0	0
2	2	0	0	2	0
3	1	0	0	0	1
4	0	1	1	0	0
5	0	2	0	2	0
6	0	4	0	0	4

*Distinct (1,0) patterns times frequencies.

Table 8.4. In other words, Table 8.4 consists of the original (1, 0) response patterns times respective frequencies. If you compare Table 8.4 with Tables 8.1 and 8.2, you can see the same interchangeability, that is, any of the three formats can be constructed from any one of the remaining two. Dual scaling of Table 8.4 provides the normed weights of Solution 1 as summarized in Table 8.5. You can see that we obtain cell scores and category weights which are identical to those in Table 8.3. Although Table 8.4 is a hybrid of the contingency and the response pattern tables, it possesses the same properties as the response pattern table so far as the number of nontrivial solutions and the corresponding squared correlation ratios are concerned.

TABLE 8.5
Normed Weights from the Hybrid Format

row	column	
1.3702	1.0801	(yes)
0.7307	-0.9258	(no)
-0.0368	1.2605	(for)
0.1959	0.1681	(unsure)
-0.4436	-1.1429	(against)
-1.2110	(η_h^2 = 0.7295)	

One final point to be mentioned here is on the relation indicated by Formula 8.4, which holds only for those solutions from the response-pattern (and the hybrid) table which correspond to the solutions from the contingency table. Since the squared correlation ratio is bounded between zero and one, a few algebraic manipulations lead to the inequality relation that

(8.6) $$1 \geq \eta_m^2 \geq 0.5 \text{ so that } 1 \geq \eta_c^2 \geq 0 .$$

This means that all the solutions from the response-pattern format which correspond to the solutions from the contingency table format have squared correlation ratios equal to or greater than 0.5. This may be a useful piece of information for checking the correspondence of the two sets of solutions. The squared correlation ratio from the hybrid format, η_h^2, is identical to η_m^2. Another inequality relation one can derive is with respect to the numbers of nontrivial solutions from the two formats as given by Equations 8.2 and 8.3,

$$m - n \succ \min(p,q) - 1 \ . \tag{8.7}$$

Thus, the response-pattern format yields more solutions than the contingency table, and the squared correlation ratios of those extra solutions from the response-pattern table are not bounded between 0.5 and 1, but can be almost as small as 0.

8.1.2 Number of Response Options

Multiple-choice data are typically characterized by more than two multiple-choice questions, or items. If there are three multiple-choice questions, the data may be summarized in a three-way or three-dimensional table of frequencies for all possible combinations of options of the three questions. However, as the number of questions increases to four, five, and so on, the tabulation of frequencies in a multiway table becomes cumbersome. It is much simpler to represent the data in the subjects-by-options table of response patterns as in Table 8.2. This format is the standard input of multiple-choice data for dual scaling.

As mentioned in the previous section, however, this format yields many solutions, often too many from a practical point of view. For instance, consider a relatively short questionnaire, consisting of 10 items, with each having 7 options, and suppose that you collect responses from 100 subjects. Then, this data set of response patterns is 100x70, and the maximum number of nontrivial solutions is given by (8.2), that is, 70 - 10 = 60. In practice, the distribution of information over so many solutions (dimensions) makes it very difficult to interpret the outcome of the analysis, and therefore it would be wise not to use too many options (for example, seven or nine), but to restrict the number to less than that.

Another problem with the number of options is on the statistic called "the total variance accounted for," that is, δ used in chapter 6. For multiple-choice data, the total variance, v_t, is given by

(8.8) $$\nu_t = \overline{m} - 1 ,$$

where m is the average number of options. Thus, in the example of seven options per item, $\nu_t = 7 - 1 = 6$. It is known that ν_t is equal to the sum of all the correlation ratios of nontrivial solutions, that is, in the case of multiple-choice data,

(8.9) $$\nu_t = \eta_1^2 + \eta_2^2 + \ . \ \ . \ \ . + \eta_{m-n}^2 .$$

As was seen in chapter 7, the statistic δ for Solution 1, for example, is given by

(8.10) $$\delta_1 = \frac{\eta_1^2 \times 100\%}{\nu_t} .$$

In the case of multiple-choice data,

(8.11) $$\delta_1 = \frac{\eta_1^2 \times 100\%}{\overline{m} - 1} .$$

Note, however, that $1 \geq \eta^2 \geq 0$. Thus, even when $\eta^2 = 1$, δ_1 for seven-option data is only 16.67%, that is, (1×100%)/6, as opposed to the common-sense expectation of 100%. Benzécri (1979) and Greenacre (1987) proposed a modified statistic which raises the value of δ. Interested readers are referred to their studies. From the user's standpoint, however, it seems that δ and modified δ's by Benzécri and Greenacre are difficult to interpret when applied to multiple-choice data. For this reason, in this book, a statistic proposed by Nishisato and Nishisato (1984), called "percent homogeneity," δ^*, will be used, which is

(8.12) $$\delta^* = \delta \nu_t = \eta^2 \times 100\% .$$

This statistic indicates the degree to which the derived solution conforms to the case of perfect internal consistency. Thus, if $\eta_1^2 = 1$, then $\delta_1^* = 100\%$, and not 16.67%. The interpretation of δ is difficult and tricky, especially for the contingency table and multiple-choice data. See a detailed discussion of δ in chapter 17.

The problems of too many solutions and too small a value of δ are peculiar to the multiple-choice data represented in the form of response patterns. Computationally at least, the problems suggest that we should not be too ambitious to capture every bit of information by using a large number of response options (e.g., completely disagree, very strongly disagree, strongly disagree, moderately disagree, slightly disagree, neutral, slightly agree, moderately agree, strongly agree, very strongly agree, completely agree). There are at least two more reasons why it may not be advisable to use such a large number of response options as seven or nine: the increased chance for outlier responses, and the possibility of overquantification. These two problems are too technical to look at right now, but are discussed in chapter 17.

8.1.3 Useful Statistics

There are a few statistics which can be used to evaluate the scaling outcome of multiple-choice data. Let us use a small example to look at those statistics. Table 8.6 presents artificial sample data and the results of Solution 1. There are three items, with three options per item, for (F), uncertain (U), and against (A).

As was the case with the contingency table, row weights (scores for subjects), y_i, and column weights (weights for response options), x_j, were determined so as to maximize the squared correlation ratio, the product--moment correlation between responses weighted by y_i and those by x_j and other criteria discussed in chapter 4. These weights satisfy a number of interesting relations:

(a) Item Difficulty. The condition that the sum of responses weighted by y_i is equal to the sum of responses weighted by x_j, both being set equal to 0, means that the sum of weighted responses of each item is 0, too (Guttman, 1941). To verify this numerically, construct a table of subjects-by-item scores (Table 8.6). This table is obtained by inserting the weights of options chosen by subjects. For instance, Subject 1 chose Option 1 of Q1 (1.67), Option 1 of Q2 (1.78), and Option 2 of Q3 (-0.25), yielding the first row of Table 8.7, that is, (1.67, 1.78, -0.25). Is this clear? Similarly, you can construct the rest of this table. See that the sum of scores of each item is 0 or close to 0, discrepancies being due to the rounding error.

The fact that all the items have means of 0, hence no mean differences among them, may appear detrimental to information retrieval, for common sense would say that different questions should have different means. Consider, for example, the following questions:

1. Do you have nightmares every night?
2. Do you catch a cold every year?

Shouldn't the means of these two items be different? Recall, however, that our task is to determine weights for response *options*, not those of items. This distinction is important to understand the property of that condition of equal item means. To explain the meaning of the condition, let us

TABLE 8.6
Multiple-Choice Data and Solution 1

Subject	Q1 F*	U	A	Q2 F	U	A	Q3 F	U	A	y_i
1	1	0	0	1	0	0	0	1	0	1.20
2	1	0	0	1	0	0	1	0	0	1.98
3	0	1	0	0	1	0	0	1	0	-0.35
4	0	0	1	0	0	1	0	0	1	-0.93
5	0	1	0	0	0	1	0	1	0	-0.63
6	0	0	1	0	0	1	0	1	0	-0.72
7	0	0	1	0	1	0	0	0	1	-0.65
8	1	0	0	0	1	0	1	0	0	1.28
9	0	1	0	0	0	1	0	1	0	-0.63
10	0	1	0	0	1	0	0	0	0	-0.55
x_j	1.67	-0.61	-0.85	1.78	-0.08	-0.81	1.83	-0.25	-0.80	

$\eta_1^2 = 0.7926$, $\delta^* = 79.26\%$, $\delta = 39.63\%$, $\nu_t = 2.00$, $\alpha = 0.8692$

Interitem Correlation

1	1.00		
2	0.80	1.00	
3	0.79	0.45	1.00

Item Statistics

Item	SS_j	r_{jt}^2	r_{jt}
1	12.04	0.9541	0.9768
2	9.04	0.7162	0.8463
3	8.93	0.7074	0.8411

consider a simple example in which one easy question and one difficult question about mathematics are asked. Furthermore, assume that each question has two response options, one correct and the other incorrect. Then, under the condition that the sum of weighted responses of each item is 0, typical dual scaling weights for the options would look like the following:

Easy Item: x_1 (correct) = 0.30, chosen by 90% of students.
x_2 (incorrect) = -2.70, chosen by 10%.

Difficult Item: x_3 (correct) = 2.00, chosen by 20%.
x_4 (incorrect) = -0.50, chosen by 80%.

Note that the sums of weighted items are 0, that is, 0.30 × 90 - 2.70 × 10 = 0 and 2.00 × 20 - 0.50 × 80 = 0. Now you can see the implication of the condition: if you miss an easy item, the penalty is great (-2.70), and if you pass a difficult item, the reward is great (2.00); likewise, if you pass an easy item, the reward is small (0.30), and if you miss a difficult item, the

TABLE 8.7
Table of Item Scores

Subject	Q1	Q2	Q3	y_i
1	1.67	1.78	-0.25	1.20
2	1.67	1.78	1.83	1.98
3	-0.61	-0.08	-0.25	-0.35
4	-0.85	-0.81	-0.80	-0.93
5	-0.61	-0.81	-0.25	-0.63
6	-0.85	-0.81	-0.25	-0.72
7	-0.85	-0.08	-0.80	-0.65
9	-0.61	-0.81	-0.25	-0.63
10	-0.61	-0.08	-0.80	-0.55
Total	0.02	0.00	0.01	

penalty is not great (-0.50). Thus, the condition of equal means of the items has the effect of looking after the effect of item difficulty or popularity.

(b) Item-Total Correlation. Look at Table 8.7. The product-moment correlation between item 1 and score y is indicated by r_{1t}. Similarly, r_{2t} and r_{3t} are product-moment correlations between item 2 and y, and item 3 and y, respectively. These are called item-total correlation. If we have two sets of scores say v_i, w_i, $i = 1, 2, \ldots, N$ (e.g., scores from N subjects on two tests v and w), the product-moment correlation is given by

$$r_{vw} = \frac{N\Sigma v_i w_i - (\Sigma v_i)(\Sigma w_i)}{\sqrt{[N\Sigma v_i^2 - (\Sigma v_i)^2]}\sqrt{[N\Sigma w_i^2 - (\Sigma w_i)^2]}} \tag{8.13}$$

This is an index of linear relationship between v and w, and r_{vw} is bounded between -1 and 1, that is, $-1 \le r_{vw} \le 1$. In dual scaling, the sum of scores on Item 1 is 0, and the sum of y_i is also 0. In addition, the sum of squares of y_i is set equal to nN (i.e., the total number of responses). Therefore, in our example of Table 8.6,

$$r_{1t} = 0.9768 \ .$$

Similarly, $r_{2t} = 0.8463$ and $r_{3t} = 0.8411$. Dual scaling determines option weights so as to maximize the average of the squares of item-total correlations. In passing, it should be mentioned that when there are only two multiple-choice items (i.e., the case of a contingency table) we have the relation $r_{1t} = r_{2t}$.

(c) Item-Total Correlation and the Squared Correlation Ratio. There exists the following simple relation between r_{jt}^2 and η^2 (Nishisato, 1980a, 1982; Saporta, 1980):

$$\eta^2 = \frac{1}{n}\sum_{j=1}^{n} r_{jt}^2$$

In other words, η^2 is the average of r_{jt}^2. In our example, 0.7926 = (0.9541 + 0.7162 + 0.7074)/3 (see Table 8.6). In the case of the contingency table, $r_{1t}^2 = r_{2t}^2 = 2\eta_c^2$.

(d) Item-Total Correlation and the Sum of Squares. The sum of squares of a set of numbers is defined as the sum of squares of differences between those numbers and their mean. Since the mean of each item scores is 0, the sum of squares of Item 1, say SS_1, is simply the sum of squares of item scores, that is,

$$SS_1 = 1.67^2 \times 2 + (-0.61)^2 + (-0.85)^2 + \ldots + (-0.61)^2 = 12.04$$

Similarly, SS_2 = 9.04, and SS_3 = 8.93. There exists the relation between SS_j and r_{jt}^2 (Nishisato, 1982) such that

(8.14) $$r_{jt}^2 = \frac{n\eta^2 SS_j}{SS_t}$$

where SS_t is the total sum of squares, that is, $SS_t = SS_1 + SS_2 + \ldots + SS_n$. In our example, $SS_t = SS_1 + SS_2 + SS_3 = 30$. In dual scaling, SS_t is set equal to the total number of responses, which can be expressed as nN provided that there are no missing responses. Thus, if $SS_t = nN$, then

(8.15) $$Nr_{jt}^2 = \eta^2 SS_j \,.$$

Since N and η^2 are constant for each solution, we can state that

(8.16) $$SS_j \propto r_{jt}^2,$$

that is, SS_j is proportional to r_{jt}^2. This is one of the most important properties of dual scaling, namely, the greater the relevance of the item to the data (i.e., the greater r_{jt}^2), the greater the contribution (i.e., the greater SS_j) of the item to explaining the data. To clarify this statement, consider typical option weights of a "good" item and a "poor" item:

Good item (r_{jt} is close to 1)	**Poor item (r_{jt} is close to 0)**
x_1 = 1.86	x_1 = 0.08
x_2 = 0.99	x_2 = -0.02
x_3 = 0.03	x_3 = -0.05
x_4 = -0.78	x_4 = 0.03
x_5 = -1.72	x_5 = -0.01

Notice that the option weights of a good item vary greatly (i.e., SS_j is large) in contrast to those of a poor item. This means that for a poor item it does not make much difference which option of the item you choose, while a good item can clearly distinguish between two people who choose different options. In other words, with a good item it is important to know which option you choose; this is not so in the case of a poor item, for your response to the item cannot convey much information about what the questionnaire intends to measure. Most subjectively derived scoring schemes (e.g., scores of 1, 2, 3, 4, 5 for five options) do not possess this property indicated by (8.16). As a consequence, they often yield an irrational way of scoring, which weights poorly correlated questions more than highly correlated ones.

(e) Internal Consistency Reliability. Lord (1958) has shown that option weighting obtained by dual scaling yields scores for the subjects which have maximal internal consistency reliability. Maximally reliable scores? Yes, that's true. But what is internal consistency reliability? To explain the concept, let us consider a simple example in which students answered 10 mathematical questions. The answers are either correct or incorrect. We arrange the questions from the easiest to the most difficult in terms of actual numbers of correct answers (i.e., the item with the largest number of correct responses is the easiest item). Suppose that we obtained data of perfect internal consistency reliability. Then we know that a student who answered six questions correctly answered the first six easiest questions correctly. Similarly, a person with three correct answers must have made correct responses to the first three questions. In contrast, when the test lacks internal consistency, there is no way to tell which three questions a person with three correct answers may have answered correctly. Thus, internal consistency means the extent to which subjects' response patterns can be predicted from their scores and item difficulty (popularity) scores.

The coefficient of internal consistency reliability is often indicated by α (alpha) and is called Cronbach's alpha (Cronbach, 1951), or generalized Kuder-Richardson reliability.[1] In dual scaling, all the item means are set equal to 0. Under this condition, internal consistency reliability α is expressed in terms of one of the optimization criteria used in dual scaling, that is, the squared correlation ratio, η^2, as:

(8.17) $$\alpha = 1 - \frac{(1-\eta^2)}{(n-1)\eta^2}, \text{ or } \eta^2 = \frac{1}{1+(n-1)(1-\alpha)},$$

---

[1] The word *generalized* is used here because it is now applied to differentially weighted multiple options, rather than dichotomously scored options.

where n is the number of multiple-choice questions. As you can see from the relationship between α and η^2 in Figure 8.1, maximization of η^2 leads to maximization of α, as Lord (1958) concluded. To illustrate this point, let $n = 11$ in (8.17). Then, $\alpha = 1 - (1 - \eta^2)/(10\eta^2) = (11\eta^2 - 1)/(10\eta^2)$. Thus, if $\eta^2 = 0.1$, $\alpha = 0.10$; if $\eta^2 = 0.2$, $\alpha = 0.60$; if $\eta^2 = 0.3$, $\alpha = 0.77$; if $\eta^2 = 0.4$, $\alpha = 0.85$; if $\eta^2 = 0.5$, $\alpha = 0.90$; and so on until we reach the absolute maximum of α being 1 when $\eta^2 = 1$.

It may be astonishing, however, to see that as η^2 decreases α becomes negative. It is known (Guttman, 1956) that α is a lower bound of reliability, which is defined in terms of "test theory" terminology as the ratio of the variance of "true scores" to the variance of the observed scores. This means that reliability is a nonnegative quantity. Thus, you may wish to consider α to be bounded between 0 and 1. If so, we find that

$$(8.18) \qquad 1 \geq \alpha \geq 0 \quad \text{means} \quad 1 \geq \eta^2 \geq \frac{1}{n},$$

where n is the number of items. Can you see in Figure 8.1 that $\alpha = 0$ when $\eta^2 = 1/n$? For instance, $\alpha = 0$ when $n = 4$, and $\eta^2 = 0.25$, which is 1/4. Similarly, if $\alpha = 0$ and $n = 10$, you know that $\eta^2 = 1/10 = 0.1$. In comparison with contingency tables, it was noted that the corresponding response--pattern format produces many more solutions than the contingency table. One of the reasonable rules for terminating extraction of further solutions is to adopt the criterion of η^2 being equal to $1/n$. In other words, one should stop the analysis when η^2 becomes smaller than $1/n$, for then α is negative.

Figure 8.1 reveals interesting features of α and η^2. One of the most remarkable relations is that for a given value of η^2 the value of α increases as the number of items increases. For instance, consider that $\eta^2 = 0.25$. Then, as n increases from 4 to 6, 10, and 20, α value changes from 0.00 to 0.40, 0.67, and 0.84, respectively. Isn't this worth noting? If you have a questionnaire with 100 multiple-choice questions, what would be the value of α? Alpha equals 0.909 when $\eta^2 = 0.100$, that is, $\delta^* = 10\%$, a case of a very heterogeneous questionnaire! This example tells us that a large value of α, such as 0.8 and 0.9, does not necessarily indicate a high degree of homogeneous responses. In contrast, η^2 is a stable quantity over the change in n because it is a ratio of two quantities, SS_b and SS_t, both of which are standardized with respect to the number of observations. These topics discussed above are important to remember in interpreting results of dual scaling. Let us now apply our knowledge to the following numerical examples.

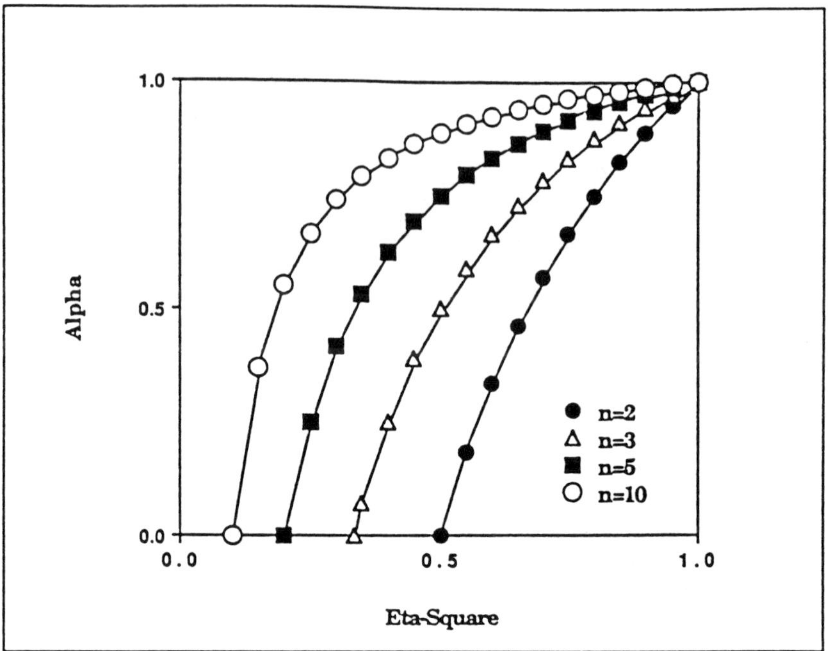

Fig. 8.1. Relation between α and η^2 as a function of the number of items, *n*

8.2 APPLICATIONS

Let us start with a simple straightforward application, to see the kinds of statistics we may look at and interpretation of the quantified results.

8.2.1 Singapore Data

One of the participants in a dual scaling workshop, held at NEC Computers Singapore Ltd. in 1985, prepared a four-item questionnaire for which data were collected from the participants. The questionnaire and the data are given in Table 8.8.

Notice that the data are presented in the condensed format rather than in the response pattern format. Therefore, the numbers in the table indicate option numbers chosen by subjects. For instance, 2 for Item 1 indicates that the subject has chosen the second option of Item 1. Similarly, (2322) of Subject 18 in the table corresponds to the response pattern (0,1,0,0,0,1,0,1,0,0,1;0). Is this clear? Remember that each item has three options, thus occupying three spaces with the pattern being one of (1,0,0),

TABLE 8.8
Singapore Data with Questionnaire

Questionnaire	Data				
		Item			
Item 1:	Subject	1	2	3	4
How old are you?					
(1) 20-29	1	3	1	2	1
(2) 30-39	2	2	1	3	2
(3) 40 or over	3	2	1	2	2
Item 2:	4	1	2	2	3
Children today are not	5	3	1	2	2
as disciplined as when I	6	1	3	1	2
was a child.	7	2	1	2	2
(1) agree	8	2	1	2	2
(2) disagree	9	1	2	3	1
(3) I cannot tell	10	3	1	2	1
Item 3:	11	1	2	2	3
Children today are not	12	2	1	1	1
as fortunate as when I	13	2	1	3	3
was a child.	14	3	1	2	1
(1) agree	15	1	1	2	3
(2) disagree	16	3	1	2	1
(3) I cannot tell	17	3	1	1	1
Item 4:	18	2	3	2	2
Religions should not be	19	3	1	2	1
taught at school.	20	2	1	2	2
(1) agree	21	1	3	3	3
(2) disagree	22	2	1	2	2
(3) indifferent	23	1	3	3	3

(0,1,0), and (0,0,1). The pattern of 12 elements is the sum of (0,1,0), (0,0,1), (0,1,0), and (0,1,0), corresponding to the condensed format of (2322).

The matrix subjected to the eigenvalue-eigenvector decomposition is what may be called the variance-covariance matrix as given in Table 8.9. This is the matrix $\mathbf{C}_1$ discussed in chapter 6.

The sum of the diagonal elements of this matrix is 2.0000, which is the total information and is equal to the average number of options minus 1. In the present example, each item has three options, hence the total information is 3 - 1 = 2. This value is also equal to the sum of the squared correlation ratios of all the solutions. The total number of solutions in the current case is 8, which is equal to the total number of options minus the number of items, that is, 12 - 4 = 8. Recall that the squared correlation ratio is the variance of quantified responses of each solution. Therefore, when we extract all eight solutions, the total information can be decomposed into individual solutions as shown in Table 8.10.

TABLE 8.9
Variance-Covariance Matrix of Options

Option	1	2	3	4	5	6	7
1	0.1739						
2	-0.0863	0.1522					
3	-0.0761	-0.0863	0.1739				
4	-0.0914	0.0362	0.0503	0.0761			
5	0.1139	-0.0565	-0.0498	-0.0753	0.2174		
6	0.0842	-0.0236	-0.0575	-0.0870	-0.0377	0.2065	
7	-0.0103	0.0181	-0.0103	0.0068	-0.0377	0.0190	0.2065
8	-0.0318	-0.0107	0.0439	0.0211	0.0067	-0.0479	-0.0813
9	0.0625	0.0016	-0.0643	-0.0413	0.0225	0.0632	-0.0486
10	-0.0479	-0.0628	0.1191	0.0317	-0.0022	-0.0615	0.0269
11	-0.0548	0.0966	-0.0548	0.0154	-0.0565	0.0181	0.0181
12	0.1224	-0.0459	-0.0704	-0.0555	0.0717	0.0488	-0.0532
	8	9	10	11	12		
8	0.0978						
9	0.909	0.1057					
10	0.0031	-0.0292	0.1630				
11	0.0116	-0.0356	-0.0922	0.1522			
12	-0.0178	0.0774	-0.0753	-0.0799	0.1848		

TABLE 8.10
Decomposition of the Total Information

Solution	η^2	δ_j	$\Sigma\delta_j$	α
1	0.6479	32.40%	32.40%	0.8189
2	0.4478	22.39%	54.78%	0.5889
3	0.2886	14.43%	69.21%	0.1782
4	0.2203	11.02%	80.23%	-0.1796
5	0.1938	9.69%	89.92%	-0.3869
6	0.1173	5.87%	95.78%	-1.5081
7	0.0553	2.77%	98.55%	-4.6939
8	0.0290	1.45%	100.00%	-10.1561

One should note that the first solution cannot account for more than [100/(average number of options minus 1)]% since the upper limit of η^2 is 1 and the total information is the average number of options minus 1. In the present case, in which the average number of options is three, the first solution cannot explain more than 50% of the total information. Notice, however, that when η^2 is less than 0.25 (i.e., $1/n$), alpha is negative. Thus, using the criterion of η^2 being $1/n$ (Nishisato, 1980a), one should retain only the first three solutions. In terms of these *admissible* solutions, the relative weights of the three solutions are 47%, 32%, and 21%. The DUAL3 output of the first three solutions is as follows:

```
                DUAL3MC Output: Singapore Data

                    DUAL SCALING OF
                  MULTIPLE-CHOICE DATA
                  (STANDARD ANALYSIS)


     TITLE OF STUDY: Singapore Data, 21 Subjects, 4 Items

            *** INPUT DATA ***

                  3 1 2 1
                  2 1 3 2
                  2 1 2 2
                  1 2 2 3
                  3 1 2 2
                  1 3 1 2
                  2 1 2 2
                  2 1 2 2
                  1 2 3 1
                  3 1 2 1
                  1 2 2 3
                  2 1 1 1
                  2 1 3 3
                  3 1 2 1
                  1 1 2 3
                  3 1 2 1
                  3 1 1 1
                  2 3 2 2
                  3 1 2 1
                  2 1 2 2
                  1 3 3 3

       *** THE TOTAL VARIANCE TO BE ACCOUNTED FOR = 2   ***


            SOLUTION  1

      SQUARED CORRELATION RATIO = 0.60526
      MAXIMUM PRODUCT-MOMENT CORRELATION = 0.77799
      PERCENTAGE HOMOGENEITY =   60.53
```

DELTA (TOTAL VARIANCE ACCOUNTED FOR): PARTIAL = 3 .26
CUMULATIVE = 30.26
RELIABILITY COEFFICIENT ALPHA = 0.78261

*** OPTIMAL SCORES AND WEIGHTS ***

NORMED	SCORE			ADJUSTED	SCORE			OPTION		NORMED	ADJUSTED
1	-0.94	11	1.73	1	-0.73	11	1.35	1	1	1.80	1.40
2	0.00	12	-0.70	2	0.00	12	-0.54	1	2	-0.31	-0.24
3	-0.50	13	0.68	3	-0.39	13	0.53	1	3	-1.19	-0.92
4	1.73	14	-0.94	4	1.35	14	-0.73	2	1	-0.65	-0.51
5	-0.79	15	0.85	5	-0.61	15	0.66	2	2	2.09	1.63
6	0.72	16	-0.94	6	0.56	16	-0.73	2	3	1.18	0.92
7	-0.50	17	-0.98	7	-0.39	17	-0.76	3	1	-0.41	-0.32
8	-0.50	18	0.09	8	-0.39	18	0.07	3	2	-0.28	-0.22
9	1.41	19	-0.94	9	1.10	19	-0.73	3	3	1.30	1.01
10	-0.9	20	-0.50	10	-0.73	20	-0.39	4	1	-0.80	-0.62
		21	1.95			21	1.52	4	2	-0.32	-0.25
								4	3	1.79	1.39

*** MATRIX OF INTER-ITEM CORRELATION ***

1	1.00			
2	0.78	1.00		
3	0.29	0.22	1.00	
4	0.67	0.43	0.30	1.00

*** ITEM STATISTICS ***

ITEM(j)	SS(j)	R^2(jt)	R(jt)
1	30.06	0.866	0.931
2	23.71	0.684	0.827
3	8.40	0.242	0.492
4	21.83	0.629	0.793

SOLUTION 2

SQUARED CORRELATION RATIO = 0.44694
MAXIMUM PRODUCT-MOMENT CORRELATION = 0.66854
PERCENTAGE HOMOGENEITY = 44.69
DELTA (TOTAL VARIANCE ACCOUNTED FOR): PARTIAL = 22.35
CUMULATIVE = 52.61
RELIABILITY COEFFICIENT ALPHA = 0.58753

*** OPTIMAL SCORES AND WEIGHTS ***

NORMED SCORE				ADJUSTED SCORE				OPTION		NORMED	ADJUSTED
1	1.04	11	1.00	1	0.69	11	0.67	1	1	0.54	0.36
2	-1.35	12	-0.11	2	-0.90	12	-0.07	1	2	-1.54	-1.03
3	-1.13	13	-0.58	3	-0.75	13	-0.39	1	3	1.30	0.87
4	1.00	14	1.04	4	0.67	14	0.69	2	1	-0.00	-0.00
5	-0.07	15	0.42	5	-0.04	15	0.28	2	2	1.56	1.04
6	-1.00	16	1.04	6	-0.67	16	0.69	2	3	-1.53	-1.03
7	-1.13	17	0.96	7	-0.75	17	0.64	3	1	-0.08	-0.05
8	-1.13	18	-1.70	8	-0.75	18	-1.14	3	2	0.14	0.10
9	1.12	19	1.04	9	0.75	19	0.69	3	3	-0.44	-0.29
10	1.04	20	-1.13	10	0.69	20	-0.75	4	1	1.34	0.89
		21	-0.37			21	-0.25	4	2	-1.61	1.08
								4	3	0.44	0.29

*** MATRIX OF INTER-ITEM CORRELATION ***

1	1.00			
2	0.15	1.00		
3	0.19	0.08	1.00	
4	0.62	0.34	-0.09	1.00

*** ITEM STATISTICS ***

ITEM(j)	SS(j)	R^2(jt)	R(jt)
1	32.53	0.692	0.832
2	14.31	0.305	0.552
3	1.08	0.023	0.151
4	36.08	0.768	0.876

SOLUTION 3

SQUARED CORRELATION RATIO = 0.32108
MAXIMUM PRODUCT-MOMENT CORRELATION = 0.56664
PERCENTAGE HOMOGENEITY = 32.11
DELTA (TOTAL VARIANCE ACCOUNTED FOR): PARTIAL = 16.05
CUMULATIVE = 68.66
RELIABILITY COEFFICIENT ALPHA = 0.29516

*** OPTIMAL SCORES AND WEIGHTS ***

NORMED SCORE				ADJUSTED SCORE				OPTION		NORMED	ADJUSTED
1	-0.11	11	-0.84	1	-0.06	11	-0.48	1	1	0.51	0.29
2	-0.52	12	1.52	2	-0.29	12	0.86	1	2	-0.57	-0.32
3	-0.83	13	-0.73	3	-0.47	13	-0.41	1	3	0.21	0.12

```
 4 -0.84 14 -0.11    4 -0.48 14 -0.06  2  1     -0.32  -0.18
 5 -0.48 15 -0.56    5 -0.27 15 -0.32  2  2     -0.96  -0.54
 6  2.90 16 -0.11    6  1.64 16 -0.06  2  3      2.58   1.46
 7 -0.83 17  1.86    7 -0.47 17  1.05  3  1      3.70   2.09
 8 -0.83 18  0.46    8 -0.47 18  0.26  3  2     -0.77  -0.44
 9  0.05 19 -0.11    9  0.03 19 -0.06  3  3     -0.08  -0.04
10 -0.11 20 -0.83   10 -0.06 20 -0.47  4  1      0.64   0.36
         21  1.03            21  0.58  4  2     -0.21  -0.12
                                       4  3     -0.69  -0.39

        *** MATRIX OF INTER-ITEM CORRELATION ***
        1    1.00
        2    0.03 1.00
        3    0.04 0.27 1.00
        4    0.06-0.23 0.24 1.00

                   *** ITEM STATISTICS ***
     ITEM(j)     SS(j)           R²(jt)          R(jt)
       1          4.46           0.068           0.261
       2         24.30           0.372           0.610
       3         49.29           0.754           0.868
       4          5.95           0.091           0.302

         ********* END OF ANALYSIS *********
```

There are a number of interesting statistics as discussed in the previous section. Let us look at those statistics before we examine weights for options and scores for subjects. In the DUAL3 output, we see a table of "item statistics" for each solution, containing the sum of squares of weighted responses of each item, SS_j, the square of item-total correlation, r_{jt}^2, and item-total correlation, r_{jt}. These statistics for the first three solutions are as shown in Table 8.11. From this table, we can observe that in the first two solutions the contribution of Item 3 is comparatively small, and that in the third solution Item 3 emerges as the dominant contributor. Although this information is interesting, we need to know more about the results to understand what they are telling us.

Since there are only four items, the present data set offers an ideal opportunity to examine the distribution of information in detail. In particular, we would like to look at the importance of response patterns in determining solutions. Table 8.12 lists the condensed form of response patterns arranged in order of their scores within each solution.

Look at Solution 1, where we can clearly see the transition of response patterns from low scores to high scores: we can find from the table which options of the items should be combined to produce low or high scores. This is a very useful piece of information for interpreting the results. We also note that high item-total correlation yields regularity in the transition of

TABLE 8.11
Item Statistics for Three Solutions

	Solution 1			Solution 2			Solution 3		
Item	SS_j	r_{jt}^2	r_{jt}	SS_j	r_{jt}^2	r_{jt}	SS_j	r_{jt}^2	r_{jt}
1	30.44	0.86	0.93	36.70	0.71	0.85	5.90	0.07	0.27
2	24.86	0.70	0.84	12.92	0.25	0.50	23.63	0.30	0.54
3	12.94	0.36	0.60	4.48	0.09	0.30	51.50	0.65	0.80
4	23.77	0.67	0.82	37.90	0.74	0.86	10.97	0.14	0.37

TABLE 8.12
Response Patterns Arranged
in Order of Their Scores

Solution 1		Solution 2		Solution 3	
1333	-1.56	2112	-0.95	1223	-0.72
1333	-1.56	2322	-0.95	1223	-0.72
1223	-1.19	2132	-0.84	2122	-0.59
1223	-1.19	1312	-0.74	2122	-0.59
1231	-1.02	2122	-0.67	2122	-0.59
1123	-0.58	2122	-0.67	2122	-0.59
1312	-0.54	2122	-0.67	1123	-0.36
2133	-0.50	2122	-0.67	3122	-0.30
2322	-0.05	2133	-0.37	2132	-0.19
2132	0.03	2111	-0.21	2133	-0.18
2122	0.48	1333	-0.16	2322	-0.06
2122	0.48	1333	-0.16	1231	0.04
2122	0.48	3122	0.07	3121	0.06
2122	0.48	1123	0.28	3121	0.06
2112	0.52	3111	0.53	3121	0.06
2111	0.61	1223	0.69	3121	0.06
3122	0.65	1223	0.69	3121	0.06
3121	0.74	1231	0.79	2112	0.41
3121	0.74	3121	0.80	1333	0.57
3121	0.74	3121	0.80	1333	0.57
3121	0.74	3121	0.80	2111	0.77
3121	0.74	3121	0.80	3111	1.06
3111	0.78	3121	0.80	1312	1.16

response-patterns. For instance, look at Item 1 of Solution 1: options show a perfect transition from low to high scores. This was achieved by the item-total correlation of 0.93. In contrast, item 3 with r_{3t} of 0.60 shows a considerable number of irregular transitions of options. This comparision offers an important point of view to evaluate the meaning of item-total correlation: how high should r_{jt} be to expect a smooth transition of options from low to high scores? The traditional test for significant correlation is not relevant here, hence cannot answer this question. The relation of item-total correlation to the smooth transition of response patterns is a topic that has never been explored.

TABLE 8.13
Adjusted Scores from Three Solutions

Subj	Sol.1	Sol.2	Sol.3	Subj	Sol.1	Sol.2	Sol.3
1	-0.74	0.80	0.06	13	0.50	-0.37	-0.18
2	-0.03	-0.84	-0.19	14	-0.74	0.80	0.06
3	-0.48	-0.67	-0.59	15	0.58	0.28	-0.36
4	1.19	0.69	-0.72	16	-0.74	0.80	0.06
5	-0.65	0.07	-0.30	17	-0.78	0.53	1.06
6	0.54	-0.74	1.16	18	0.05	-0.95	-0.06
7	-0.48	-0.67	-0.59	19	-0.74	0.80	0.06
8	-0.52	-0.95	0.41	20	-0.48	-0.67	-0.59
9	1.02	0.79	0.04	21	1.56	-0.16	0.57
10	-0.74	0.80	0.06	22	-0.48	-0.67	-0.59
11	1.19	0.69	-0.72	23	1.56	-0.16	0.57
12	-0.61	-0.21	0.77				

The fact that response patterns contain the full information of the data presents an exciting possibility of its use in graphical display: plot subjects, label them by their response patterns, and interpret clusters of subjects in terms of their common response patterns.[2] Table 8.13 lists adjusted scores of subjects from the three solutions, and two-dimensional plots of subjects are as shown in Figure 8.2. As you see, response patterns of subjects are used as labels. Look at the second quadrant of Figure 8.2(a), where seven subjects lie with response patterns (3121), (3111), and (3122). The common pattern is (31**), which indicates that those subjects are all 40 years of age or over, and regard children today as not disciplined as when they were children. Six subjects share the pattern (31*1), indicating that they further agree to the teaching of religions at school. In Solution 3, however, one of the six subjects with the pattern (3111) moves away from the other five who disagree with the statement of items, as shown in their response-pattern (3121). In this way, we can partition the space into regions, for instance, the region (2**2), that is, age 30-39, and against religions taught at school, and the region (1***), that is, the age 20-29. It is important and interesting to note that an item which has a high value of r_{jt} divides the

[2] See Nishisato (1990, 1992a).

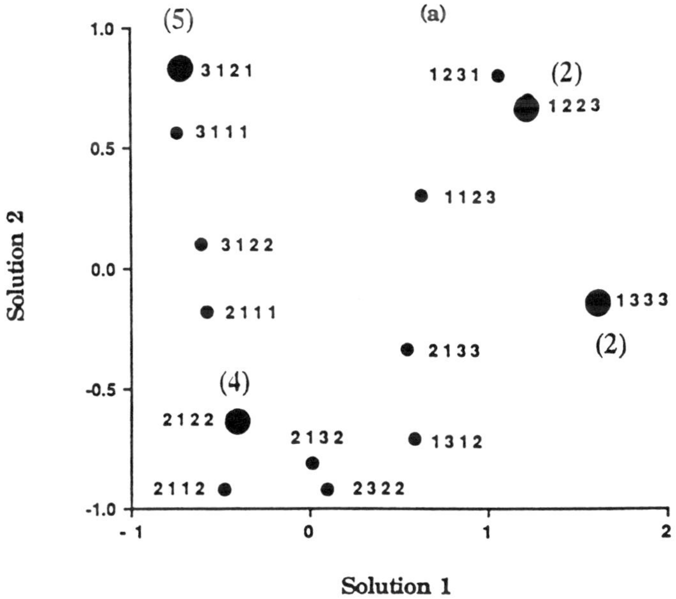

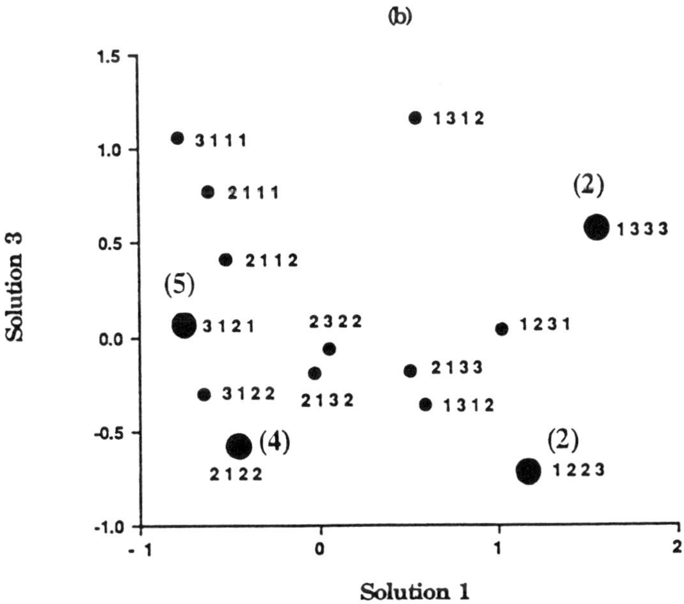

Fig. 8.2(a) and (b). Distribution of subjects and their5 response patterns.

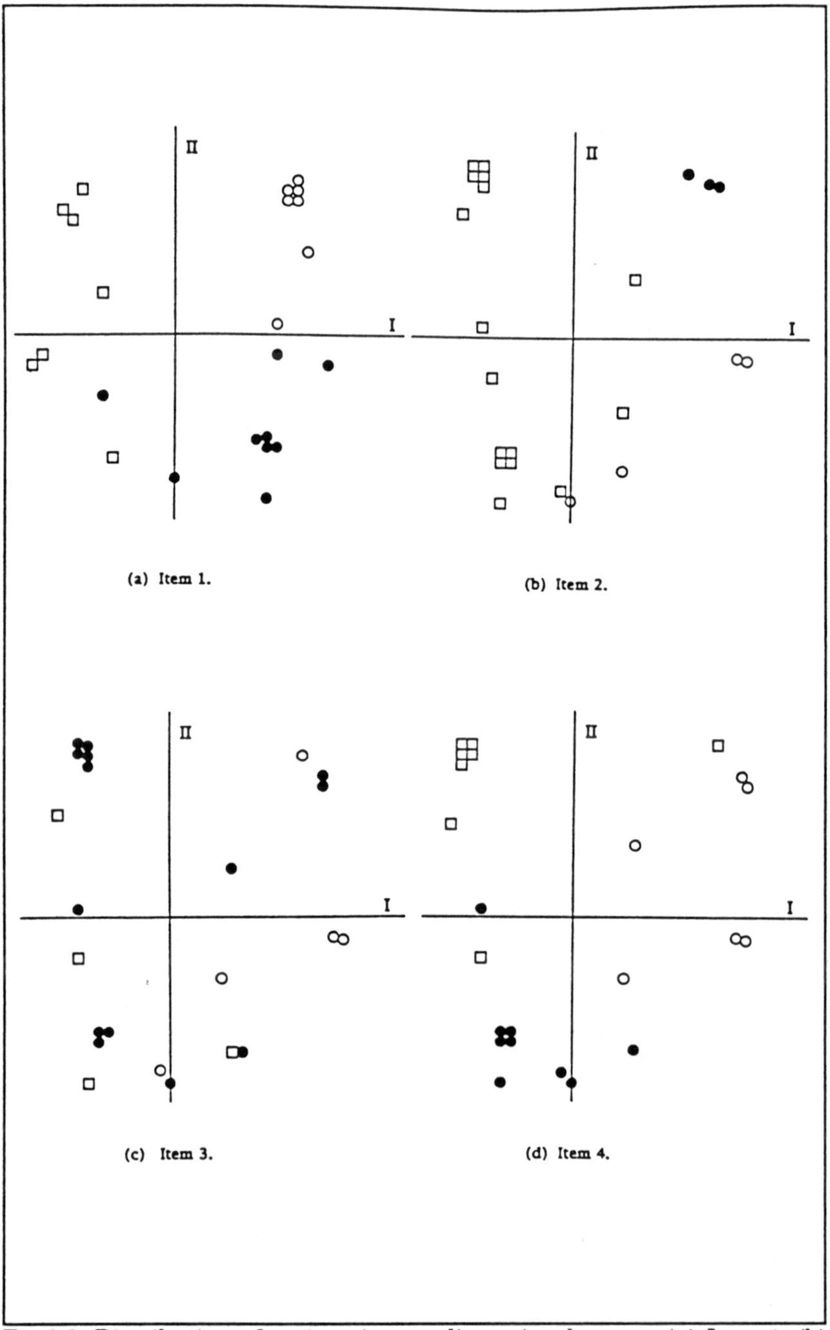

Fig.8.3. Distribution of options in two-dimensional spaces (a) Item 1, (b) Item 2, (c) Item 3, (d) Item 4.

space cleanly by their options. For instance, look at item 1 in Figure 8.2(a): we can partition the space by patterns (1***), (2***), and (3***), where the asterisk indicates that we ignore the response to that item. In terms of Solutions 1 and 2 [Figure 8.2(a)], the space is relatively neatly partitioned by age (a***), discipline (*d**), and religion (***r). Figure 8.3 shows item-by-item distributions of subjects who chose particular options in the space of the first two solutions. We can see from the graphs how important it is to have high item-total correlation in order for the item to contribute to classification of subjects into clusters, that is, for options to be distributed in separate regions. Out of the four items, it looks as though Items 1, 2, and 4 are "good" items for the purpose of classification; Item 3, however, does not cleanly divide the space into regions by its options.

Before we leave this section, let us look at more mathematical aspects of this data analysis. As we recall, eight solutions exhaust the information in the data set. In the eight-dimensional space, the contribution of an item with m_j options, as measured by the sum of squares, can be expressed numerically as shown in Table 8.14.

TABLE 8.14
Contribution of Each Item to the Total Space

Dimension	Sum of Squares				
Solution	Item 1	Item 2	Item 3	Item 4	Total
1	30.44	24.96	12.94	23.77	92
2	36.70	12.92	4.48	37.90	92
3	5.90	23.50	51.50	10.97	92
4	8.39	33.51	47.72	2.38	92
5	14.55	28.57	41.57	7.46	92
6	0.21	25.86	13.28	52.65	92
7	34.34	13.65	9.79	34.24	92
8	53.48	21.16	2.73	14.63	92
Sum	184	184	184	184	

Notice that all the marginals of the table are constant. The sum of squares of Item 1 for Solution 1, for example, is calculated as

$$f_1x_1^2+f_2x_2^2+f_3x_3^3 = (7\times1.6830^2)+(9\times0.4298^2)+(7\times1.1304^2) = 30.44$$

If we express by $\mathbf{x}_{jk}$ the weight of m_j options of Item j for Solution k and by $\mathbf{D}_{cj}$ the diagonal matrix of the frequencies of m_j options of Item j, the sum of squares of Item j for Solution k, SS_{jk}, can be expressed as

(8.19) $$SS_{jk} = \mathbf{x}_{jk}'\mathbf{D}_{cj}\mathbf{x}_{jk} \; .$$

The total contribution of Item j to the total space $SS_{T(j)}$, is

(8.20) $$SS_{T(j)} = \sum_{k=1}^{K} SS_{jk} = nN(m_j - 1) \; .$$

In our example, $SS_{T(j)}$ = 4x23x22(2-3) = 184, j = 1, 2, 3, 4. The sum of squares of all the items for Solution k, SS_k, is given by

(8.21) $$SS_k = \sum_{j=k}^{n} SS_{jk} = nN \; .$$

In our example, SS_k = 4 × 34 = 92, k = 1, 2, ..., 8.

TABLE 8.15
Option Contributions of Item 2, x_{jpk}^2

Solution	Option 1 $f_{21} = 16$	Option 2 $f_{22} = 3$	Option 3 $f_{23} = 4$
1	0.4695	3.0615	2.0400
2	0.0006	2.6332	1.2511
3	0.0331	2.6429	3.7916
4	0.0575	8.7492	1.5866
5	0.1893	0.7762	5.7653
6	0.4554	4.6126	1.1846
7	0.1880	0.0209	2.6445
8	0.3566	4.1702	0.7350
$SS_{T(jp)}$	28.00	80.00	76.00

Let us now look at contributions of individual options to the total space. Three option frequencies and the squared option weights of Item 2 are as shown in Table 8.15. The column total $SS_{T(jp)}$ is the contribution of Option p of Item j, given by

(8.22) $$SS_{T(jp)} = \sum_{k=1}^{K} f_{jp}x_{jpk}^2 = n(N - f_{jp}) \; ,$$

where f_{jp} is the frequency of Option p of Item j, and x_{jpk} is the weight of Option p of Item j for Solution k. Note that, for example, $SS_{T(21)}$ = 4(23 - 16) = 28. It is important to keep in mind that the total contribution of an option is inversely related to its popularity. In other words, those options chosen by fewer people have greater contributions to the total space. These marginal constraints imposed on the outcome are discussed again in chapter 17.

8.2.2 As Compared with an Average Person

As you know by now, dual scaling yields not only option weights but also scores for subjects in association with option weights. Let us now analyze a slightly larger data set than that of Example 1. Recall that we briefly looked at a questionnaire called "As Compared with an Average Person" listed in Table 2.4 as one of the illustrative examples of dual scaling. Let us analyze the data in Table 8.16 to see what information we can extract about personality traits. Since all the items have two options, the total information, or association, to be analyzed is 1 (i.e., 2 - 1). Table 8.17 shows some statistics for the first 11 solutions. As you see, the value of η_{11}^2 is 0.0304, which is smaller than 0.0333 (i.e., 1/30), resulting in a negative value of the internal consistency reliability coefficient α. In contrast, η_{10}^2 is 0.0392 with α being equal to 0.1556. Therefore, we will look at only the first 10 solutions. Let us first look at main contributors to each of the 10 solutions in terms of the item-total correlation, r_{jt}. To make it easier to look at, only those values which are greater than 0.35 are listed in Table 8.18.[3] Tables 8.19—8.28 summarize dichotomous responses of those questions in Table 8.17 in the descending order of r_{jt}. As you can see in these tables, each item does in general contribute to more than one solution. On the basis of these tables, our task typically includes the interpretation of the solutions. Does the questionnaire tap any categories of personality traits as included in Freud's psychoanalytic theory, Jung's analytic theory, Murray's personology, Roger's self theory, or what not? As is usually the case, the first solution tends to be a general division, and as we go down the list of the subsequent solutions, we typically see more idiosyncratic traits revealed by a small number of subjects. Therefore, if we are interested in Solution 2, for example, we may first look at the main characteristics of the solution such as high-handed, devouring, unpleasant, and clumsy, and then look for subjects whose scores on this solution are high. In this way, it is easy to identify those subjects who posses those traits captured by the solution. As is clear from these tables, which list only main contributors, the number of items listed becomes smaller as we move from Solution 1 down to Solution 10. This decrease also implies that the number of subjects who contribute to these solutions get smaller, too. Look at Solutions 8, 9, and 10. After major contributors of personality traits

[3] This value of 0.35 is chosen for convenience and is arbitrary.

TABLE 8.16
As Compared with an Average Person. . .

SS	1	2	3	4	5	6	7	8	9	10	11	12	13	14	15	16	17	18	19	20	21	22	23	24	25	26	27	28	29	30
1	2	1	2	2	2	1	1	2	2	2	2	2	1	2	2	2	1	2	2	2	2	2	2	2	2	2	2	2	2	1
2	1	1	1	2	2	2	1	2	2	1	2	1	1	2	2	1	1	1	2	1	2	1	2	1	2	2	1	2	1	1
3	1	2	1	2	2	2	1	2	1	1	1	1	2	1	2	1	1	2	2	1	1	1	1	1	2	2	1	2	1	1
4	2	1	1	2	1	2	1	1	2	2	1	2	1	1	1	1	2	1	1	2	1	2	1	2	2	2	2	2	1	1
5	1	1	2	2	1	2	1	2	1	1	2	1	1	2	1	2	2	2	2	1	1	1	2	1	2	2	2	1	1	1
6	2	1	1	2	1	2	1	1	2	1	2	2	1	1	2	1	1	1	2	1	1	2	1	1	2	2	2	1	1	1
7	2	1	2	1	1	2	1	1	2	2	1	2	2	2	2	1	2	1	1	1	2	2	1	2	1	1	1	1	1	1
8	1	2	1	2	1	2	1	1	1	1	1	1	2	1	2	1	2	2	1	1	1	1	1	1	2	1	1	2	1	2
9	1	2	1	2	1	2	2	1	1	1	1	1	1	1	1	1	2	2	2	1	1	1	1	1	2	1	1	1	2	1
10	2	1	2	1	1	2	2	1	2	2	2	2	2	1	1	2	1	2	2	1	2	2	2	2	2	2	2	2	2	2
11	1	1	1	2	1	2	1	2	2	1	1	2	1	1	2	1	2	2	2	1	2	2	1	1	2	2	1	2	2	1
12	1	2	1	2	1	2	1	1	1	1	2	2	1	1	1	2	2	1	1	1	2	1	1	1	1	2	2	2	1	2
13	2	2	0	2	1	1	1	2	2	2	1	2	1	1	2	1	2	2	2	1	1	2	2	1	2	2	2	1	2	2
14	1	1	1	2	2	2	1	2	1	1	1	1	1	1	1	2	2	2	1	1	1	1	1	1	2	2	1	2	1	2
15	2	1	1	2	2	1	1	1	2	1	1	2	1	2	2	2	1	1	2	1	2	2	1	1	2	2	2	1	1	1
16	1	1	1	2	1	2	2	2	1	2	2	1	2	1	2	2	2	2	2	1	2	1	2	1	2	2	2	1	1	1
17	1	2	1	2	1	2	1	2	1	1	2	1	1	2	2	1	2	2	2	1	1	1	1	1	1	2	1	1	2	1
18	2	1	2	2	2	1	1	2	2	2	2	1	1	1	2	0	1	2	1	1	1	2	2	2	2	2	2	1	1	1
19	1	2	1	2	1	2	2	1	1	2	1	2	1	1	1	1	2	2	1	1	1	1	2	2	2	2	1	1	1	2
20	2	2	2	2	1	2	2	2	1	1	2	2	1	2	2	2	2	2	2	1	1	2	1	1	2	2	1	2	1	2
21	1	1	1	2	2	2	1	1	2	1	2	1	2	1	2	1	2	2	2	1	1	1	1	1	2	2	2	1	2	2
22	1	1	1	2	2	2	0	2	2	1	2	2	1	1	2	2	1	2	2	1	2	2	2	1	2	2	1	2	2	1
23	2	1	1	2	2	2	1	0	2	0	0	0	0	2	0	2	2	1	2	1	2	2	1	1	2	0	0	2	1	2
24	2	1	1	2	1	2	2	1	2	1	1	1	1	1	1	1	2	2	1	1	1	1	1	1	1	1	1	1	1	2
25	1	2	1	2	1	2	1	2	2	1	2	1	2	1	2	1	2	2	2	1	1	1	1	1	1	2	1	1	1	2
26	1	2	1	2	1	1	1	2	1	1	2	1	1	1	2	1	2	2	2	1	1	1	1	1	2	2	1	1	1	1
27	2	1	1	2	2	2	2	2	1	1	1	1	1	1	2	1	2	2	2	1	2	2	1	1	2	2	1	1	1	2
28	1	2	1	2	1	2	1	2	2	2	1	2	1	1	2	1	2	2	2	1	2	2	1	1	2	2	1	1	2	1
29	2	2	1	2	2	2	1	2	2	2	2	2	1	1	2	2	2	2	2	1	2	2	1	1	2	2	2	2	2	1

TABLE 8.17
Some Statistics on the First 11 Solutions

Solution	η^2	δ	δ^*	α
1	0.1935	18.89%	18.89%	0.8563
2	0.1300	12.69%	31.58%	0.7692
3	0.0876	8.55%	40.13%	0.6407
4	0.0741	7.23%	47.37%	0.5692
5	0.0674	6.58%	53.94%	0.5227
6	0.0649	6.34%	60.29%	0.5036
7	0.0513	5.01%	65.30%	0.3629
8	0.0483	4.71%	70.01%	0.3204
9	0.0447	4.37%	74.38%	0.2636
10	0.0392	3.83%	78.21%	0.1556
11	0.0304	2.96%	81.17%	-0.1011

TABLE 8.18
Coefficients of Item-Total Correlation Greater than 0.35

Item	Solution 1	2	3	4	5	6	7	8	9	10
friendly	65	--	--	--	--	--	--	48	--	--
quiet	51	--	--	--	--	--	39	--	--	--
mature	52	--	41	--	--	--	--	--	--	--
clumsy	--	59	--	00	55	--	--	--	--	--
outspoken	37	46	--	--	--	--	45	--	--	--
miserly	42	--	--	--	40	--	--	--	--	--
self-conscious	--	--	38	--	--	--	--	46	--	--
high-handed	--	75	--	--	--	--	--	--	--	--
chatting	63	--	--	36	--	--	--	--	--	--
courteous	56	36	43	--	--	--	--	--	--	--
impatient	--	40	--	44	--	--	--	--	--	--
optimistic	60	--	--	--	--	41	--	--	--	--
meticulous	--	--	--	--	48	--	38	--	--	39
inquisitive	--	--	--	--	--	36	49	--	--	--
presumptuous	--	54	--	36	--	41	--	--	--	--
creative	47	--	--	63	--	--	--	--	--	--
timid	57	--	--	--	--	--	40	--	--	--
inflexible	--	--	75	--	--	--	--	--	--	--
eat more	--	70	--	--	42	--	--	--	--	--
ambitious	46	--	--	--	43	--	--	--	--	--
humorous	53	--	--	--	47	--	--	--	--	--
cheerful	77	--	--	35	--	--	--	--	--	--
tasteful	45	--	62	--	--	--	--	--	--	--
pleasant	53	60	35	--	--	--	--	--	--	--
flamboyant	--	38	--	--	--	35	38	--	--	--
wasteful	36	58	--	--	--	--	--	--	--	--
generous	59	--	--	--	--	--	--	--	58	--
vocal	--	--	--	--	--	49	--	--	40	--
fortunate	--	--	--	40	44	--	--	--	--	--
easily hurt	--	--	--	--	--	35	--	--	--	48

Note. The decimal point is omitted (e.g., 35 is 0.35).

were captured by the first seven solutions, the data must still have retained information on such specific traits, or combinations of traits, captured by the last three solutions.

TABLE 8.19
Solution 1

One End of Scale	r_{jt}	The Other End of Scale
More cheerful	0.77	Less cheerful
More friendly	0.65	Less friendly
More chatting	0.63	Less chatting
More optimistic	0.60	Less optimistic
More generous	0.59	Less generous
Less timid	0.57	More timid
More courteous	0.56	Less courteous
More humorous	0.53	Less humorous
More pleasant	0.53	Less pleasant
More mature	0.52	Less mature
Less quiet	0.51	More quiet
More creative	0.47	Less creative
More ambitious	0.46	Less ambitious
More tasteful	0.45	Less tasteful
Less miserly	0.42	More miserly
More outspoken	0.37	Less outspoken
More wasteful	0.36	Less wasteful

TABLE 8.20
Solution 2

One End of Scale	r_{jt}	The Other End of Scale
More high-handed	0.75	Less high-handed
Eat more	0.70	Eat less
Less pleasant	0.60	More pleasant
More clumsy	0.59	Less clumsy
More wasteful	0.58	Less wasteful
More presumptuous	0.54	Less presumptuous
More outspoken	0.46	Less outspoken
Less patient	0.40	More patient
More flamboyant	0.38	Less flamboyant
Less courteous	0.36	More courteous

TABLE 8.21
Solution 3

One End of Scale	r_{jt}	The Other End of Scale
Less flexible	0.75	More flexible
More tasteful	0.62	Less tasteful
More courteous	0.43	Less courteous
More mature	0.41	Less mature
More self-conscious	0.38	Less self-conscious
More pleasant	0.35	Less pleasant

TABLE 8.22
Solution 4

One End of Scale	r_{jt}	The Other End of Scale
More creative	0.63	Less creative
Less patient	0.44	More patient
Less fortunate	0.40	More fortunate
Less chatting	0.36	More chatting
Less presumptuous	0.36	More presumptuous
Less cheerful	0.35	More cheerful

TABLE 8.23
Solution 5

One End of Scale	r_{jt}	The Other End of Scale
More clumsy	0.55	Less clumsy
Less meticulous	0.48	More meticulous
Less humorous	0.47	More humorous
Less fortunate	0.44	More fortunate
Less ambitious	0.43	More ambitious
Eat more	0.42	Eat less
More miserly	0.40	Less miserly

TABLE 8.24
Solution 6

One End of Scale	r_{jt}	The Other End of Scale
More vocal	0.49	Less vocal
More optimistic	0.41	Less optimistic
Less presumptuous	0.41	More presumptuous
Less inquisitive	0.36	More inquisitive
More flamboyant	0.35	Less flamboyant
More easily hurt	0.35	Less easily hurt

TABLE 8.25
Solution 7

One End of Scale	r_{jt}	The Other End of Scale
More inquisitive	0.49	Less inquisitive
Less outspoken	0.45	More outspoken
More timid	0.40	Less timid
More quiet	0.39	Less quiet
Less meticulous	0.38	More meticulous
Less flamboyant	0.38	More flamboyant

TABLE 8.26
Solution 8

One End of Scale	r_{jx}	The Other End of Scale
More friendly	0.58	Less friendly
More outspoken	0.46	Less outspoken

TABLE 8.27
Solution 9

One End of Scale	r_{jx}	The Other End of Scale
More generous	0.58	Less generous
Less vocal	0.40	More vocal

TABLE 8.28
Solution 10

One End of Scale	r_{jx}	The Other End of Scale
More easily hurt	0.48	Less easily hurt
More meticulous	0.39	Less meticulous

CHAPTER 9
Sorting Data

9.1. SPECIAL PROBLEMS

The application of dual scaling to sorting data is due to Takane (1980a), who discovered that one of his preferred methods of analyzing sorting data is exactly the same as dual scaling applied to a transposed variant of the response-pattern matrix. This format has already been looked at in chapter 3.

To facilitate our discussion, let us consider a small example of sorting data in the condensed format (Table 9.1) and the response-pattern format (Table 9.2). In this example, each of four subjects sorted six fruits, apple, orange, pineapple, banana, strawberry, and mango, into a number of piles. Subject 3, for example, placed the apple and orange together in the first pile, the pineapple, banana, and mango in the second pile, and the strawberry in the third pile.

TABLE 9.1
Sorting Data (Condensed Format)

	Subject			
	1	2	3	4
Apple	1	1	1	1
Orange	1	1	1	2
Pineapple	2	1	2	2
Banana	2	2	2	3
Strawberry	1	3	3	2
Mango	2	2	2	3

As was mentioned in chapter 3, the input data for dual scaling (Table 8.2) look like multiple-choice data, except that the rows and the columns are interchanged. More fundamentally, however, sorting data are quite different from multiple-choice data. In multiple-choice data individual differences show up only in terms of responses over a limited number of choices, called response options, the number of options of each item being

TABLE 9.2
Sorting Data (Response-Pattern Format)

	Subject											
	1		2			3			4			Weight
Apple	1	0	1	0	0	1	0	0	1	0	0	y_1
Orange	1	0	1	0	0	1	0	0	0	1	0	y_2
Pineapple	0	1	1	0	0	0	1	0	0	1	0	y_3
Banana	0	1	0	1	0	0	1	0	0	0	1	y_4
Strawberry	1	0	0	0	1	0	0	1	0	1	0	y_5
Mango	0	1	0	1	0	0	1	0	0	0	1	y_6
Weight	x_1	x_2	x_3	x_4	x_5	x_6	x_7	x_8	x_9	x_{10}	x_{11}	

determined solely by the researcher; in contrast, in sorting data, individual differences are revealed through subjects' unrestricted or free choices of piles, rather than the researcher's imposing decision on the number of piles. This distinction appears to explain the fact that dual scaling of sorting data often yields too many solutions to interpret. Therefore, the problem of how many solutions to adopt becomes more difficult with sorting data than with multiple-choice data or contingency tables.

The comparison of sorting data with multiple-choice data suggests an interesting idea of what one may call a "restricted sorting task" as opposed to free sorting. What if we impose the condition that each subject should sort n objects into k piles, k being a constant, such as 2, 3, or, 5? What if we ask the subjects to sort n objects into at most k piles? While these possibilities might merit serious consideration, they are not pursued any further in this book.

Under free sorting, the number of piles each subject uses may be interesting to look at. Keep in mind that the following correspondence exists between multiple-choice data and sorting data:

Multiple-choice data	Sorting data
Rows = Subjects	Rows = Objects to be sorted
Columns = Options of items	Columns =Piles of subjects

Thus, the item statistics (e.g., SS_j and r_{jt}) for multiple-choice data are "subject statistics" for sorting data. Therefore, dual scaling of sorting data provides a matrix of intersubject correlations, the sum of squares of each subject (SS_j), and the correlation between each subject and the group of subjects (r_{jt}). Some of these statistics may be influenced by the number of piles each subject uses. These subject statistics present an interesting opportunity to investigate individual differences. For instance, the investigator may have an idea about possible criteria for sorting, say, criteria A,

B and C. If so, augment the objects-by-subjects data matrix of response patterns with response patterns of the three prototype (theory-derived) subjects A, B, and C. Dual scaling of this augmented matrix will reveal, through the intersubject correlation matrix, which subjects are likely to have used criterion A, B, or C, and, through r_{jt}, which of A, B, and C may have contributed more to a particular solution.

You will recall that there was no way to plot multiple-choice items because all the items have zero means, and that there tend to be too many options to plot. In contrast, sorting data provide a table of coordinates of objects, and it is easy, at least operationally, to plot them. The only possible problem may be that there are too many graphs to draw.

One of the important advantages of sorting data lies in the fact that the sorting task can deal with a large number of objects or stimuli. It is difficult to appreciate this advantage right now, but you will see in Part 3 that other data types have a serious limitation in the number of objects they can handle (see discussions on paired comparison and rank-order data in chapters 10 and 11, respectively).

Once the similarities and differences between multiple-choice data and sorting data are understood, dual scaling of sorting data is straightforward. At the time of interpreting the results, those differences must be kept in mind. This may not always be a simple task. Let us therefore look at simple applications and learn how sorting data are analyzed.

9.2. APPLICATIONS

Two examples are presented in this section. The first example was obtained from Covert (1990) and the second from Nishisato and Nishisato (1984).

9.2.1 Sorting 51 Car Models

For a course on applied scaling at the Ontario Institute for Studies in Education, Dwayne Covert conducted a survey on the similarity of 51 car models. At his place of work, he randomly chose 17 respondents and asked them to group the models in terms of their similarity. Table 9.3 shows the data. Whenever sorting data are analyzed, you are likely to observe a large number of solutions. In other words, the information tends to distribute widely over many dimensions. This is probably due to the fact that respondents have the freedom of adopting any number of piles, unlike multiple--choice data in which each subject must choose one out of a fixed number of options per item. Table 9.4 shows some statistics of the first 20 solutions.

TABLE 9.3
Sorting Data on 51 Car Models

Car Model	Respondent 1	2	3	4	5	6	7	8	9	10	11	12	13	14	15	16	17
Acura Integra	4	1	1	3	6	1	1	1	2	3	7	3	5	7	2	1	2
Acura Legend	5	1	7	5	2	2	2	10	2	1	5	2	5	4	1	3	3
Audi 5000	3	2	7	5	8	3	2	2	2	1	5	2	4	4	1	3	3
BMW 535	3	2	7	5	8	3	2	2	2	1	5	1	4	4	1	3	3
Buick Le Sabre	6	4	6	2	3	6	6	9	3	1	3	13	2	5	10	4	2
Buick Riviera	6	4	6	2	12	2	5	11	3	4	5	17	3	5	11	4	5
Buick Regal	6	4	6	2	8	3	2	8	4	1	3	12	2	5	4	4	3
Cadillac Eldorado	3	4	8	5	2	3	2	8	4	1	3	11	2	6	4	4	3
Chevrolet Beretta	5	7	3	3	13	1	1	13	5	4	7	16	3	5	12	2	2
Chevrolet Camaro	2	6	5	4	4	5	3	16	5	3	7	7	4	3	13	2	1
Chevrolet Lumina	6	7	6	2	1	4	5	13	3	4	6	9	3	5	9	7	5
Chevrolet Sprint	1	5	4	1	11	4	4	15	1	5	10	6	1	1	6	5	4
Dodge Colt	1	5	4	1	11	4	4	15	1	5	10	6	1	1	6	5	4
Dodge Shadow	5	5	3	1	14	4	5	13	8	4	4	16	3	1	11	7	2
Ford Escort	1	5	4	1	14	4	4	15	1	5	4	4	1	1	6	5	4
Ford Mustang	2	6	5	3	4	2	3	16	5	5	7	18	4	3	12	1	1
Ford Probe GT	4	6	1	3	7	1	1	5	5	5	1	3	3	2	5	1	2
Honda Accord	6	1	3	2	6	6	2	1	3	3	8	15	5	7	2	1	2
Honda Civic	6	5	4	6	11	4	4	15	1	5	4	4	3	1	6	6	5
Honda CRX	4	5	2	6	15	4	7	17	7	2	10	15	4	2	3	6	1
Honda Prelude	5	3	1	3	10	2	3	5	2	2	8	3	5	7	5	1	2
Hyundai Excel	1	5	4	2	11	4	5	15	1	5	4	6	3	1	6	5	4
Lincoln Continental	3	4	8	5	2	3	2	8	4	1	3	11	2	6	4	3	3
Mazda 626	5	1	3	2	6	5	2	7	3	4	8	15	5	4	8	2	4
Mazda Miata	4	3	2	4	15	7	7	3	7	2	9	10	4	2	3	6	4
Mazda MX-6 GT	4	3	1	3	10	2	1	14	6	3	1	3	5	2	5	8	2
Mazda RX-7	4	3	2	4	9	1	3	17	6	2	1	5	4	2	7	6	1
Mercury Cougar	5	4	6	3	1	2	6	11	3	4	5	20	2	3	13	8	2
Mercury Sable	5	3	6	2	12	2	2	9	3	4	5	9	3	5	9	8	3
Nissan 240 SX	4	3	2	3	7	2	1	14	7	3	1	3	4	2	5	1	1
Nissan 300 ZX	4	3	1	4	9	1	3	6	6	2	2	5	4	2	7	3	1
Nissan Maxima	6	7	6	2	6	2	2	10	2	1	6	8	5	4	8	8	2
Nissan Micra	1	5	4	1	11	5	4	15	1	5	10	6	1	1	6	5	5
Nissan Sentra	5	7	3	1	14	6	5	12	8	5	8	15	3	4	2	1	4
Oldsmobile Cutlass	6	4	6	2	1	6	5	11	3	4	5	13	2	6	9	4	2
Oldsmobile Delta 88	6	4	6	5	3	6	2	9	4	1	3	12	2	5	10	4	3
Pontiac Grand Am	6	7	6	3	13	2	5	14	3	4	7	18	5	5	12	2	2
Pontiac Grand Prix	2	7	6	3	1	3	6	11	5	4	5	17	4	5	5	1	2
Pontiac Tempest	6	7	3	1	13	4	5	13	5	4	7	16	3	5	14	7	5
Pontiac Trans AM	2	6	5	4	4	5	3	16	5	3	1	7	4	3	15	2	1
Plymouth Daytona	4	7	5	3	7	5	3	5	5	5	1	7	3	7	12	2	2
Plymouth Reliant	6	7	3	1	14	4	5	12	1	5	4	16	1	5	1	5	4
Saab 900 S	3	2	7	5	8	3	2	4	2	1	2	1	5	4	1	3	3
Toyota MR2	4	3	2	4	7	7	7	17	7	2	2	14	4	2	3	6	1
Toyota Celica	4	3	1	4	10	6	1	14	6	2	1	3	3	7	5	1	2
Toyota Corolla	6	5	3	6	14	4	4	7	8	5	8	4	5	1	2	7	4
Toyota Cressida	5	1	4	5	13	2	2	10	8	1	6	8	5	4	8	8	5
Toyota Supra	4	3	5	4	9	1	3	6	6	2	2	5	4	2	7	1	1
Volvo 740	3	2	7	5	8	3	2	2	2	1	5	2	5	4	1	3	5
VW Fox	5	5	3	1	14	4	1	12	8	5	4	6	1	1	6	5	5
VW Golf	1	5	4	1	13	4	4	7	8	4	6	19	3	4	2	5	4

Notice the lack of a typical concentration of distinctly large values of η^2 in the first few solutions.

TABLE 9.4
Distribution of the 21 η_j^2

Solution	η_j^2	δ_j	$\Sigma\delta_j$	Solution	η_j^2	δ_j	$\Sigma\delta_j$
1	0.8006	9.86%	9.86%	11	0.2405	2.96%	63.18%
2	0.7899	9.73%	19.59%	12	0.1915	2.36%	65.54%
3	0.6109	7.53%	27.12%	13	0.1839	2.27%	67.80%
4	0.5247	6.46%	33.58%	14	0.1773	2.18%	69.99%
5	0.4864	5.99%	39.57%	15	0.1719	2.12%	72.10%
6	0.4088	5.04%	44.61%	16	0.1523	1.88%	73.98%
7	0.3545	4.37%	48.98%	17	0.1447	1.78%	75.76%
8	0.3259	4.01%	52.99%	18	0.1428	1.76%	77.52%
9	0.3159	3.89%	56.88%	19	0.1385	1.71%	79.23%
10	0.2704	3.33%	60.21%	20	0.1298	1.60%	80.83%

If we were to decide how many solutions to look at only by examining the distribution of η_j^2, sorting data are unlikely to make the decision task very simple. In practice, we should try to interpret as many solutions as possible. In other words, the interpretability of a solution seems to be a more practical way to make the decision than the distribution of η_j^2.

Let us look at the first six solutions to see if we can interpret them, and in the course of this analysis we may decide whether or not we should examine more solutions. The adjusted weights of the six solutions are shown in Table 9.5. It seems sensible to identify within each solution those car models with extreme weights, that is, those with large weights (positive weights of large values) and those with small weights (negative weights of large absolute values), and examine what characteristics those cars at the two extreme ends have. You may expect that the two extreme groups should represent bipolar attributes such as expensive versus inexpensive or large versus small. As you will see shortly, each solution is not necessarily characterized by a clearly discernable single attribute, but may be more complicated than that, showing, for example, expensive cars on one extreme end and small cars on the other extreme end. Since there is no built-in response model for dual scaling, the results of dual scaling can reveal strange ways of subjects' thinking or judgment.

Table 9.6 is based on Solution 1, and shows those car models with comparatively large weights and those with negative weights of large absolute values. Recall that the sign of weights, negative or positive, has no substantive implications. Even if the signs are reversed, the maximized

TABLE 9.5
Car Data: Six Solutions

Car	Solution 1	2	3	4	5	6
Acura Integra	0.30	0.52	-0.48	-0.54	-0.95	-0.99
Acura Legend	-1.19	0.30	0.64	-0.93	-0.54	0.22
Audi 5000	-1.42	0.46	1.19	-1.23	-0.17	0.51
BMW 535	-1.43	0.46	1.22	-1.24	-0.16	0.51
Buick Le Sabre	-1.08	0.14	-0.78	1.49	0.37	-0.20
Buick Riviera	-0.59	-0.14	-0.93	0.84	-0.22	0.61
Buick Regal	-1.53	0.24	0.10	0.94	0.82	-0.46
Cadillac Eldorado	-1.89	0.40	0.61	0.83	1.33	-0.98
Chevrolet Beretta	0.23	0.02	-1.03	-0.24	-0.23	0.31
Chevrolet Camaro	0.84	0.90	-0.99	-1.17	1.86	0.78
Chevrolet Lumina	-0.31	-0.42	-0.96	0.63	-0.51	0.84
Chevrolet Sprint	0.65	-1.85	0.68	-0.09	0.61	-0.41
Dodge Colt	0.65	-1.85	0.68	-0.09	0.61	-0.41
Dodge Shadow	0.19	-0.97	-0.45	0.13	-0.42	0.42
Ford Escort	0.59	-1.80	0.56	-0.10	0.44	-0.33
Ford Mustang	0.71	0.71	-0.98	-1.02	1.37	0.54
Ford Probe GT	0.75	0.65	-0.45	-0.31	-0.22	-0.97
Honda Accord	-0.13	0.08	-0.60	-0.11	-1.00	-0.43
Honda Civic	0.55	-1.42	0.52	0.13	0.32	-0.08
Honda CRX	1.15	0.45	1.27	1.01	-0.14	0.93
Honda Prelude	0.40	0.64	-0.35	-0.36	-0.82	-1.06
Hyundai Excel	0.49	-1.61	0.38	0.02	0.40	-0.26
Lincoln Continental	-1.88	0.43	0.74	0.63	1.27	-0.94
Mazda 626	-0.21	-0.24	-0.39	-0.24	-0.93	0.17
Mazda Miata	1.36	1.02	1.69	1.58	-0.33	1.64
Mazda MX-6 GT	0.59	0.80	-0.32	-0.15	-0.78	-1.23
Mazda RX-7	1.23	1.23	0.83	0.43	0.06	-0.36
Mercury Couger	-0.47	0.26	-1.12	0.47	0.11	0.50
Mercury Sable	-0.60	0.07	-0.68	0.63	-0.48	0.49
Nissan 240 SX	0.87	0.94	0.05	0.10	-0.38	-0.58
Nissan 300 ZX	0.98	1.26	0.77	0.07	0.03	-0.77
Nissan Maxima	-0.67	0.05	-0.32	-0.31	-1.01	0.27
Nissan Micra	0.62	-1.73	0.58	-0.19	0.69	-0.37
Nissan Sentra	0.13	-0.65	-0.30	-0.11	-0.67	0.04
Oldsmobile Cutlass	-0.85	0.03	-0.91	1.20	0.06	0.22
Oldsmobile Delta 88	-1.42	0.21	-0.24	1.29	0.68	-0.43
Pontiac Grand AM	-0.02	0.13	-1.11	-0.05	-0.21	0.36
Pontiac Grand Prix	-0.19	0.36	-0.92	0.17	0.07	0.39
Pontiac Tempest	0.07	-0.58	-0.87	0.07	-0.31	0.78
Pontiac Trans Am	0.93	0.97	-0.92	-1.25	1.96	0.65
Plymouth Daytona	0.68	0.51	-0.81	-0.65	0.50	-0.19
Plymouth Reliant	0.18	-1.14	-0.04	-0.05	-0.07	0.20
Saab 900 S	-1.39	0.48	1.29	-1.27	-0.27	0.36
Toyota MR2	1.30	1.19	1.40	1.11	-0.15	0.93
Toyota Celica	0.71	0.80	-0.23	0.05	-0.69	-1.36
Toyota Corolla	0.31	-1.11	0.05	-0.04	-0.42	0.05
Toyota Cressida	-0.59	-0.20	-0.04	-0.54	-0.96	0.31
Toyota Supra	1.12	1.27	0.60	0.06	0.20	-0.66
Volvo 740	-1.40	0.33	1.09	-1.27	-0.35	0.50
VW Fox	0.43	-1.39	0.18	-0.15	-0.01	-0.20
VW Golf	0.23	-1.18	0.07	-0.18	-0.36	0.17

TABLE 9.6
Cars with Extreme Weights (Solution 1)

Negative Weights		Positive Weights	
Car	Weight	Car	Weight
Cadillac Eldorado	-1.89	Mazda Miata	1.36
Lincoln Continental	-1.88	Toyota MR2	1.30
Buick Regal	-1.53	Mazda RX-7	1.23
BMW 535	-1.43	Honda CRX	1.15
Oldsmobile Delta 88	-1.42	Toyota Supra	1.12
Audi 5000	-1.42	Nissan 300 ZX	0.98
Volvo 740	-1.40		
Saab 900S	-1.39		
Acura Legend	-1.19		
Buick Le Sabre	-1.08		

squared correlation ratio η^2 remains the same, meaning that there is no ordinary connotation of their signs. Looking at those cars with negative weights, we can see some commonality that may be called "luxuriousness" in contrast to those cars with positive weights that may be characterized by "sportiness." When luxuriousness is on one end, one may be tempted to call the other end cheap or nonluxurious. However, the cluster of cars with positive weights are more suited to be called sporty than cheap. This is a subjective interpretation, and you may differ from it. For instance, an equally plausible interpretation is in terms of their size, "large or mid-sized" versus "compact or subcompact."

TABLE 9.7
Cars with Extreme Weights (Solution 2)

Negative Weights		Positive Weights	
Car	Weight	Car	Weight
Chevrolet Sprint	-1.80	Toyota Supra	1.27
Dodge Colt	-1.85	Nissan 300 ZX	1.26
Ford Escort	-1.80	Mazda RX-7	1.23
Nissan Micra	-1.73	Toyota MR2	1.19
Hyundai Excel	-1.61	Mazda Miata	1.02
Honda Civic	-1.42	Pontiac Trans Am	0.97
V.W. Fox	-1.39	Nissan 240 SX	0.94
V.W. Golf	-1.18	Chevrolet Camaro	0.90
Plymouth Reliant	-1.14		
Toyota Corolla	-1.11		

Let us now look at Solution 2 as shown in Table 9.7. Those cars with negative weights are relatively inexpensive, no frills cars. What about those cars with positive weights? They look like larger, sporty cars. What about Solution 3 (Table 9.8)? It looks as though this solution has picked up subjects' classification of cars in terms of American cars and foreign cars. As we continue to interpret solutions one by one in order, the task becomes increasingly more difficult. Out of the remaining three solutions (Tables 9.9—9.11), it seems that only Solution 5 is easily interpretable, that is, Japanese cars versus American cars. However, a car maniac may immediately see common characteristics of clusters captured by Solutions 4 and 6. So we will leave the interpretation of the remaining solutions to the imagination of the reader.

TABLE 9.8
Cars with Extreme Weights (Solution 3)

Negative Weight		Positive Weight	
Car	Weight	Car	Weight
Mercury Couger	-1.12	Mazda Miata	1.69
Pontiac Grand AM	-1.11	Toyota MR2	1.40
Chevrolet Beretta	-1.03	Saab 900S	1.29
Chevrolet Camaro	-0.99	Honda CRX	1.27
Ford Mustang	-0.98	BMW 535	1.22
Chevrolet Lumina	-0.96	Audi 5000	1.19
Buick Riviera	-0.93	Volvo 740	1.09
Pontiac Grand Prix	-0.92		
Pontiac trans Am	-0.92		
Oldsmobile Cutlass	-0.91		

TABLE 9.9
Cars with Extreme Weights (Solution 4)

Negative Weight		Positive Weight	
Car	Weight	Car	Weight
Saab 900S	-1.27	Mazda Miata	1.58
Volvo 740	-1.27	Buick Le Sabre	1.49
Pontiac TransAm	-1.25	Oldsmobile Delta 88	1.37
BMW 535	-1.24	Oldsmobile Cutlass	1.20
Audi 5000	-1.23	Toyota MR2	1.11
Chevrolet Camaro	-1.17	Honda CRX	1.01
Ford Mustang	-1.02	Buick Regal	0.94

TABLE 9.10
Cars with Extreme Weights (Solution 5)

Negative Weight		Positive Weight	
Car	Weight	Car	Weight
Nissan Maxima	-1.01	Pontiac Trans Am	1.96
Honda Accord	-1.00	Chevrolet Camaro	1.86
Toyota Cressida	-0.96	Ford Mustang	1.37
Acura Integra	-0.95	Cadillac Eldorado	1.33
Mazda 626	-0.93	Lincoln Continental	1.27
Honda Prelude	-0.82	Buick Regal	0.82

TABLE 9.11
Cars with Extreme Weights (Solution 6)

Negative Weight		Positive Weight	
Car	Weight	Car	Weight
Toyota Celica	-1.36	Mazda Miata	1.64
Mazda MX-6 GT	-1.23	Honda CRX	0.93
Honda Prelude	-1.06	Toyota MR2	0.93
Acura Integra	-0.99	Chevrolet Lumina	0.84
Cadillac Eldorado	-0.98		
Ford Probe GT	-0.97		

9.2.2 Classifying 39 Countries

Twenty-one people who participated in a two-day workshop were asked to sort 39 countries into piles. Table 9.12 is the questionnaire used, and Table 9.13 is the 39x21 condensed data matrix obtained from them. In this data set, the largest number, that is, the maximal number of piles used by subjects, is 9, while you recall that in the Covert data, the largest number was 20, and this contributed to generating a large number of solutions. Even 9, which is much smaller than 20, is still very large if you compare it with the number of options in multiple-choice data. Thus, what you would expect is again a large number of solutions, and perhaps squared correlation ratios diminishing only gradually in successive solutions, unlike the ordinary case of multiple-choice data, in which a small number of them are dominant.

Some statistics on the first 10 solutions are listed in Table 9.14. As expected, the squared correlation ratio diminishes only gradually in the

TABLE 9.12
Questionnaire for Sorting 39 Countries

Please sort the following countries into as many groups as you like in terms of their similarity and dissimilarity. Pick one country and assign "1" to the country. Then, assign "1" to those countries (any number of countries, including 0) which are similar to this country. Then, pick another country from the remaining countries, and assign "2" to this country. Assign "2" to all the countries which are similar to this country. And, continue in this way by assigning 3, 4, . . . and so on until all the countries are sorted into groups.		
( ) Brazil	( ) India	( ) Romania
( ) Britain	( ) Iran	( ) USSR
( ) Canada	( ) Italy	( ) Singapore
( ) China	( ) Japan	( ) South Africa
( ) Cuba	( ) Lebanon	( ) Spain
( ) Denmark	( ) Libya	( ) Sweden
( ) Egypt	( ) Netherlands	( ) Switzerland
( ) Ethiopia	( ) New Zealand	( ) Tanzania
( ) Finland	( ) Nigeria	( ) Thailand
( ) France	( ) Norway	( ) USA
( ) Greece	( ) Peru	( ) Venezuela
( ) Guatemala	( ) Pakistan	( ) West Germany
( ) Hungary	( ) Poland	( ) Yugoslavia

successive solutions. Let us look at only the first five solutions. Table 9.15 lists several countries from each of the extreme ends (i.e., those countries with large extreme weights, positive and negative) of each solution.

Examine clusters of countries in each solution, and see if you can interpret them. This task is probably easier than that of the previous example of cars because most of us are more familiar with countries than with cars. There are a few points of interest. China in Solution 3 is there because of political similarity and possibly geographical proximity as well. China in Solution 4 is there definitely because of geographical closeness. It is also interesting to note that those countries in the Middle East tend to be clustered with African countries. Solution 5 has countries from diverse areas in one cluster, that is, USA, Canada, New Zealand, Britain, and South Africa. These countries, however, can be placed into a single group in terms of the language, for these countries are primarily English-speaking.

In interpreting results from sorting data, it would be interesting to introduce a few "ideal" subjects who, for example, sort the countries in terms of geographical proximity, political similarity, or economic prosperity.

TABLE 9.13
Condensed Data Matrix of Assigned Group Numbers

Country	Subject 1	2	3	4	5	6	7	8	9	10	11	12	13	14	15	16	17	18	19	20	21
Brazil	2	1	2	1	1	1	1	1	1	2	1	1	1	1	1	1	1	1	2	1	1
Britain	1	2	1	2	2	2	2	2	2	3	2	2	3	2	5	2	2	2	1	2	2
Canada	1	2	1	2	2	2	2	2	3	3	2	3	3	2	5	2	2	2	1	2	2
China	4	3	4	3	8	3	3	3	4	1	3	4	7	3	3	3	3	3	4	3	3
Cuba	4	4	4	1	1	3	3	4	1	1	1	3	1	3	2	1	1	1	4	3	3
Denmark	1	5	1	4	4	4	2	5	2	3	2	2	5	2	1	4	2	2	2	4	4
Egypt	3	6	2	1	3	5	4	6	5	4	2	5	6	4	3	5	4	4	3	5	5
Ethiopia	3	6	3	5	5	5	4	4	5	5	1	5	6	4	1	5	5	4	4	5	5
Finland	1	5	1	4	4	4	2	5	2	3	2	2	5	5	1	4	4	2	3	4	4
France	1	7	2	4	2	6	2	5	2	3	2	2	5	2	5	6	2	2	1	4	2
Greece	4	7	1	4	7	6	2	6	5	3	2	2	1	6	3	6	4	2	2	7	5
Guatemala	2	1	2	5	1	1	1	4	1	4	1	1	1	1	4	1	1	1	3	1	3
Hungary	4	7	4	4	6	3	3	4	2	2	1	2	4	3	1	7	2	2	4	3	3
India	3	1	1	6	8	2	2	1	4	4	1	4	2	1	4	8	6	5	2	6	5
Iran	3	2	3	6	3	7	4	4	6	4	1	5	2	2	4	5	5	4	3	8	5
Italy	1	3	1	4	7	6	2	5	2	3	2	2	1	3	3	6	4	2	1	7	5
Japan	1	4	1	3	8	8	2	5	4	3	2	4	7	4	3	3	6	3	1	6	2
Lebanon	3	2	1	1	3	7	4	4	6	4	3	5	2	2	4	8	5	4	3	8	5
Lybia	3	2	3	1	3	5	4	4	6	5	3	5	2	2	4	5	5	4	3	5	5
Netherlands	1	5	1	4	4	6	2	5	2	3	2	2	5	5	1	6	2	2	2	4	4
New Zealand	1	6	1	2	9	2	2	2	2	3	4	6	3	6	5	2	2	2	2	2	6
Nigeria	3	7	2	5	5	2	4	6	5	2	3	5	6	7	4	5	1	6	3	5	5
Norway	1	4	1	4	4	4	2	5	2	3	2	2	5	5	1	4	4	2	2	4	4
Peru	2	8	2	1	1	1	1	1	1	2	2	1	1	7	1	1	1	1	3	1	1
Pakistan	3	1	3	6	8	2	2	4	4	4	1	4	2	1	4	8	5	4	3	6	5
Poland	4	3	4	1	6	3	3	4	2	1	2	2	4	8	2	7	2	2	4	3	3
Romania	4	1	4	1	6	3	3	4	2	1	1	2	4	1	3	7	5	2	4	3	3
USSR	4	1	4	1	6	3	3	3	2	1	1	2	4	1	2	7	6	2	4	3	3
Singapore	3	2	2	3	8	2	2	5	4	4	3	4	7	2	3	3	3	3	3	6	6
S. Africa	1	3	1	5	9	5	4	7	5	4	4	5	6	3	5	2	2	7	2	2	2
Spain	1	4	1	4	7	6	2	5	2	4	2	2	5	4	5	1	4	2	2	7	5
Sweden	1	1	1	4	4	4	2	5	2	3	2	2	5	5	5	4	4	2	2	4	4
Switzerland	1	3	2	4	4	6	2	5	2	3	2	2	5	5	5	6	2	2	2	4	4
Tanzania	3	2	2	5	5	5	4	1	5	5	4	5	6	6	1	5	4	6	3	5	5
Thailand	3	2	1	3	8	8	2	1	4	4	3	4	7	2	1	3	1	3	3	6	6
U.S.A.	1	5	2	2	2	9	2	3	3	3	2	3	3	7	5	2	2	2	1	2	2
Venezuela	2	4	1	1	1	1	1	1	1	2	4	1	1	2	4	1	1	1	2	1	1
W.Germany	1	3	1	2	6	6	2	5	2	3	2	2	5	5	2	6	2	2	1	4	2
Yugoslavia	2	1	4	4	6	3	3	1	2	1	1	2	4	1	1	7	4	2	4	3	3

TABLE 9.14
Distribution of the First 10 η_j^2

Solution	η_j^2	δ_j	$\Sigma\delta_j$	Solution	η_j^2	δ_j	$\Sigma\delta_j$
1	0.6901	13.18%	13.18%	6	0.3291	6.28%	57.85%
2	0.6148	11.74%	24.91%	7	0.2255	4.31%	62.15%
3	0.5310	10.14%	35.05%	8	0.2186	4.17%	66.32%
4	0.4564	8.71%	43.76%	9	0.2004	3.83%	70.15%
5	0.4087	7.80%	51.56%	10	0.1644	3.14%	73.29%

TABLE 9.15
Two Clusters of Countries from Each Solution

Solution 1		Solution 2	
Netherlands	Iran	Libya	Brazil
Sweden	Libya	Iran	Peru
Switzerland	Lebannon	Lebanon	Venezuela
Denmark	Ethiopia	Ethiopia	Guatemala
Norway	Nigeria	Egypt	Cuba
West Germany	Tanzania	Tanzania	
Finland			
Solution 3		**Solution 4**	
China	Peru	Ethiopia	Thailand
USSR	Venezuela	Poland	Singapore
Poland	Brazil	Hungary	Japan
Romania	Guatemala	Romania	India
Hungary		Tanzania	China
Yugoslavia		Yugoslavia	
		Egypt	
		USSR	
Solution 5			
Norway	USA		
Finland	Canada		
Sweden	New Zealand		
Denmark	Britain		
Netherlands	South Africa		

Once the results are obtained, we can then find from intersubject correlation those subjects who must have used a particular sorting criterion, geographical, political, or economic (see an example of this type of analysis discussed in Chapter 13). It would also be interesting to collect sorting data on many countries before and after a series of lectures on geography, economics, cultures, and religions. The main question then is to see if the lectures have influenced the subjects on their judgment of similarity.

Likewise, data can be collected at the beginning of a geography course and at the last class, to compare the difference between the two data sets.

In terms of methodology of sorting data, it is often the case that some subjects create too many piles to be manageable: the total number of piles that subjects use has a direct implication for the number of possible solutions. Thus, it is a worthwhile topic to investigate whether or not limiting the number of piles subjects are allowed to use would affect the nature of each of the first few solutions. It may be difficult to assess the exact effects of the limiting of the number of piles one can use on the outcome, but such a study is needed before we consider the restricted sorting task for practical use.

PART III
Dominance Data

Part 2 dealt with quantification problems typically handled by the well-known formulations of correspondence analysis, multiple correspondence analysis, dual scaling, homogenity analysis, optimal scaling and reciprocal averaging. Greenacre (1978, p.xi) stated that although the assumptions underlying correspondence analysis are minimal, "perhaps the only fully-fledged assumption is that data elements be nonnegative!" This statement appears to have created the false impression among some researchers that we must limit the application of the method only to data matrices of nonnegative elements such as those data types dealt with in Part 2. These data are absolute measurements given in the forms of absence/presence and frequencies, and indeed they are what correspondence analysis was formulated for.

Part 3 contains applications of dual scaling to dominance data that are ordinal measurements. As shown in the next three chapters, input data can involve negative elements. More important than the mixed signs of data elements, however, are special properties of the data which are common to all three data types to be discussed in Part 3, that is, row-conditionality and the absence of the trivial solution that incidence data have (see chapter 6).

CHAPTER 10
Paired Comparison Data

10.1 SPECIAL PROBLEMS

The three data types discussed so far (i.e., contingency tables, multiple--choice data and sorting data) can be subjected, without modifications, to the method of reciprocal averages. This is so because data elements are nonnegative, and the sum of a row (a column) represents the number of responses in that row (column). Once negative elements are involved, however, the sum of the elements of a row no longer tells us how many responses are involved in that row. This is exactly the problem with paired comparison data, and two other types to be discussed in the ensuing chapters. Although it may not be well known, the extension of the method to data matrices containing negative elements was successfully formulated into a practical method by Guttman more than 40 years ago, in 1946. His extension was not obvious because he dealt with the difference between two matrices of nonnegative elements.

Let us look at the basic idea of this extension, following Nishisato's formulation (1978a). Consider paired comparisons of Coke (X_1), Pepsi (X_2), and 7-UP (X_3). Each subject is asked to indicate his or her preference for X_j or X_k when pair (X_j, X_k) is presented. For pair (X_j, X_k), the response of subject i, ${}_if_{jk}$, is coded as follows:

$$(10.1)\qquad {}_if_{jk} = \begin{cases} 1, & \text{if Subject } i \text{ prefers } X_j \text{ to } X_k,\ X_j \succ X_k . \\ 0, & \text{if Subject } i \text{ provides equality judgment, } X_j = X_k . \\ -1, & \text{if Subject } i \text{ prefers } X_k \text{ to } X_j,\ X_j \prec X_k . \end{cases}$$

Suppose that the first two subjects' responses, ${}_if_{jk}$, are as given in Table 10.1. Let x_1, x_2, and x_3 be unknown scale values of X_1, X_2, and X_3, respectively. In terms of the scale values, Table 10.1 indicates that $x_1 > x_2$, $x_1 > x_3$, $x_2 < x_3$ for Subject 1, and that $x_1 > x_2$, $x_1 < x_3$, $x_2 < x_3$ for Subject 2. The coding of 1, 0, and -1 indicates the direction of pairwise contrasts of scale values. In other words, Table 10.1 can be expressed in terms of x_1, x_2, and x_3 as Table 10.2. This table can be rearranged into a subjects-by-stimuli table (Table 10.3).

TABLE 10.1
Paired Comparison Data

	Pairs		
Subjects	(X_1,X_2)	(X_1,X_3)	(X_2,X_3)
1	1	1	-1
2	1	-1	-1

TABLE 10.2
Weighted Paired Response

Subjects	Contrast of Scale Values		
1	$(x_1 - x_2)$	$(x_1 - x_3)$	$-(x_2 - x_3)$
2	$(x_1 - x_2)$	$-(x_1 - x_3)$	$-(x_2 - x_3)$

TABLE 10.3
Weighted Subject-by-Stimulus Table

	Stimuli		
Subjects	1	2	3
1	x_1, x_1	$-x_2, -x_2$	$-x_3, x_3$
2	$x_1, -x_1$	$-x_2, -x_2$	x_3, x_3

Note that Table 10.3 was obtained from Table 10.2 by removing parentheses and putting signed scale values stimulus by stimulus. Notice and keep in mind that each cell of the table contains two responses, that is, the total number of stimuli minus one. If we consider the table of cell totals, we obtain Table 10.4. The table of coefficients is therefore given by Table 10.5. This is called the *dominance table*. Each of n stimuli ($n = 3$ in this example) is compared with $(n - 1)$ stimuli, and each element of the dominance table indicates "the number of times the stimulus is preferred to the remaining $(n - 1)$ stimuli minus the number of times the other stimuli are preferred to it." Therefore, in Table 10.5, each stimulus was comapred with the two remaining stimuli, and elements 2, -2, and 0 indicate, respectively, twice preferred, twice not preferred, and once preferred and once not preferred.

The dominance table such as Table 10.5 is the input data matrix for dual scaling of paired comparison data. Note not only that there are negative elements, but also that the sum of dominance numbers of the stimuli is zero for each subject (e.g., Subject 1 provided 2, -2, and 0 for X_1, X_2, and X_3; 2-2+0 = 0),

$$\Sigma e_{ij} = 0 \text{ for each } i,\ i = 1, 2, \ldots, n .$$

TABLE 10.4
Summary Table of Weighted Responses

Subjects	Stimuli 1	2	3
1	$2x_1$	$-2x_2$	0
2	0	$-2x_2$	$2x_3$

TABLE 10.5
Dominance Table

Subjects	Stimuli 1	2	3
1	2	-2	0
2	0	-2	2

Or, in matrix notation,

$$\mathbf{E1} = \mathbf{0} .$$

Therefore, when there are N subjects and n stimuli and N is much larger than n, the maximal number of solutions is $n - 1$.[1] There are three points that one should keep in mind about the dominance table:

(a) Number of Responses Involved. Recall the method of reciprocal averages, that is, one of the methods for finding dual scaling weights for rows and columns of a data table. You calculated row averages of responses weighted by column weights, and column averages of responses weighted by row weights. The total number of responses in each row (column) was simply the sum of responses in the row (column). In the case of the dominance table, the sum of elements of each row is zero, which is no longer the total number of responses needed in the calculation of the row average. Instead, the total number of each row and that of each column of the dominance table must be calculated as follows. Suppose that N subjects are involved in paired comparisons of n stimuli. Then, each element of the dominance table is, as we have just seen in the above example, a summary statistic based on $(n - 1)$ comparisons. Thus, the total number of responses of each row and that of each column of the $N \times n$ dominance table are respectively equal to $n(n - 1)$ and $N(n - 1)$, the total number of responses in the table, that is, f_t, being now $Nn(n - 1)$. One of the unique aspects of dominance table is the fact that each cell has the same number of responses, that is, $(n - 1)$. This is particularly important, considering that quantification of incidence

[1] This is related to the fact that the rank of the $N \times n$ dominance matrix $\mathbf{E}$ is $n - 1$. This is obvious because elements of each row of $\mathbf{E}$ sums to zero, that is, linear dependence among n elements of each row of $\mathbf{E}$.

data, as discussed in Part 2, is much influenced by uneven distributions of responses in different cells. In other words, dominance data do not have this powerful source of influence on the outcome of quantification. Implications of equal cell frequency versus nonequal are discussed in chapter 17.

(b) Ipsativity (Special Marginal Constraint). In data analysis, we often encounter a data table in which either row marginals or column marginals are constant and all equal. For instance, consider job descriptions of college professors A, B, and C in terms of their time allocations over teaching, research, student advising, and administration, expressed in percentages. The sum of those percentage figures of each professor is 100. This set of measurements is called *ipsative*, and ipsative data are not always easy to analyze. Suppose you collect from 10 professors data on their time allocations over teaching, research, and administrative work, all expressed in terms of percentage. Suppose further that you suspect positive correlation between teaching and research, but negative correlation between research and administrative work. Can the data support your common-sense relationships? From your 10×3 ipsative data (i.e., row marginals are all 100), the ordinary (product-moment) correlation of any pair of the three variables is likely to be negative, contrary to the common-sense expectation. Try to generate a 10×3 table, with row marginals being all equal to 100, and verify numerically that all the correlation coefficients between pairs of the three variables are negative. Another example of ipsative data is obtained when each subject is asked to rank five world leaders according to the order of their preference. If the rank numbers are used as data, each subject's total is 1+2+3+4+5, that is, 15, and the N×5 table of rank numbers is ipsative. In this instance, our common sense is good enough to convince us that these rank numbers are meaningful only within each subject. In other words, different subjects' rank numbers are probably not comparable, for rank 3 of one subject may be quite different from rank 3 of another subject in terms of the strength of their support. In this context, data with identical row marginals are sometimes called *row-conditional*, and data with identical column marginals are said to be *column-conditional*. Dual scaling of paired comparison data therefore means an extension of this quantification method from unconditional data to conditional data.

(c) Relationship Between Row Weights and Column Weights. In the previous chapters, the dual relations were looked at, and the discrepancy between the row space and the column space was noted. It was nevertheless emphasized that overlaying of the two discrepant spaces into a common graph could not be dispensed with from the data analysis routine, for such practice would typically provide the researcher with insight into interpretations of dual scaling outcome. There are problems with row-column joint display of dominance data, similar to that of incidence data. At the same time, there are some fundamentally different problems, peculiar to dominance data. To convince you on this point, let us consider two data tables, one contingency table and one dominance table, which have almost perfect internal consistency. As we have already seen, we would expect that the row weights of the contigency table would be vastly different among themselves, the same being true with respect to the column weights. In contrast, the weights for the rows of the row-conditional dominance table would in this case be almost equal, that is, the variance of the absolute values of row weights is nearly zero, rather than a maximum. The column weights of the dominance table would show large variation

among themselves. To be more precise about the row weights of a row-conditional table, dual scaling as formulated by Nishisato (1978a, 1980a) will provide row weights being either 1 or -1 when data possess perfect internal consistency, or close to 1 or -1 when the consistency is nearly perfect. Table 10.6 shows examples of a contingency table and a dominance table which have high internal consistency. Observe the similarity and the difference in weights between the two tables, especially the difference between the two sets of row weights, indicated by *. The row weights show only how similar the row *patterns* of dominance numbers are to the *pattern* of the derived scale values (i.e., weights 1.39, 0.25, -0.25, -1.39 in Table 10.6). The same applies to dual scaling of rank-order data (chapter 11) and successive categories data (chapter 12). Graphical display of dominance data is examined in detail in chapter 14.

TABLE 10.6
Two Data Types with High Internal Consistency

	Contingency Table						Dominance Table				
	1	2	3	4	Weight		1	2	3	4	Weight
1	33	2	0	1	1.53*	1	3	1	-1	-3	1.03*
2	1	28	1	0	.54*	2	3	-1	1	-3	.91*
3	0	2	34	3	-.66*	3	3	2	-1	-3	1.03*
4	2	1	2	47	-1.06*	4	3	1	-1	-3	1.03*
Weight	1.54	.56	-.62	-1.04		Weight	1.39	.25	-.25	-1.39	

The dominance matrix, say **E**, can easily be obtained from the subject-by-pair matrix of responses coded by equation (10.1). Nishisato (1978a) proposed the following formula:

$$\mathbf{E} = \mathbf{F}\mathbf{A} \qquad (10.2)$$

where **F** is the input matrix of responses coded by (10.1), with rows being subjects and columns being pairs arranged in the specific order (1,2), (1,3), . . . (1,n), (2,3), . . . (2,n), . . . (n-1,n), and **A** is called the deisgn matrix (e.g., Bock and Jones, 1968; Nishisato, 1975, 1978a). For those who are interested in formulas, a simple numerical example of (10.2) is given as follows: Let us augment Table 10.1 by adding two more subjects and indicate this 4×3 matrix by **F**, that is,

$$\mathbf{F} = \begin{bmatrix} 1 & 1 & -1 \\ 1 & -1 & -1 \\ 1 & 1 & -1 \\ -1 & -1 & 1 \end{bmatrix}$$

Note that these are artificial numbers, and that important points are:

1. The four rows represent responses, coded by (10.1), of four subjects
2. Columns 1, 2, 3 correspond to pairs (1, 2), (1, 3), and (2,3)
3. There are three stimuli.

The corresponding design matrix is of order "pairs-by-stimuli," that is, 3×3 in the current example, or, in general, $n(n-1)/2 \times n$, where N is the number of subjects and n is the number of stimuli. The design matrix for the current example is given by

$$\mathbf{A} = \begin{bmatrix} 1 & -1 & 0 \\ 1 & 0 & -1 \\ 0 & 1 & -1 \end{bmatrix}$$

The three rows of this matrix show comparisons (1,2), (1,3), and (2,3), respectively. Thus, the dominance matrix is obtained by (10.2) as follows:

$$\mathbf{E} = \begin{bmatrix} 1 & 1 & -1 \\ 1 & -1 & -1 \\ 1 & 1 & -1 \\ -1 & -1 & 1 \end{bmatrix} \begin{bmatrix} 1 & -1 & 0 \\ 1 & 0 & -1 \\ 0 & 1 & -1 \end{bmatrix} = \begin{bmatrix} 2 & -2 & 0 \\ 0 & -2 & 2 \\ 2 & -2 & 0 \\ -2 & 2 & 0 \end{bmatrix}$$

This is another way to construct a dominance matrix. Check if these dominance numbers are correct by simply counting, that is, the total number of times a particular stimulus was preferred to the remaining stimuli, minus the number of times it was not.

In summary, we have noted the following:

1. Coding of paired comparison by 1, 0, and -1
2. Transforming paired comparisons to dominance numbers
3. Row-conditionality of the dominance table
4. The total number of responses in each row being $n(n-1)$, that in each column $N(n-1)$, and the total number of responses $Nn(n-1)$
5. Distinct characteristics of row (subject) weights and column (stimulus) weights, in contrast to those of incidence data.

Let us now look at two examples to see how dual scaling of dominance tables can be interpreted.

10.2 APPLICATIONS

10.2.1 Eight Christmas Party Plans

The first example is obtained from Nishisato and Nishisato (1984, pp. 54-69). Data were collected by Ian Wiggins as a part of an assignment for a course in scaling at the Ontario Institute for Studies in Education. The object of the data collection was to find people's preference patterns for eight plans for a Christmas party:

> X_1 = a pot-luck at someone's home in the evening
> X_2 = a pot-luck in the group room
> X_3 = a pub/restaurant crawl after work
> X_4 = a reasonably priced lunch in an area restaurant
> X_5 = keep to one's self
> X_6 = an evening banquet at a restaurant
> X_7 = a pot-luck at someone's home after work
> X_8 = a ritzy lunch at a good restaurant (tablecloths)

Twenty-eight pairs of party plans (X_j, X_k) were presented in a random order to 14 behavioral science researchers at a research institute, and the subjects were asked to indicate which one of the plans in each pair they would prefer. Table 10.7 contains the responses of the 14 subjects to the 28 pairs. As indicated earlier, paired comparision responses to (X_j, X_k) should be coded 1, -1, and 0, indicating, respectively X_j being preferred to X_k, X_k preferred to X_j, and a tied (equal preference) response. However, to simplify the table and to avoid typing errors, -1 is replaced with 2 in Table 10.7. In other words, responses $X_j > X_k$, $X_j < X_k$, and $X_j = X_k$ are coded for convenience 1, 2, and 0, respectively.

TABLE 10.7
Preferences for Christmas Party Plans

Pair (X_j,X_k)

SSj=	1	1	1	1	1	1	1	2	2	2	2	2	2	3	3	3	3	3	4	4	4	4	5	5	5	6	6	7
k=	2	3	4	5	6	7	8	3	4	5	6	7	8	4	5	6	7	8	5	6	7	8	6	7	8	7	8	8
1	1	1	2	1	1	2	1	2	2	2	2	2	2	2	1	1	2	1	1	1	2	1	1	2	1	2	1	2
2	2	2	2	1	2	1	2	1	2	1	2	1	2	2	1	1	1	2	1	1	1	2	2	2	2	1	2	2
3	1	1	1	1	1	2	1	1	1	1	1	2	1	1	1	1	2	1	1	1	2	1	2	2	2	2	1	1
4	2	1	2	1	1	1	2	1	1	1	1	1	2	2	1	2	2	2	1	1	1	2	2	2	2	2	2	2
5	2	2	2	1	2	1	2	2	2	1	2	2	2	2	1	2	1	2	1	1	1	1	2	2	2	1	2	2
6	1	1	1	1	1	1	1	2	2	1	2	2	2	2	1	2	2	2	1	1	1	1	2	2	2	2	2	1
7	1	1	1	1	1	2	1	1	2	1	1	2	1	2	1	1	2	1	1	1	2	1	2	2	2	2	2	1
8	1	1	1	1	1	2	1	1	2	1	2	2	1	2	1	2	2	1	1	2	2	1	2	2	1	2	1	1
9	1	2	2	1	1	2	1	2	2	1	1	2	2	1	1	1	2	1	1	1	2	1	2	2	2	2	2	1
10	1	2	1	1	2	2	2	2	2	1	2	2	2	1	1	1	1	1	1	2	2	2	2	2	2	1	1	2
11	1	2	1	1	1	1	1	2	2	2	2	2	2	1	1	1	1	1	1	1	1	1	2	2	2	2	2	2
12	2	2	2	2	1	2	2	1	2	1	1	1	1	2	1	1	1	1	1	1	1	1	1	1	1	2	2	1
13	1	2	1	1	2	1	2	2	2	2	2	2	2	1	1	1	1	1	1	2	1	2	2	2	2	1	1	2
14	2	2	2	2	1	2	1	2	1	1	1	1	1	1	1	1	1	1	2	1	2	1	1	2	1	2	1	1

Note. In this table, -1 is coded 2 for simplicity.

Generally speaking, people are different in their preference for a type of party, and dual scaling can identify such individual differences. In contrast, the traditional method of paired comparisons (see Bock & Jones, 1968) ignores such individual differences by averaging them out. As we will see, however, there are occasions in which individual differences are important in interpreting data, or more interesting than the average preference pattern. Dual scaling incorporates individual differences in determining scale values of stimuli (i.e., Christmas party plans in the current example). Noting that 2 in Table 10.7 is substituted for -1, the 14×8 dominance table can be obtained by counting for each subject the number of times a particular party plan is preferred to the other stimuli minus the number of times it is not preferred. Table 10.8 is the result of this transformation.

TABLE 10.8
Subject-by-Stimulus Table of Dominance Numbers

	Stimulus							
Subject	1	2	3	4	5	6	7	8
1	3	-7	1	5	-1	-3	5	-3
2	-3	1	1	5	-7	1	-5	7
3	5	3	1	-1	-7	-3	7	-5
4	1	5	-5	3	-7	-3	-1	7
5	-3	-3	1	7	-7	3	-3	5
6	7	-5	-3	5	-7	-1	3	1
7	5	1	-1	3	-7	-5	7	-3
8	5	-1	-3	1	-5	3	7	-7
9	1	-3	5	3	-7	-5	7	-1
10	-1	-5	7	-3	-7	5	1	3
11	5	-7	7	3	-5	-3	-1	1
12	-5	5	3	7	1	-7	-1	-3
13	1	-7	7	-1	-5	5	-3	3
14	-3	5	7	-1	1	-5	3	-7

Notice that some subjects have all the dominance numbers distinct (e.g., Subject 3), while others have some dominance numbers repeated (e.g., Subject 2), indicating various degrees of subjects' contributions to distinguishing party plans. For instance, the dominance numbers of Subject 2 indicate that Party Plans 2, 3, and 6 cannot be distinguished, for they are all equal to 1. Thus, weights for the subjects are determined by two factors, the discriminability (distinctiveness) of the dominance numbers and the similarity of the pattern of dominance numbers to that of the entire dominance table. Each element in the dominance table is based on seven comparisons. Thus, each row marginal and each column marginal of the table are 56 (7×8) and 98 (7×14), respectively.

As was mentioned in relation to (10.2), the total number of solutions from the $N \times n$ table is typically equal to n - 1. In the present example, this number is 8 - 1 = 7. Table 10.9 shows the relevant statistics. From the distribution of η^2 over seven solutions, it looks as though we may consider two solutions or four solutions because there are two relatively large drops in the distribution, one between Solution 2 and

TABLE 10.9
Distribution of Information over Seven Solutions

Solution	η^2	δ	$\Sigma\delta$
1	0.1411	33.72%	33.72%
2	0.1098	26.24%	59.96%
3	0.0652	15.59%	75.55%
4	0.0551	13.18%	88.73%
5	0.0284	6.80%	95.53%
6	0.0131	3.12%	98.65%
7	0.0057	1.35%	100.00%

Solution 3, and the other between Solution 4 and Solution 5. This data set has been analyzed, and it is known that the first three solutions are interpretable, but that Solution 4 is not. Therefore, we will look at only the first three solutions. Table 10.10 lists adjusted weights for the eight party plans of the three solutions.

As discussed earlier in this chapter, plotting row weights and column weights in the same graph required a special interpretation. Thus, before examining weights for the subjects, let us plot weights of stimuli in Table 10.10 to see what the data can tell us about people's choices for the Christmas party. Figures 10.1 and 10.2 are plots of the plans based on the weights in Table 10.10.

TABLE 10.10
Weights for the Party Plans on the First Three Dimensions

Party Plans	Solution 1	Solution 2	Solution 3
Pot-luck in the evening	0.39	0.14	0.13
Pot-luck (group room)	-0.35	0.32	-0.38
Pub/restaurant crawl	0.19	-0.12	0.26
Reasonably priced lunch	0.30	-0.08	-0.35
Keep to one's self	-0.73	0.24	0.18
Evening banquet	-0.15	-0.36	0.28
Pot-luck after work	0.41	0.46	0.09
Ritzy lunch	-0.06	-0.59	-0.21

Solution 1 has Plans 7, 1 and 4 on one side and Plans 5 and 2 on the other side. Noting that Plan 5 is "Keep to one's self" and Plan 2 "a pot-luck in the group room," we see a rather negative cluster toward a party. In contrast, the other end consists of "a pot-luck at someone's home in the evening," "a pot-luck at someone's home after work" and "a pub/restaurant crawl after work," all of which seem to suggest informal and convivial parties. Can we therefore name the first solution the dimension of *conviviality*? Solution 2 has on one side "a pot-luck at someone's home after work," "a pot-luck in the group room," and "keep to one's self" and on the other side "a ritzy lunch at a good restaurant (tablecloths)" and "an evening banquet at a restaurant." There is a clear contrast of inexpensive versus expensive plans. Can we therefore name this the dimension of *cost*? What about Solution 3? From Table

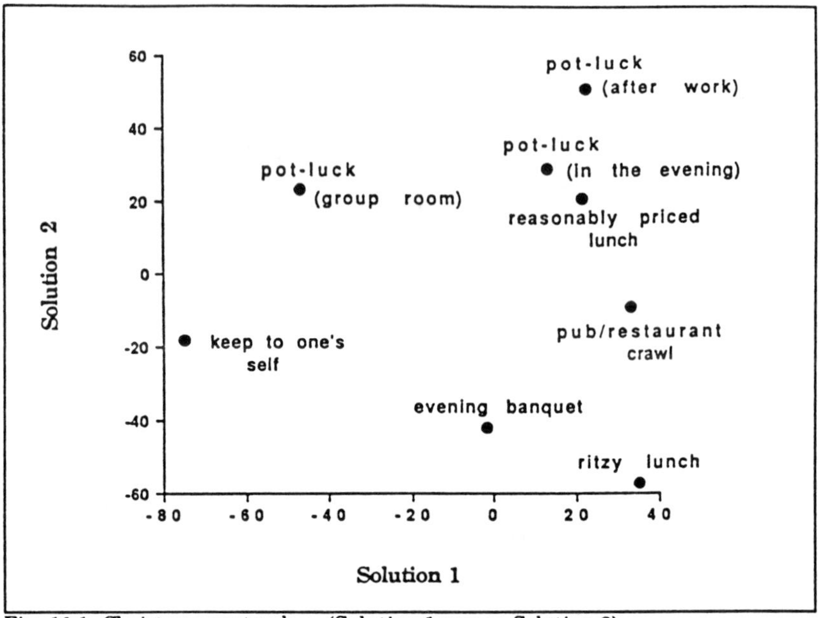

Fig. 10.1. Christmas party plans (Solution 1 versus Solution 2)

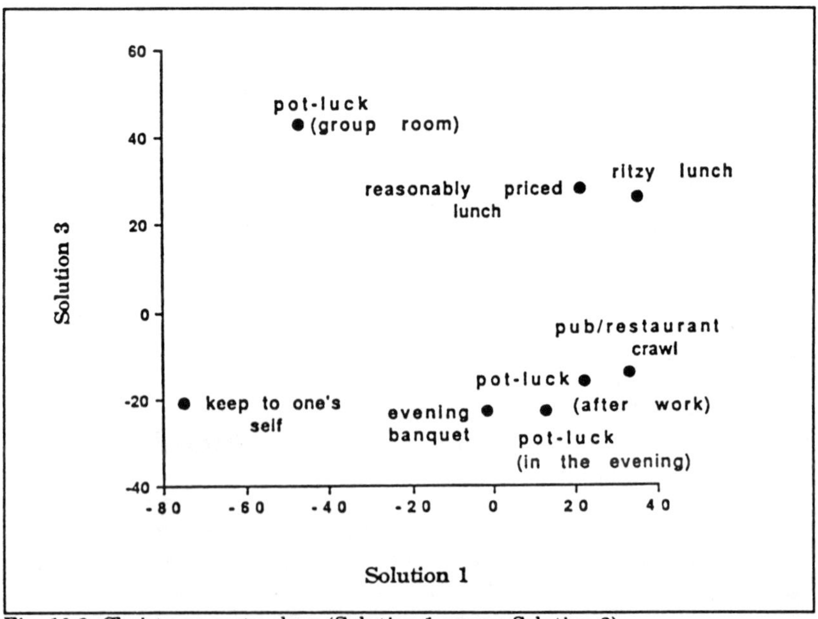

Fig. 10.2. Christmas party plans (Solution 1 versus Solution 3)

10.1 and Figure 10.2, we can see the two contrasting clusters, "an evening banquet at a restaurant," "a pub/restaurant crawl after work," "Keep to one's self," "a pot-luck at someone's home in the evening," and "a pot-luck at someone's home after work" on one side, and "a reasonably priced lunch in an area restaurant," "a pot-luck in the group room," and "a ritzy lunch at a good restaurant." Since the data were collected from workers in a research institute, we can interpret "Keep to one's self" to be an evening activity, and "a pot-luck in the group room" to be a daytime party. Then it becomes clear that Solution 3 identifies the contrast between daytime parties and evening parties. Therefore, we can call it the dimension of *time*.

The results are quite interesting. But how can we relate the findings to choices of individual respondents? We have earlier noted that the weights for subjects in dual scaling of dominance data indicate the extent to which individuals contribute to a particular solution, or the similarity of individuals' choice patterns to the pattern captured by a particular solution. Therefore, the above question can be answered by examining the weights for subjects. Table 10.11 lists the weights from three solutions.

TABLE 10.11
Weights for Subjects

Subject	Solution 1	Solution 2	Solution 3
1	0.41	0.14	0.11
2	0.12	-0.49	-0.36
3	0.45	0.38	-0.03
4	0.16	-0.17	-0.55
5	0.25	-0.49	-0.22
6	0.56	-0.06	-0.09
7	0.53	0.32	-0.18
8	0.41	0.31	0.13
9	0.57	0.12	-0.02
10	0.30	-0.38	0.30
11	0.49	-0.20	0.14
12	-0.05	0.22	-0.39
13	0.25	-0.48	0.32
14	-0.01	0.43	0.02

Notice that 12 subjects out of 14 have positive weights. This means a heavy inclination towards convivial parties. When the first solution is very dominant in the entire set of solutions, the weights for the columns (i.e., party plans) are similar to the column averages of the dominance matrix. Thus, even when the first solution is not very dominant, we can nevertheless see some discernable relationship between the column weights of Solution 1 and column averages of dominance numbers. Therefore, we expect that dominance numbers of subjects in the fifth column ("Keep to one's self") are likely to be negative. It is interesting to note that only those two subjects who obtained negative weights on Solution 1, that is, Subjects 12 and 14, have positive dominance numbers for Party Plan 5. Although Party Plan 5 is not their first choice, their responses show slight inclinations towards Party Plan 5. More specifically, their dominance numbers tell us the following preference orders:

Subject 12: $X_4 > X_2 > X_3 > X_5 > X_7 > X_8 > X_1 > X_6$.
Subject 14: $X_3 > X_2 > X_7 > X_5 > X_4 > X_1 > X_6 > X_8$.

Both chose Plan 2 (a pot-luck in the group room) as their second most preferred plan. Recall that Plan 2 was second least preferred one in the first solution.

What about the other two solutions? It is clear that the distributions of weights of the two solutions are quite different from that of Solution 1. In Solutions 2 and 3, positive weights and negative weights are more evenly distributed. When a subject's weight is positive and close to the largest scale value of stimuli, it indicates that the subject's choice pattern is similar to that of the solution. When it is negative and close to the end of the scale, it shows that the reversed pattern of the subject's choice is similar to that of the solution. This explanation may sound complicated, but the basic point is that subjects rank stimuli according to the order of their distances to the stimuli, the closest stimulus first and the furthest last. Thus, we can now relate the previous results on the party plans to the choice patterns of subjects. Let us look at the cost dimension, that is, Solution 2. Noting those inexpensive plans are on the positive side (see Figure 10.1), we see that Subjects 14, 3, 7, 8, and 12, for example, preferred inexpensive plans, while Subjects 5, 2, 10, and 11 tended to choose expensive party plans. Similarly, we can interpret the time dimension, that is, Solution 3. Subjects 13 and 10 preferred evening party plans, while Subjects 4, 12, 2, and 5 chose daytime party plans over evening ones. It is important to note here that there is nothing to prevent the emergence of all possible positive-negative patterns of subjects' weights. In other words, the analysis can accommodate such preference patterns as (inexpensive, daytime), (inexpensive, evening), (expensive, daytime), and (expensive, evening). As may be inferred from Solution 1, it is also possible that all the weights for subjects are positive. This happens when all the subjects have similar preference patterns. Keep in mind, however, that the weights for the columns (party plans) of the dominance patterns are very different from those for subjects. The column weights always have positive and negative weights evenly distributed. This difference between row weights and column weights of dominance data comes from the fact that the dominance matrix has the property Equation 10.2 only with respect to the rows. Let us now look at another example of paired comparisons.

10.2.2 Seriousness of Eight Crimes

The seriousness of eight criminal acts will be judged by paired comparison, and dual scaling will be used to assess similarities and differences among them. The eight crimes are as follows:

(A) Arson;	(B) Burglary;	(C) Counterfeiting;
(F) Forgery;	(H) Homicide;	(K) Kidnapping;
(M) Mugging;	(R) Receiving stolen goods.	

TABLE 10.12
Seriousness of Eight Crimes

	Pair(X_j, X_k)																											
SSj=	A	A	A	A	A	A	A	B	B	B	B	B	B	C	C	C	C	C	F	F	F	F	H	H	H	K	K	M
k=	B	C	F	H	K	M	R	C	F	H	K	M	R	F	H	K	M	R	H	K	M	R	K	M	R	M	R	R
1	1	1	1	2	2	2	1	1	1	2	2	2	1	1	2	2	2	2	2	2	2	1	1	1	1	1	1	1
2	1	1	1	2	0	1	1	1	1	2	2	0	1	0	2	2	2	2	2	2	2	1	1	1	1	1	1	1
3	1	1	1	2	2	1	1	0	1	2	2	1	1	0	2	2	2	1	2	2	1	1	1	1	1	1	1	1
4	1	1	1	2	2	1	1	0	1	2	2	2	1	0	2	2	2	1	2	2	2	0	1	1	1	2	1	1
5	1	1	1	2	2	1	1	1	1	2	1	1	1	1	2	2	1	1	2	2	1	1	1	1	1	1	1	1
6	1	1	1	2	2	1	1	1	1	2	2	2	1	2	2	2	2	1	2	2	2	1	1	1	1	0	1	1
7	1	1	1	1	1	2	1	1	1	2	2	2	1	2	2	2	2	1	2	2	2	1	1	2	1	2	1	1
8	1	1	1	2	2	1	1	1	1	2	2	2	1	2	2	2	2	2	2	2	2	2	1	1	1	1	1	1
9	1	1	1	2	2	1	1	1	1	2	2	1	1	1	2	2	1	1	2	2	2	1	1	1	1	1	1	1
10	0	2	1	2	2	2	1	1	1	2	2	1	1	0	2	2	2	1	2	2	2	1	1	1	1	2	1	1
11	1	2	1	2	2	2	1	2	1	2	2	2	1	1	2	2	1	1	2	2	2	2	1	2	1	1	1	1
12	1	1	1	0	1	1	1	1	1	2	2	0	1	0	2	2	2	0	2	2	2	0	1	1	1	0	1	1
13	1	1	1	0	1	1	1	2	1	2	2	2	1	1	2	2	2	1	2	2	2	1	1	1	1	1	1	1
14	1	1	1	2	0	1	1	1	1	2	2	2	1	2	2	2	2	1	2	2	2	1	1	1	1	1	1	1
15	1	1	1	0	0	0	1	0	0	2	2	0	1	0	2	2	2	1	2	2	2	1	1	1	1	0	1	1
16	1	1	1	2	0	1	1	1	1	2	2	0	1	0	2	2	2	2	2	2	2	1	1	1	1	1	1	1
17	1	1	1	2	1	2	1	1	1	2	2	2	1	2	2	2	2	1	2	2	2	1	1	1	1	2	1	1
18	1	0	1	1	1	0	1	1	0	2	2	0	2	0	2	2	0	0	2	2	0	0	1	1	1	1	1	0
19	1	1	0	0	0	0	1	1	1	0	0	0	1	0	2	2	2	2	2	2	2	0	0	0	1	0	1	1
20	0	1	1	2	2	2	1	1	0	2	2	2	0	0	2	2	2	0	2	2	2	0	1	1	1	1	1	1
21	1	1	1	1	1	1	1	2	2	2	2	2	0	2	2	2	0	1	2	2	1	1	1	1	1	1	1	1
22	1	1	1	2	1	1	1	1	1	2	2	2	1	2	2	2	2	1	2	2	2	1	1	1	1	1	1	1
23	1	1	0	1	1	2	2	2	0	2	2	1	2	0	1	2	1	1	2	1	1	1	2	2	2	1	1	1

Since there are eight stimuli (crimes), the total number of combinations of two out of eight is 8×(8 - 1)/2 = 28. These pairs were presented to 23 students, enrolled in a course on applied scaling at the Ontario Institute for Studies in Education. The 23×28 data matrix is shown in Table 10.12.

TABLE 10.13
Distribution of the Squared Correlation Ratios

Solution (j)	η_j^2	δ_j	$\Sigma\delta_j$
1	0.3084	81.39%	81.39%
2	0.0232	6.13%	87.52%
3	0.0178	4.70%	92.22%
4	0.0143	3.78%	96.00%
5	0.0079	2.10%	98.10%
6	0.0045	1.20%	99.29%
7	0.0027	0.71%	100.0%

The total number of solutions is seven, and the distribution of the seven squared correlation ratios is shown in Table 10.13. This is a case in which the first solution is very dominant, and the remaining solutions are by comparison marginal.

Therefore, we will look at only the first solution. Let us first look at the dominance matrix and normed weights for subjects on Solution 1, as shown in Table 10.14.

TABLE 10.14
Subject-by-Crime Table of Dominance Numbers

Subject	1	2	3	4	5	6	7	8	y_i
1	1	-1	-5	-5	7	5	3	-5	1.097
2	4	0	-6	-4	7	4	0	-5	1.115
3	3	0	-3	-2	7	5	-3	-7	1.030
4	3	-2	-3	-5	7	3	3	-6	1.086
5	3	3	01	-3	7	3	-5	-7	0.903
6	3	-1	-5	-3	7	4	2	-7	1.146
7	5	-1	-5	-3	3	1	7	-7	0.957
8	3	-1	-7	-5	7	5	1	-3	1.099
9	3	1	-1	-5	7	5	-3	-7	1.052
10	-2	2	-2	-4	7	3	3	-7	0.910
11	-1	-3	3	-7	5	5	3	-5	0.769
12	6	0	-5	-5	6	2	1	-5	1.069
13	6	-3	-1	-5	6	3	1	-7	1.089
14	4	-1	-5	-3	7	4	1	-7	1.164
15	4	-2	-3	-3	6	3	2	-7	1.039
16	4	0	-6	-4	7	4	0	-5	1.115
17	3	-1	-5	-3	7	1	5	-7	1.089
18	5	-3	-3	-3	5	3	-2	-2	0.751
19	3	2	-6	-4	3	3	3	-4	0.823
20	0	-2	-5	-4	7	5	3	-4	1.000
21	7	-6	-2	1	5	3	-2	-6	0.862
22	5	-1	-5	-3	7	3	1	-7	1.162
23	2	-4	2	2	-3	3	-1	-1	-0.071

In Table 10.14, the normed weights were used since we are looking at only one solution, that is, no comparisons are made between solutions. Normed weights are expected to be all 1 when the data set involves only one solution. In other words, when the first solution accounts for 100% of the total information, all y_i are equal to 1. From this viewpoint, we can look at the current results. As you see in the table, weights of many subjects are very close to 1, except Subject 23, whose weight is -0.071. The fact that this subject's weight is close to 0 means two things. First, his or her response pattern on different crimes does not resemble the pattern that this data set as a whole indicates. Second, therefore, this subject's response pattern was nearly ignored in calculating scale values of the eight crimes. If you extract more than one solution, however, this subject's response pattern will be captured by some of the solutions. What then is so unique about this response pattern? According to the dominance numbers of Subject 23, the crimes can be arranged in order from the most serious to the least serious as follows:

1. Kidnapping
2. Arson, counterfeiting, forgery

3. Mugging, receiving stolen goods
4. Homicide
5. Burglary

Our common sense would not agree with this being a typical pattern of judgment. Who in their right mind would say that homicide is less serious than forgery or receiving stolen goods? Did the subject fail to understand this judgmental task? Did the subject try to sabotage this scaling project? Did the subject answer the task in a random fashion? No matter what the explanation, it is comforting to know that dual scaling is capable of differentially weighting responses from different subjects. In this particular case, this subject's responses were almost totally ignored in determining the first solution. From Solution 1, we obtain the eight crimes arranged in order of their seriousness as shown in Figure 10.3.

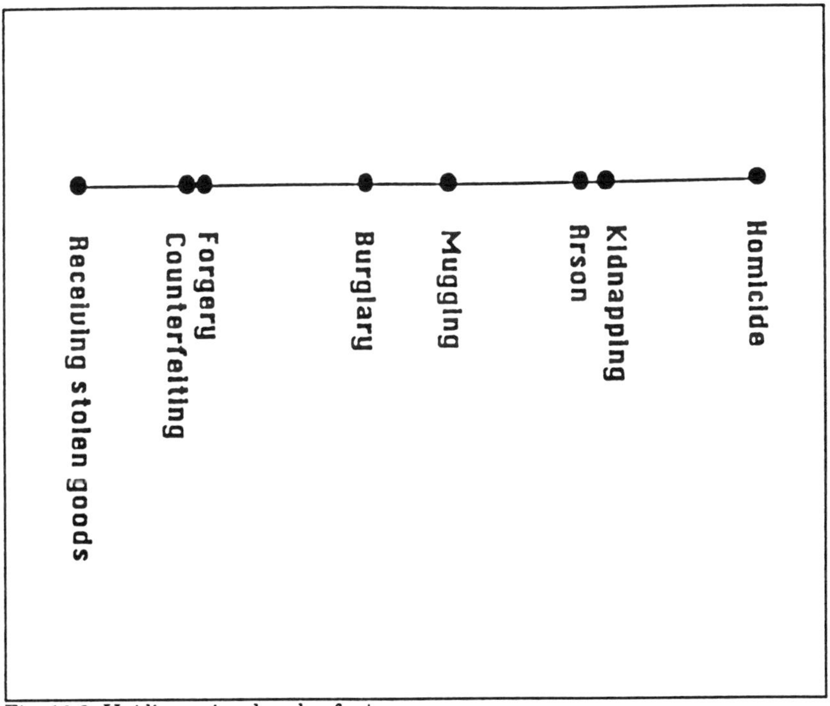

Fig. 10.3. Unidimensional scale of crimes.

As our common sense dictates, forgery and counterfeiting are very closely located on this continuum. Wouldn't you agree with this arrangement of the crimes?

We will look at only the first solution of the DUAL3 output here. As you will see, the signs of the weights in this output are reversed completely, which we can tell from the dominance of the minus signs in the subject weights and columns of the dominance table. When this should happen, we must reverse the signs of both subject and stimulus weights in reporting the results.

DUAL SCALING OF
PAIRED COMPARISON DATA
(STANDARD ANALYSIS)

TITLE OF STUDY: Analysis of Seriousness of Eight Crimes

*** INPUT DATA ***

1	1	1	2	2	2	1	1	1	2	2	2	1	1	2	2	2	2	2	2	2	1	1	1	1	1	1	1
1	1	1	2	0	1	1	1	1	2	2	0	1	0	2	2	2	2	2	2	2	1	1	1	1	1	1	1
1	1	1	2	2	1	1	0	1	2	2	1	1	0	2	2	2	1	2	2	1	1	1	1	1	1	1	1
1	1	1	2	2	1	1	0	1	2	2	2	1	0	2	2	2	1	2	2	2	0	1	1	1	2	1	1
1	1	1	2	2	1	1	1	1	2	1	1	1	1	2	2	1	1	2	2	1	1	1	1	1	1	1	1
1	1	1	2	2	1	1	1	1	2	2	2	1	2	2	2	2	1	2	2	2	1	1	1	1	0	1	1
1	1	1	1	1	2	1	1	1	2	2	2	1	2	2	2	2	1	2	2	2	1	1	2	1	2	1	1
1	1	1	2	2	1	1	1	1	2	2	2	1	2	2	2	2	2	2	2	2	2	1	1	1	1	1	1
1	1	1	2	2	1	1	1	1	2	2	1	1	1	2	2	1	1	2	2	2	1	1	1	1	1	1	1
0	2	1	2	2	2	1	1	1	2	2	1	1	0	2	2	2	1	2	2	2	1	1	1	1	2	1	1
1	2	1	2	2	2	1	2	1	2	2	2	1	1	2	2	1	1	2	2	2	2	1	2	1	1	1	1
1	1	1	0	1	1	1	1	1	2	2	0	1	0	2	2	2	0	2	2	2	0	1	1	1	0	1	1
1	1	1	0	1	1	1	2	1	2	2	2	1	1	2	2	2	1	2	2	2	1	1	1	1	1	1	1
1	1	1	2	0	1	1	1	1	2	2	2	1	2	2	2	2	1	2	2	2	1	1	1	1	1	1	1
1	1	1	0	0	0	1	0	0	2	2	0	1	0	2	2	2	1	2	2	2	1	1	1	1	0	1	1
1	1	1	2	0	1	1	1	1	2	2	0	1	0	2	2	2	2	2	2	2	1	1	1	1	1	1	1
1	1	1	2	1	2	1	1	1	2	2	2	1	2	2	2	2	1	2	2	2	1	1	1	1	2	1	1
1	0	1	1	1	0	1	1	0	2	2	0	2	0	2	2	0	0	2	2	0	0	1	1	1	1	1	0
1	1	0	0	0	0	1	1	1	0	0	0	1	0	2	2	2	2	2	2	2	0	0	0	1	0	1	1
0	1	1	2	2	2	1	1	0	2	2	2	0	0	2	2	2	0	2	2	2	0	1	1	1	1	1	1
1	1	1	1	1	1	1	2	2	2	2	2	0	2	2	2	0	1	2	2	1	1	1	1	1	1	1	1
1	1	1	2	1	1	1	1	1	2	2	2	1	2	2	2	2	1	2	2	2	1	1	1	1	1	1	1
1	1	0	1	1	2	2	2	0	2	2	1	2	0	1	2	1	1	2	1	1	1	2	2	2	1	1	1

*** CHOICE FREQUENCY TABLE (MATRIX E) ***

1	-1	-5	-5	7	5	3	-5
4	0	-6	-4	7	4	0	-5
3	0	-3	-2	7	5	-3	-7
3	-2	-3	-5	7	3	3	-6
3	3	-1	-3	7	3	-5	-7
3	-1	-5	-3	7	4	2	-7
5	-1	-5	-3	3	1	7	-7
3	-1	-7	-5	7	5	1	-3
3	1	-1	-5	7	5	-3	-7
-2	2	-2	-4	7	3	3	-7
-1	-3	3	-7	5	5	3	-5
6	0	-5	-5	6	2	1	-5
6	-3	-1	-5	6	3	1	-7
4	-1	-5	-3	7	4	1	-7

4	-2	-3	-3	6	3	2	7
4	0	-6	-4	7	4	0	-5
3	-1	-5	-3	7	1	5	-7
5	-3	-3	-3	5	3	-2	-2
3	2	-6	-4	3	3	3	-4
0	-2	-5	-4	7	5	3	-4
7	-6	-2	1	5	3	-2	-6
5	-1	-5	-3	7	3	1	-7
2	-4	2	2	-3	3	-1	-1

*** THE TOTAL VARIANCE TO BE ACCOUNTED FOR = .378882 ***

SOLUTION 1

SQUARED CORRELATION RATIO = 0.30836
MAXIMUM PRODUCT-MOMENT CORRELATION = 0.55530
PERCENTAGE HOMOGENEITY = 30.84
DELTA (TOTAL VARIANCE ACCOUNTED FOR): PARTIAL = 81.39
CUMULATIVE = 81.39

*** OPTIMAL WEIGHT VECTORS ***

*** ROWS ***

	NORMED	WEIGHTED
1	-1.0965	-0.6089
2	-1.1148	-0.6190
3	-1.0303	-0.5721
4	-1.0858	-0.6030
5	-0.9034	-0.5016
6	-1.1462	-0.6365
7	-0.9569	-0.5314
8	-1.0995	-0.6105
9	-1.0520	-0.5842
10	-0.9105	-0.5056
11	-0.7694	-0.4272
12	-1.0693	-0.5938
13	-1.0891	-0.6048
14	-1.1637	-0.6462
15	-1.0392	-0.5771
16	-1.1148	-0.6190
17	-1.0889	-0.6047
18	-0.7513	-0.4172
19	-0.8228	-0.4569
20	-0.9999	-0.5552
21	-0.8620	-0.4787
22	-1.1622	-0.6454
23	0.0708	0.0393

*** COLUMNS ***

	NORMED	WEIGHTED
1	-0.8290	-0.4604
2	0.2176	0.1208
3	0.9571	0.5315
4	0.9359	0.5197
5	-1.5819	-0.8784
6	-0.8752	-0.4860
7	-0.2822	-0.1567
8	1.4578	0.8095

********* END OF ANALYSIS *********

We have looked at two examples of paired comparison data as analyzed by dual scaling. You would probably agree that dual scaling offers very detailed analysis. The first example demonstrated the legitimacy of differentially weighting subjects' responses in order to extract different preference patterns. The second example revealed the wisdom of differentially weighting subjects to identify a dominant dimension through avoiding unwanted effects of an idiosyncratic subject. Dual scaling of paired comparison data offers a good way to look into individual differences in preference judgment. Let us now look at another type of dominance data, that is, rank-order data.

CHAPTER 11
Rank-Order Data

11.1 SPECIAL PROBLEMS

Ranking is a very popular way of collecting data. For instance, members of a panel are asked to rank eligible baseball players for the Hall of Fame. Typically, the best one is given rank 1, which is worth as many points as the panel may decide, and the worst one the last rank. Unlike paired comparisons, ranking becomes increasingly more difficult as the number of stimuli or objects to be ranked increases. For example, imagine asking a group of people to rank 20 flower arrangements in order of excellence at a garden show. Wouldn't that be a difficult task? Can you rank 30 politicians in terms of the confidence you have in them? You may be certain about your ranking of the first 5 or so, but you would not be able to swear to the accuracy of your ranking of, for example, the last 10. In contrast, paired comparison is easy no matter how many stimuli (objects) are involved, since the task is always to choose one out of the two in a pair. But as the number of stimuli, n, increases, the number of pairs, $n(n\text{-}1)/2$, may soon become hopelessly large. For instance, the total number of pairs for n of 10, 20, and 30 are respectively 45, 190, and 435, suggesting that some subjects may get tired of judging pairs of stimuli even when n is only 20. Therefore, it looks as though both ranking and paired comparison are limited to a small number of stimuli, say up to 15.

In terms of the information collected by ranking and paired comparison, the two methods have an important difference: ranking requires that responses be "transitive," or unidirectionally streamlined, while paired comparison responses can be intransitive.[1] This regularity in rank-order data simplifies the computation of dominance numbers from rank numbers. Let n be the number of stimuli, and K_{ij} the rank that Subject i gave to stimulus j. Then the dominance number for Subject i and stimulus j, e_{ij}, is given by

---

[1] If A>B and B>C, transitivity means that A>C. If, however, A>B, B>C, and C>A, this is an example of intransitive relations among A, B, and C.

(11.1)
$$e_{ij} = n + 1 - 2K_{ij} \,.$$

Consider rank-order data in Table 11.1, where $n = 5$. Using Equation 11.1, we calculate the dominance number of Subject i for stimulus j by $e_{ij} = n + 1 - 2K_{ij} = 6 - 2K_{ij}$, as shown in Table 11.2. This formula applies to the case in which no missing responses are involved.

TABLE 11.1
Rank-Order Data (K_{ij})

	Stimulus				
Subject	1	2	3	4	5
1	4	3	1	5	2
2	5	3	2	4	1
3	4	2	3	5	1

TABLE 11.2
Dominance Table (e_{ij})

	Stimulus					
Subject	1	2	3	4	5	Total
1	-2	0	4	-4	2	0
2	-4	0	2	-2	4	0
3	-2	2	0	-4	4	0

11.1.1 Treatment of Incomplete Ranking and Missing Responses

When subjects rank only the first n^* stimuli out of n $(n^*<n)$, we assume that those unranked stimuli would have been ranked after the ranked ones. If this is the case, then the average of the unused rank numbers should be assigned to the $(n\text{-}n^*)$ stimuli. Although this procedure was discussed in chapter 3, let us look at a small numerical example. Table 11.3 shows rank numbers of five stimuli, obtained from three subjects, where an asterisk indicates an unranked stimulus. Table 11.4 shows a complete table, obtained by using the procedure discussed previously, and Table 11.5 the corresponding dominance table. Notice that this procedure shows reduced discriminability among rank numbers when subjects rank only the first n^* stimuli. See the dominance numbers of Subjects 3 and 1. Therefore, as long as the number of stimuli is not too large, one should ask the subjects to rank all stimuli.

TABLE 11.3
Rank-Order Data with Missing Responses

Subject	Stimulus 1	2	3	4	5
1	3	*	1	*	2
2	1	*	2	3	4
3	1	*	*	*	2

TABLE 11.4
Complete Rank-Order Data by Imputation

Subject	Stimulus 1	2	3	4	5
1	3	4.5	1	4.5	2
2	1	5	2	3	4
3	1	4	4	4	2

TABLE 11.5
Dominance Table Obtained from Table 11.4

Subject	Stimulus 1	2	3	4	5	Total
1	0	-3	4	-3	2	0
2	4	-4	2	0	-2	0
3	4	-2	-2	-2	2	0

Consider a slightly different situation in which missing responses occur in a random fashion. This type of missing responses could occur, for example, in worm-eaten data sheets. In such a case, we cannot justify the procedure, which is based on the assumption that the subject judged the first (best, most preferred) n^* stimuli. Instead, we code the dominance number for a missing response as 0, and dominance numbers for observed responses are given by

(11.2) $$e_{ij} = n_i + 1 - 2K_{jp} ,$$

where n_i is the total number of responses of Subject i, that is, n minus the number of missing responses, and K_{ij} are down-adjusted.[2] Table 11.6 shows rank numbers with missing responses, Table 11.7 has down-adjusted rank numbers, and Table 11.8 is the table of dominance numbers. Notice that in both cases of missing responses, the dominance table is still row-conditional and each row sum is 0.

TABLE 11.6
Rank-Order Data with Randomly Missing Responses

			Stimulus			
Subject	1	2	3	4	5	n_i^*
1	*	3	1	*	*	2
2	1	5	*	3	*	3
3	1	5	*	4	2	4

TABLE 11.7
Down-Adjusted Rank-Order Data

			Stimulus			
Subject	1	2	3	4	5	n_i^*
1	*	2	1	*	*	2
2	1	3	*	2	*	3
3	1	4	*	3	2	4

TABLE 11.8
Dominance Table

			Stimulus			
Subject	1	2	3	4	5	Sum
1	0	-1	1	0	0	0
2	2	-2	0	0	0	0
3	3	-3	0	-1	1	0

The two cases are quite different, however, when analysis is carried out. The dominance matrix obtained in the first case (e.g., Table 11.4) can be handled in the standard way, using the number of responses in each

[2] If rank numbers 1, 3, 4, and 8 are used, they are changed to 1, 2, 3, and 4, respectively.

row, the number in each column and the total number, respectively, $n(n-1)$, $N(n-1)$, and $nN(n-1)$. These numbers are exactly the same as the corresponding numbers used in paired comparisons of n stimuli by N subjects. The dominance matrix obtained from the second case (e.g., Table 11.8) is quite different because some responses are permanently missing. Let us use the data in Table 11.7 as an example. Noting that those cells with the asterisks have no response, we can conclude that the other cells contain responses based on (n_i-1) comparisons. Therefore, the numbers of responses for cells in Table 11.7 are as shown in Table 11.9. As you can tell from this example, this case of missing responses makes the analysis somewhat complicated. Luckily, however, most missing data sets in practice do not fall into this category of random missing responses, but into the first case of partial ranking of stimuli. Let us now look at two examples of rank-order data.

TABLE 11.9

	Stimulus					
Subject	1	2	3	4	5	Total
1	0	1	1	0	0	2
2	2	2	0	2	0	6
3	3	3	0	3	3	12
Total	5	6	1	5	3	

11.2 APPLICATIONS

Analysis of rank-order data is essentially identical to that of paired comparison data. Computationally, however, rank-order data are much simpler to handle than paired comparison data. The first example is obtained from Nishisato and Nishisato (1984), and the second example is the data collected by Wiggins.

11.2.1 Ranking of 10 Government Services

The following 10 government services and facilities were ranked by 31 adults according to the order of "being satisfactory."

A. Public transit system
B. Postal service

C. Medical care, including hospitals and clinics
D. Sports, recreational facilities
E. Police protection
F. Public libraries
G. Cleaning streets
H. Restaurants
I. Theaters
J. Overall planning and development

TABLE 11.10

	Service									
Subj	A	B	C	D	E	F	G	H	I	J
1	1	7	9	10	2	6	3	8	5	4
2	6	10	9	5	3	1	7	2	4	8
3	9	8	4	3	5	6	10	2	1	7
4	2	10	5	6	3	1	4	8	7	9
5	2	10	6	7	4	1	5	3	9	8
6	1	3	5	6	7	8	2	4	10	9
7	7	10	1	6	5	3	8	4	2	9
8	2	10	6	7	4	1	5	3	9	8
9	2	10	5	8	4	1	6	3	7	9
10	2	10	5	9	8	7	4	1	3	6
11	9	10	7	6	5	1	4	2	3	8
12	6	10	7	4	2	1	3	9	8	5
13	1	10	3	9	6	4	5	2	7	8
14	8	6	5	3	10	7	9	2	1	4
15	8	10	9	6	4	1	3	2	5	7
16	3	5	10	4	6	9	8	2	1	7
17	1	10	8	9	3	5	2	6	7	4
18	5	4	9	3	10	8	7	2	1	6
19	2	10	6	7	8	1	5	4	3	9
20	1	4	2	10	9	7	6	3	5	8
21	2	10	5	7	3	1	4	6	8	9
22	6	3	9	4	10	8	7	2	1	5
23	6	9	10	4	8	7	5	2	1	3
24	5	2	1	9	10	4	8	6	3	7
25	2	10	6	7	9	1	3	4	5	8
26	7	10	9	5	2	6	3	1	4	8
27	8	7	10	3	5	9	4	2	1	6
28	3	8	6	7	5	10	9	2	4	1
29	2	10	7	9	4	1	5	3	6	8
30	2	10	9	1	4	7	5	3	6	8
31	4	10	9	7	5	1	3	2	6	8

The "most satisfactory" was given rank 1, the "second most satisfactory" rank 2, and so on until the "least satisfactory" was given rank 10. The data are shown in Table 11.10. There are no missing responses. The total number of possible solutions of this 31×10 rank-order data is 9, that is, 10 - 1. The distribution of the squared correlation ratio is as shown in Table 11.11.

TABLE 11.11
Distribution of Squared Correlation Ratio

Solution	η_j^2	δ_j	$\Sigma\delta_j$
1	0.1543	37.87%	37.87%
2	0.0911	22.35%	60.22%
3	0.0544	13.35%	73.57%
4	0.0432	10.62%	84.18%
5	0.0198	4.85%	89.04%
6	0.0169	4.16%	93.19%
7	0.0110	2.70%	95.90%
8	0.0088	2.16%	98.06%
9	0.0079	1.94%	100.00%

The first two solutions are comparatively dominant in this set of solutions. If we go further down the list, we notice a distinct gap between Solution 4 and Solution 5. We should therefore examine the first four solutions. Table 11.12 shows adjusted weights for the 10 government services and service facilities of the first 4 solutions.

TABLE 11.12

Service	Solution 1	Solution 2	Solution 3	Solution 4
A	0.40	-0.14	0.36	0.35
B	-0.76	-0.26	0.11	0.03
C	-0.14	-0.33	0.31	-0.33
D	-0.26	0.19	-0.30	-0.06
E	0.25	-0.19	-0.38	0.06
F	0.60	-0.15	-0.08	-0.34
G	0.25	-0.15	-0.13	0.21
H	0.23	0.55	0.17	0.03
I	-0.19	0.54	0.06	-0.11
J	-0.38	-0.50	-0.14	0.16

In interpreting rank-order data, there is an important issue to keep in mind. It involves the direction or sign of the solution. Suppose you asked a number of people to rank (a) baseball, (b) tennis, and (c) wrestling according to the order of their preference. In dual scaling, the signs of weights can be completely reversed because the maximization of η^2 is not affected by reversing the signs of the weights. In other words, if x_1 and x_2 maximize η^2, so do $-x_1$ and $-x_2$. The direction or the sign of weights of Solution 1, however, can ususally be inferred from column averages of the dominance matrix **E**. Suppose that dual scaling of the ranking of (a), (b), and (c) yielded $x_a<x_b<x_c$, meaning wrestling first and baseball last, while the column averages of **E** showed $e_a>e_b>e_c$. Then we could conclude that the signs of x's must be reversed so that $x_a>x_b>x_c$.

Another problem related to this involves the nature of the scale, or the solution. Suppose that we rectified the signs of the weights—if the signs of the weights x_j were reversed, the signs of y_i must also be reversed. Let us examine the signs of subject weights y_i. If y_i are all positive, and are all located closest to the largest value of x_j, then we can say that scale values x_j constitute a "preference scale" or a directional scale. Otherwise, we cannot talk about a directional scale. Why not? Well, this is not an easy question, but it is one of the essential problems that need to be clarified. No matter what data types we may analyze, the basic requirement of the first solution is to approximate the information contained in the data as closely as possible, and the second solution should be such that together with the first solution it must reproduce the input data as much as possible. Thus, if we can assume that subjects and objects occupy the same space (i.e., if weights for rows and those for columns can be plotted in the same space), the position of a subject must be such that he or she chooses the most closely located object first, the second most closely located object second, and so on. As hinted earlier, of course, the space for rows and the space for columns do not match exactly, but are discrepant. However, the interpretation of the scaled outcome is viewed from this ideal standpoint of common space: the subject's ranking is determined according to the order of object proximities to his or her location in the solution space. Thus, if a subject has a positive weight which is closest to x_a in the above example, the subject's ranking is a first, b second, and c third; if a subject's weight is closest to x_b, then to x_a, and then to x_c, the order of the choice is b, a, and c. Even if the weights of both of these subjects are positive, there are two preference scales, one for each of them, which makes it difficult to talk about a "preference scale" for the solution. This interpretation of row and column weights of rank-order data is the same as that used in Coombs' *unfolding technique.*[3] In dual scaling, the problem of discrepant spaces for rows and columns makes the intrepretation of the derived quantities difficult, and this problem is discussed again in chapter 14. Although the interpretation of row and column weights in paired comparision data was

---

[3] See Coombs (1964).

not fully discussed here, there exists the same expectation as for rank-order data: the first solution must best approximate paired comparisons in the input data.

TABLE 11.13
Weights for Subjects

	Normed Weights				Adjusted Weights			
Subj	Sol.1	Sol.2	Sol.3	Sol.4	Sol.1	Sol.2	Sol.3	Sol.4
1	0.65	-0.51	-0.19	2.02	0.25	-0.15	-0.04	0.42
2	1.15	1.08	-0.81	-0.69	0.45	0.33	-0.19	-0.14
3	-0.16	1.51	-0.14	-1.68	-0.06	0.46	-0.03	-0.35
4	1.39	-0.73	-0.18	-0.31	0.54	-0.22	-0.04	-0.06
5	1.54	-0.21	0.25	0.02	0.60	-0.06	0.06	0.00
6	0.21	-0.69	1.22	1.52	0.08	-0.21	0.28	0.32
7	0.63	0.63	0.56	-2.22	0.25	0.19	0.13	-0.46
8	1.54	-0.21	0.25	0.02	0.60	-0.06	0.06	0.00
9	1.52	-0.03	0.66	-0.42	0.60	-0.01	0.15	-0.09
10	0.75	1.02	1.56	0.80	0.29	0.31	0.36	0.17
11	0.99	1.08	-0.71	-1.33	0.39	0.33	-0.17	-0.28
12	0.97	-0.74	-1.80	-0.24	0.38	-0.22	-0.42	-0.05
13	1.25	0.02	1.55	0.12	0.49	0.01	0.36	0.02
14	-0.70	1.63	0.56	-0.94	-0.28	0.49	0.13	-0.20
15	1.16	0.85	-1.12	-0.49	0.45	0.26	-0.26	-0.10
16	-0.22	1.65	0.40	1.19	-0.09	0.50	0.09	0.25
17	1.13	-0.38	-0.17	1.81	0.44	-0.11	-0.04	0.38
18	-0.58	1.72	0.64	0.53	-0.23	0.52	0.15	0.11
19	1.25	0.65	1.02	-0.54	0.49	0.20	0.24	-0.11
20	0.12	-0.07	2.66	0.28	0.05	-0.02	0.62	0.06
21	1.52	-0.64	0.05	-0.19	0.60	-0.19	0.01	-0.04
22	-0.76	1.63	0.60	0.48	-0.30	0.49	0.14	0.10
23	0.03	1.86	-0.36	0.81	0.01	0.56	-0.08	0.17
24	-0.43	-0.23	2.12	-1.44	-0.17	-0.07	0.49	-0.30
25	1.28	0.34	0.96	-0.17	0.50	0.10	0.22	-0.03
26	0.87	1.23	-1.1-	0.48	0.34	0.37	-.26	0.10
27	-0.20	1.77	-0.97	0.77	-0.08	0.54	-0.23	0.16
28	-0.19	0.93	0.65	1.31	-0.08	0.28	0.15	0.27
29	1.55	0.17	0.45	0.08	0.61	0.05	0.10	0.02
30	0.76	0.92	-0.59	1.23	0.30	0.28	-0.14	0.26
31	1.47	0.60	-0.22	0.19	0.58	0.18	-0.05	0.05

Table 11.13 shows the normed and adjusted weights of subjects. In rank-order and paired comparision data, normed weights are expected to be all 1s if no individual differences in judgment exist. For this reason, normed weights are often more interesting to look at than adjusted weights, which are typically very small. As you can see, each solution

contains mixed signs, indicating that there is no fixed rank order of government services that is endorsed by everyone in the group.

You can see considerable individual differences, which would convince you that each solution is far from a common preference scale for the entire group of subjects. It is also the case that normed weights are easier to look at than adjusted weights if we want to see the magnitute of individual differences, for adjusted weights are often too small to see what the expected value is like in each solution. In chapter 17, we see that η^2 from dominance data are comparatively smaller than those from incidence data, and may never reach the absolute maximum of 1 even when the responses reveal perfect internal consistency.

To interpret the results, let us plot adjusted weights for the 10 services. In this case, adjusted weights are used so that different solutions reflect their relative importance to the total space (see Figure 11.1). Solution 1 indicates a strong tendency in the data for high regard for public libraries and the transit system, followed by police protection and cleaning streets, but a low opinion of the postal service, followed by overall planning/development and sports/recreational facilities. Solution 2 shows strong support from some subjects for theaters and restaurants, followed by sports/recreational facilities. This is understandable, for Toronto has good theaters and restaurants if you can afford them. If you belong to a private club, you have access to first-rate facilities for sports such as golf and tennis. The opposite side is represented by medical care and postal service. How about Solution 3? The contribution of this solution is much less than the first two. There are very strong supporters, Subjects 20 and 24, who indicate a high regard for the transit system and medical care, and a low opinion of police protection and sports/recreational facilities. We should also note, however, that there are many supporters of the opposite pattern, that is, a high regard for police protection and sports/recreational facilities, and a low opinion for the transit system and medical care. Solution 4 shows a cluster of library and medical care on one side, and transit system and cleaning streets on the other. If we want to identify supporters of all of these contrasts, it is relatively easy to do so by looking at the weights of subjects shown in Table 11.13.

11.2.2 Ranking Eight Christmas Party Plans

Recall the eight Christmas party plans that were discussed in chapter 10. The same plans were used, and the same 14 subjects were asked to rank the 8 plans according to the order of their preference, 1 being the first choice and 8 the last. The data were collected by Wiggins. The eight plans are:

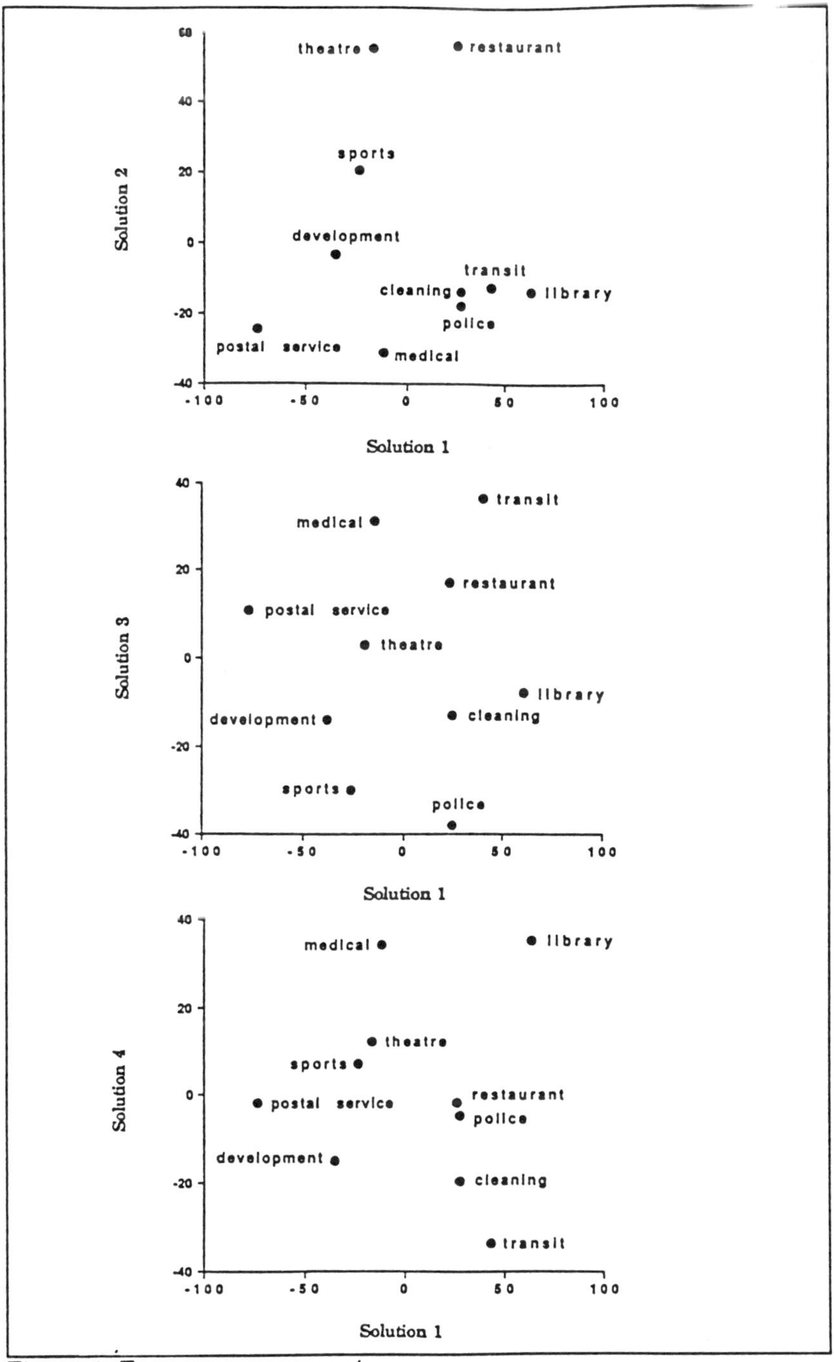

Fig. 11.1. Ten government services.

X_1: A pot-luck at someone's home in the evening
X_2: A pot-luck in the group room
X_3: A pub/restaurant crawl after work
X_4: A reasonably priced lunch in an area restaurant
X_5: Keep to one's self
X_6: An evening banquet at a restaurant
X_7: A pot-luck at someone's home after work
X_8: A ritzy lunch at a good restaurant (tablecloths)

TABLE 11.14

Subj	X_1	X_2	X_3	X_4	X_5	X_6	X_7	X_8
1	6	1	5	3	2	8	4	7
2	3	8	1	6	7	5	4	2
3	5	7	1	6	8	2	4	3
4	4	6	2	3	8	7	1	5
5	2	4	6	3	7	5	1	8
6	2	4	5	3	8	7	1	6
7	1	7	6	3	8	5	2	4
8	7	5	3	1	8	4	6	2
9	4	2	7	3	8	6	5	1
10	5	1	2	4	7	6	3	8
11	6	4	3	2	8	7	5	1
12	3	8	4	2	5	6	1	7
13	3	2	1	6	4	7	5	8
14	5	8	1	4	7	3	6	2

TABLE 11.15

Solution	η^2	δ_j	$\Sigma\delta_j$
1	0.1415	33.01%	33.01%
2	0.1227	28.62%	61.63%
3	0.0658	15.36%	76.99%
4	0.0551	12.86%	89.86%
5	0.0225	5.26%	95.12%
6	0.0160	3.74%	98.85%
7	0.0049	1.15%	100.00%

Rank-order data collected are shown in Table 11.14. Seven solutions can exhaustively analyze the corresponding dominance matrix. Statistics for those seven solutions are shown in Table 11.15. In terms of the statistic

δ_j, we should perhaps examine four solutions. However, recalling that the paired comparison data on Christmas party plans produced only three interpretable solutions, we should look at the first three solutions.[4] Table 11.16 shows adjusted weights for the eight Christmas party plans on three solutions. From these weights, it is clear that the three solutions roughly correspond to the three solutions obtained from paired comparison data. Thus, instead of examining the results further, let us look at the DUAL3 output. Observe, for instance, that there are two subjects whose weights on Solution 1 are negative, and that Subject 1 has the weight of -1.0833. As expected from this weight, Subject 1 ranked X_5 ("Keep to one's self") second, and X_2 ("Pot-luck in the group room") first, that is, preferring these unpopular plans over the others. Isn't this interesting and amazing? Dual scaling certainly picks up otherwise hidden patterns of responses, doesn't it?

TABLE 11.16

Party Plan	Solution 1	Solution 2	Solution 3
X_1	0.13	0.29	-0.23
X_2	-0.47	0.23	0.43
X_3	0.33	-0.09	-0.14
X_4	0.21	0.21	0.28
X_5	-0.75	-0.18	-0.21
X_6	-0.02	-0.42	-0.23
X_7	0.22	0.51	-0.16
X_8	0.35	-0.57	0.26

```
                DUAL SCALING
              RANK-ORDER DATA
            (STANDARD ANALYSIS)


TITLE OF STUDY: Ranking Eight Christmas Party Plans


            *** INPUT DATA ***

         6  1  5  3  2  8  4  7
         3  8  1  6  7  5  4  2
         5  7  1  6  8  2  4  3
```

[4] According to Weingardern and Nishisato (1986), rank-order data produce a smaller number of interpretable solutions, due to the transitivity of judgment, than paired comparision data.

4	6	2	3	8	7	1	5
2	4	6	3	7	5	1	8
2	4	5	3	8	7	1	6
1	7	6	3	8	5	2	4
7	5	3	1	8	4	6	2
4	2	7	3	8	6	5	1
5	1	2	4	7	6	3	8
6	4	3	2	8	7	5	1
3	8	4	2	5	6	1	7
3	2	1	6	4	7	5	8
5	8	1	4	7	3	6	2

*** CHOICE FREQUENCY (DOMINANCE) TABLE (MATRIX E) ***

-3	7	-1	3	5	-7	1	-5
3	-7	7	-3	-5	-1	1	5
-1	-5	7	-3	-7	5	1	3
1	-3	5	3	-7	-5	7	-1
5	1	-3	3	-5	-1	7	-7
5	1	-1	3	-7	-5	7	-3
7	-5	-3	3	-7	-1	5	1
-5	-1	3	7	-7	1	-3	5
1	5	-5	3	-7	-3	-1	7
-1	7	5	1	-5	-3	3	-7
-3	1	3	5	-7	-5	-1	7
3	-7	1	5	-1	-3	7	-5
3	5	7	-3	1	-5	-1	-7
-1	-7	7	1	-5	3	-3	5

*** THE TOTAL VARIANCE TO BE ACCOUNTED FOR = .4285714

SOLUTION 1

SQUARED CORRELATION RATIO = 0.14147
MAXIMUM PRODUCT-MOMENT CORRELATION = 0.37613
PERCENTAGE HOMOGENEITY = 14.15
DELTA (TOTAL VARIANCE ACCOUNTED FOR): PARTIAL = 33.01
CUMULATIVE = 33.01

*** OPTIMAL WEIGHT VECTORS ***

*** ROWS ***

	NORMED	WEIGHTED
1	-1.0833	-0.4074
2	1.4014	0.5271
3	1.3057	0.4911
4	1.3085	0.4922
5	0.3442	0.1295

*** COLUMNS ***

	NORMED	WEIGHTED
1	0.3506	0.1319
2	-1.2558	-0.4723
3	0.8659	0.3257
4	0.5645	0.2123
5	-2.0040	-0.7538

6	0.8015	0.3015
7	1.2235	0.4602
8	1.0887	0.4095
9	0.5469	0.2057
10	0.0541	0.0203
11	1.1052	0.4157
12	0.7177	0.2700
13	-0.4637	-0.1744
14	1.3225	0.4974

6	-0.0411	-0.0155
7	0.5883	0.2213
8	0.9317	0.3504

SOLUTION 2

SQUARED CORRELATION RATIO = 0.12267
MAXIMUM PRODUCT-MOMENT CORRELATION = 0.35024
PERCENTAGE HOMOGENEITY = 12.27
DELTA (TOTAL VARIANCE ACCOUNTED FOR): PARTIAL = 28.62
CUMULATIVE = 61.63

*** OPTIMAL WEIGHT VECTORS ***

*** ROWS ***

	NORMED	WEIGHTED
1	-0.9948	-0.3484
2	0.4373	0.1532
3	0.6864	0.2404
4	-1.0582	-0.3706
5	-1.6679	-0.5842
6	-1.6059	-0.5624
7	-0.7914	-0.2772
8	0.5885	0.2061
9	-0.0764	-0.0268
10	-1.2754	-0.4467
11	0.1493	0.0523
12	-1.1737	-0.4111
13	-0.8987	-0.3148
14	1.0269	0.3597

*** COLUMNS ***

	NORMED	WEIGHTED
1	-0.8389	-0.2938
2	-0.6647	-0.2328
3	0.2475	0.0867
4	-0.5975	-0.2093
5	0.5084	0.1781
6	1.1867	0.4156
7	-1.4630	-0.5124
8	1.6215	0.5679

SOLUTION 3

SQUARED CORRELATION RATIO = 0.06583
MAXIMUM PRODUCT-MOMENT CORRELATION = 0.25658
PERCENTAGE HOMOGENEITY = 6.58
DELTA (TOTAL VARIANCE ACCOUNTED FOR): PARTIAL = 15.36
CUMULATIVE = 76.99

*** OPTIMAL WEIGHT VECTORS ***

*** ROWS ***

	NORMED	WEIGHTED
1	-1.0101	-0.2592
2	0.8342	0.2140
3	0.7463	0.1915
4	0.0460	0.0118
5	0.3222	0.0827
6	-0.2479	-0.0636
7	0.3696	0.0948
8	-1.4482	-0.3716
9	-2.0617	-0.5290
10	-0.5977	-0.1534
11	-1.8241	-0.4680
12	1.0838	0.2781
13	0.3104	0.0796
14	0.3453	0.0886

*** COLUMNS ***

	NORMED	WEIGHTED
1	0.9098	0.2334
2	-1.6653	-0.4273
3	0.5430	0.1393
4	-1.0941	-0.2807
5	0.8192	0.2102
6	0.8816	0.2262
7	0.6339	0.1626
8	-1.0281	-0.2638

********* END OF ANALYSIS *********

CHAPTER 12
Successive Categories (Rating) Data

12.1 SPECIAL PROBLEMS

Successive categories data are different from any of the data types discussed so far, and present a few topics that require attention. This section introduces a unique formulation of dual scaling, treatment of missing responses and the problem of multidimensional decomposition.

12.1.1 Basic Formulation

Dual scaling of successive categories is based on recent studies (Nishisato, 1980b; Nishisato and Sheu, 1984). Although two distinct but equivalent approaches are available, the simpler one will be presented here. As discussed earlier, a group of subjects are asked to rate n objects in terms of a *common set* of ordered categories (e.g., never, sometimes, often, always). The responses are summarized in the subjects-by-objects table with elements being the chosen category codes. Suppose that four TV programs (A, B, C, and D) were rated by three people in terms of 1=poor, 2=mediocre, 3=fair, 4=good, 5=excellent, and that responses were summarized as in Table 12.1.

TABLE 12.1
Rating Four TV Programs

Subject	A	B	C	D
1	4	5	1	3
2	3	4	2	2
3	5	5	2	1

Subject 1, for instance, placed A into category good (4), B into excellent (5), C into poor (1), and D into fair (3). Let us consider a continuum, along which the four objects are located together with category boundaries, say τ_1

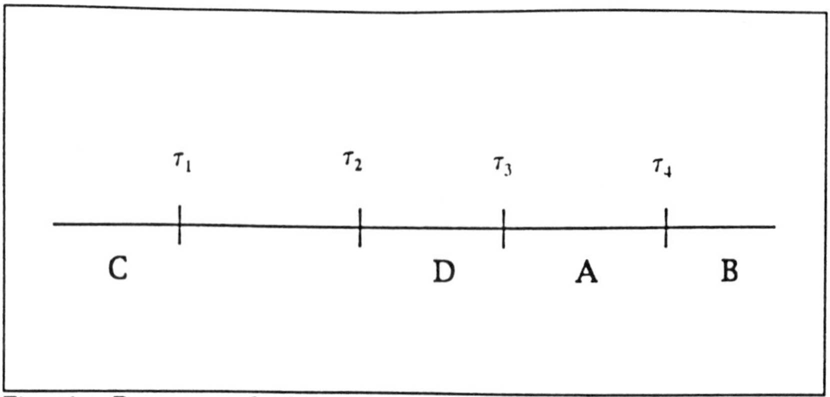

Fig. 12.1. Responses of subject 1 interpreted on the continuum.

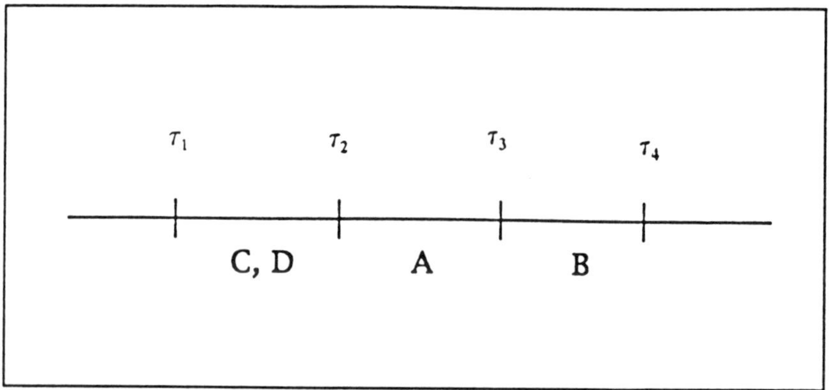

Fig. 12.2. Responses of subject 2 interpreted on the contimuum.

(the boundary between categories poor and mediocre), τ_2 (between mediocre and fair), τ_3 (between fair and good), and τ_4 (between good and excellent). On this continuum, Subject 1 classified objects A, B, C, and D as shown in Figure 12.1. In other words, Subject 1 placed C in category 1 (poor), that is, before categories boundary τ_1, D in category 3 (fair), that is, between boundaries τ_2 and τ_3, A in category 4 (good), that is between τ_3 and τ_4, and B in category 5 (excellent), that is, above τ_4. Therefore, we can assign numbers to category boundaries and objects according to the order in which they appear on the continuum, namely, rank 1 to C, rank 2 to τ_1, rank 3 to τ_2, rank 4 to D, rank 5 to τ_3, rank 6 to A, rank 7 to τ_4, and rank 8 to B. Similarly, the responses of Subject 2 can be expressed as in Figure 12.2. To do so, give rank 1 to τ_1, rank 2.5 (the average of rank 2 and rank 3) to C and D, rank 4 to τ_2, rank 5 to A, rank 6 to τ_3, rank 7 to B, and rank 8 to τ_4. Responses of Subject 3 yield rank 1 for D, rank 2 for τ_1, rank 3 for C, rank 4 for τ_2, rank 5 for τ_3, rank 6 for τ_4, and rank 7.5 (the average of ranks 7 and 8) for A and B (if you feel lost, examine Figures 12.1 and 12.2 once again). On the basis of the above orderings, we can now construct rank--order data, in which category boundaries are ranked together with stimuli (see Table 12.2).

TABLE 12.2
Rank Order Data Constructed from Table 12.1

Subject	τ_1	τ_2	τ_3	τ_4	A	B	C	D
1	2	3	5	7	6	8	1	4
2	1	4	6	8	5	7	2.5	2.5
3	2	4	5	6	7.5	7.5	3	1

When the total number of category boundaries and stimuli is relatively large, the conversion of the input data into rank orders may not be easy. This conversion, however, can be expressed in a systematic process as follows:

1. Let n and m be respectively the total number of stimuli and the total number of category boundaries (i.e., m + 1 successive categories).
2. Let k_0 be the total number of stimuli which are classified into Category 1, that is, the lowest category. Assign each of those k_0 stimuli the average rank $(k_0+1)/2$.
3. Assign the rank k_0+1 to the category boundary τ_1.
4. Let k_1 be the total number of stimuli which are classified into Category 2. Assign each of those k_1 stimuli the average rank $(k_1+1)/2 + (k_0+1)$.
5. Assign the rank $(k_0 1)+(k_1+1)$ to the category boundary τ_2.
6. Let k_2 be the total number of stimuli which are classified into category 3. Assign each of those k_2 stimuli the average rank $(k_2+1)/2 + (k_0+1) + (k_1+1)$.
7. Assign the rank $(k_0+1)+(k_1+1)+(k_2+1)$ to the category boundary τ_3.

The above process continues until we assign rank numbers to all $(m+n)$ quantities. Let us apply the procedure to the data in Table 12.1, in which n = 4, and m = 4. For Subject 1:

Step 1: $k_0 = 1$ (Stimulus C is rated 1), and C is given rank $(k_0+1)/2 = 1$,
Step 2: The rank given to τ_1 is $(k_0+1) = 2$,
Step 3: $k_1 = 0$,
Step 4: The rank given to τ_2 is $(k_0+1)+(k_1+1) = 3$,
Step 5: $k_2 = 1$ (Stimulus D is rated 3), and D is given rank $(k_2+1)/2+(k_0+1)+(k_1+1) = 4$,

Step 6: The rank given to τ_3 is $(k_0+1)+(k_1+1)+(k_2+1) = 5$,
Step 7: $k_3 = 1$ (A is rated 4), and A is given rank $(k_3+1)/2+(k_0+1)+(k_1+1)+(k_2+1) = 6$,
Step 8: The rank given to τ_4 is $(k_0+1)+(k_1+1)+(k_2+1)+(k_3+1) = 7$,
Step 9: $k_4 = 1$ (Stimulus B is rated 5), and B is given rank $(k_4+1)/2 +(k_0+1) +(k_1+1) +(k_2+1) +(k_3+1) = 8$.

In the same way, successive category judgments of Subjects 2 and 3 can be converted to rank-order data. This process looks complicated, but once it is programmed the computation is quite straightforward.

In chapter 11, we saw how simple it was to derive dominance numbers from rank numbers, namely, the number of objects or stimuli (n) plus 1, minus twice the rank number. In the case of successive categories data, the derived rank-order table is not subjects-by-objects but, as you see in Table 12.2, the subjects-by-(category boundaries plus objects). Let N be the number of subjects. Then the dominance number of Subject i, for the jth column of the rank-order table, is given by

$$e_{ij} = 2K_{ij} - (n + m + 1) \text{ ,} \qquad (12.1)$$

where K_{ij} is the rank number that subject i gave to object j, namely, the element of cell (i,j) of the rank-order table. Notice that in successive categories data, rank number 1 is given to the object or category boundary which is located at the lowest (left-most) position, and therefore that Formula 12.1 was obtained by changing the direction of the formula used for rank-order data, that is, $e_{ij} = (n + 1) - 2K_{ij}$. In other words, rank 1 in rank-order data typically indicates the "best," the "highest," or the "top," while in successive categories data the meaning of K_{ij} is reversed. Let us now apply Formula 12.1 to the data in Table 12.2. The results are given in Table 12.3. Can you see how Table 12.3 was obtained? Recall that the sum of each row of a dominance table is zero. Indeed, all the row sums of Table 12.3 are 0. If the sum of a row of a dominance table is different from 0, it will be important to check the computation or the presence of missing responses (see the next section). Once the dominance table is obtained from a complete data set, dual scaling of the table requires the specifications of the numbers of responses in rows and columns. Assuming that the data are complete, we may follow the logic used for paired comparison and rank-order data, except that successive categories data involve m category boundaries in addition to n objects. Given the $N\times(n+m)$ dominance table, each cell is the outcome of $(n + m - 1)$ comparisons. In other words, each subject compares an object or a category boundary with the remaining $(n + m - 1)$ objects and category boundaries. Therefore, the total number of comparisons (responses) for each subject (i.e., each row total) is

$(n + m)(n + m - 1)$. Similarly, the total number of responses for each column (the column total) is $N(n + m - 1)$. The total number of responses in the dominance table is therefore given by $N(n + m)(n + m - 1)$.

TABLE 12.3
Dominance Table Obtained from Table 12.2 by Equation 12.1

Subject	τ_1	τ_2	τ_3	τ_4	A	B	C	D	Total
1	-5	-3	1	5	3	7	-7	-1	0
2	-7	-1	3	7	1	5	-4	-4	0
3	-5	-1	1	3	6	6	-3	-7	0

Now that the dominance table, row totals, and column totals are specified, one can carry out, for example, the method of reciprocal averages to derive weights for subjects and scale values for category boundaries (the first m columns) and objects (the remaining n columns).

12.1.2 Treatment of Missing Responses

In practice, missing responses do occur. In such cases, the following procedure is recommended:

1. Code misssing responses as "0" in the rating table (e.g., Table 12.1).

2. Rank category boundaries and objects as before, but without those objects which were not rated (i.e., missing responses).

3. Let n_i^* be the number of objects, minus the number of objects that Subject i did not rate.

4. Calculate dominance numbers for rated objects by the formula:

(12.2)
$$e_{ij} = 2K_{ij} - (n_i^* + m + 1)$$

As for those objects which were not rated, set $e_{ij} = 0$.

Let us look at an example. Suppose that we obtained rating data in which six responses were missing (Table 12.4). The value of n is four, but n_i^* are 3, 2, and 1 for Subjects 1, 2, and 3, respectively. In other words, the numbers of missing responses for the three subjects, $n - n_i^*$, are respectively 1, 2, and 3. Rankings of the four category boundaries and objects are given in Table 12.5, where omitted objects are coded 0. Notice that missing responses do not affect category boundaries, but only stimuli.

Using Formula 12.2 and entries in Table 12.5, we obtain the dominance table, Table 12.6.

TABLE 12.4
Rating Four TV Programs with Six Missing Responses

Subject	A	B	C	D	n_i^*
1	4	0	1	3	3
2	3	4	0	0	2
3	0	5	0	0	1

TABLE 12.5
Rank-Order Data Constructed from Table 12.4

Subject	τ_1	τ_2	τ_3	τ_4	A	B	C	D	$m + n_i^* + 1$
1	2	3	5	7	7	0	1	4	8
2	1	2	4	6	3	5	0	0	7
3	1	2	3	4	0	5	0	0	6

TABLE 12.6
Dominance Table When Some Responses Are Missing

Subject	τ_1	τ_2	τ_3	τ_4	A	B	C	D	Total
1	-4	-2	2	6	4	0	-6	0	0
2	-5	-3	1	5	-1	3	0	0	0
3	-4	-2	0	2	0	4	0	0	0

When missing responses are involved in data, we must modify the total numbers of responses for rows and columns, discussed in the previous section. To arrive at the correct numbers, let us use Table 12.6, where Subjects 1, 2, and 3 omitted, respectively, one, two, and three objects from rating. Table 12.7 shows the number of responses for each cell, as well as row totals, column totals, and the grand total. Notice that omitted objects have 0s, and that the remaining cells of each subject have the identical number, 6 for Subject 1, 5 for Subject 2, and 4 for Subject 3. Thus, row totals are relatively easy to calculate, but this is not the case with column totals, which depend on the locations of missing responses, as demonstrated in Table 12.7. Once the row totals and the column totals are specified in this way, the corresponding dominance table can be subjected, for example, to the method of reciprocal averages, or the table may be transformed, using those totals, for the eigenvalue decomposition.

TABLE 12.7
Numbers of Responses (Comparisons) Involved in Table 12.6

Subject	τ_1	τ_2	τ_3	τ_4	A	B	C	D	Total
1	6	6	6	6	6	0	6	6	42
2	5	5	5	5	5	5	0	0	30
3	4	4	4	4	0	4	0	0	20
Total	15	15	15	15	11	9	6	6	92

12.1.3 Multidimensionality

The current formulation does not use the order constraint on category boundaries. Nevertheless, it is almost always the case that the first solution provides correctly ordered category boundaries, that is,

(12.3) $$\tau_1 \leq \tau_2 \leq \tau_3 \ . \ \ . \ \ . \leq \tau_m \ .$$

Solution 1, however, is unlikely to explain the data exhaustively. Thus, the investigator would want to proceed with the analysis to obtain more than one solution. As is known empirically, it is improbable that the subsequent solutions would reveal correctly ordered category boundaries. For this reason, the current practice is to extract only one solution, even when multidimensionality is a definite possibility. How can we handle multidimensionality of the data and the order constraint on τ_k in more than one dimension?

Nishisato (1986c) proposed the point-of-view analysis. He noted that category boundaries are always ordered correctly within each subject in the dominance matrix (see, e.g., Table 12.2), and therefore that the first solution from a dominance matrix consisting of any subset of rows (subjects) of the entire dominance table is likely to maintain the correct order of τ_k. Thus, he proposed three methods for clustering subjects (rows of the dominance matrix) and to submit each dominance submatrix to dual scaling. The idea is that each cluster of subjects would tap a relatively distinct dimension, while satisfying the order constraint on the category boundaries. The three procedures are:

(a) Nonoverlapping clustering. Obtain p solutions ($p>0$) without the order constraint, and categorize Subject i into one of the p groups in which the subject has the largest contribution (i.e., the largest absolute score).

(b) Overlapping clustering. With respect to each of the p solutions, choose q subjects ($q>0$, q being a preset value) of the q largest absolute scores.

(c) Step-wise nonoverlapping clustering. Obtain the first solution, and choose those subjects whose scores exceed, for instance, 0.90, to form the first group. Remove those subjects in the first group from the data matrix, and subject the data of the remaining subjects to dual scaling. Select those subjects whose weights exceed 0.90, who then constitute the second group. Remove those subjects in this group from the matrix that has just been analyzed, and subject the remaining data to dual scaling. This process can be continued until the number of the remaining subjects becomes too small to be called a group. Each group then is submitted to dual scaling to obtain stimulus scale values and correctly ordered category boundaries.

Solutions obtained from different subgroups of subjects by any one of these procedures are likely to be correlated. Furthermore, stimulus scale values and category boundaries are calculated only from a subgroup of subjects, not from the entire group. Nishisato (1986c) presented an example to show that several solutions obtained from (c) were almost orthogonal to one another. Although his further empirical work is encouraging, more work is needed to improve these procedures so that the optimal multidimensional decomposition of the data may be obtained.

Another method of multidimensional decomposition is simply to ignore category boundaries. Suppose that the investigator is not interested in obtaining category boundaries from the data. One can then convert successive category data into rank-order data of only stimuli by ignoring category boundaries. For instance, Table 12.1 can be transformed into rank-order data as in Table 12.8.

TABLE 12.8
Rank-Order Transformation of Rating Data in Table 12.1

Subject	A	B	C	D
1	2	1	4	3
2	2	1	3.5	3.5
3	1.5	1.5	3	4

Table 12.8 can then be subjected to dual scaling for multidimensional decomposition of stimuli. Notice, however, that under this scheme category boundaries cannot be determined, and further that n stimuli are ranked in terms of only $(m+1)$ successive categories, resulting almost inevitably in many tied ranks. Imagine, for instance, 20 stimuli rated in terms of 4 categories. In the rank-order conversion of the rating data, there are at

most only 4 distinct rank numbers for the 20 stimuli judged by each subject. This example shows one of the hidden aspects[1] of rating data: successive category rating does not produce much discriminative data. It is empirically known, however, that tied ranks have relatively few effects on the derived data structure (Nishisato, 1988c).

12.2 APPLICATIONS

The main procedural difference between successive categories data and rank-order data is that the input data in the former format must be converted into ranking of both category boundaries and stimuli. Since this conversion does not present any new aspect of dual scaling, we will look at only one numerical example, and devote another section for a description of a situation in which successive categories data may present an interesting application.

12.2.1 Seriousness of Criminal Acts

You will recall from chapter 10 that 23 subjects provided paired comparisons of 8 criminal acts. Now a different group of 17 subjects rated the same 8 criminal acts with respect to their seriousness in terms of 4 successive categories (Table 12.9).

As mentioned earlier, the current formulation of dual scaling for successive categories data generally provides only one solution that satisfies the natural order of category boundaries. Therefore, for this current example, only the first solution is obtained by the DUAL3 package.

```
            DUAL SCALING OF
       SUCCESSIVE CATEGORIES DATA

  TITLE OF STUDY: Seriousness of Crimes

          *** INPUT DATA ***
        4  2  2  2  4  3  3  1
        4  2  2  2  4  4  3  1
```

[1] In chapter 14, we will see another aspect of successive categories data, that is, some difficulty in joint graphical display of row weights and column weights which is unique to this data type.

TABLE 12.9
Seriousness of Crimes Judged by Successive Categories by 17 Subjects

Subject	A	B	C	Crimes F	H	K	M	R
1	4	2	2	2	4	3	3	1
2	4	2	2	2	4	4	3	1
3	3	2	2	2	4	3	3	1
4	4	3	2	2	4	4	4	3
5	4	3	2	2	4	4	3	2
6	4	3	3	2	4	4	3	2
7	4	1	2	2	4	4	2	1
8	4	4	2	2	4	4	3	2
9	3	2	1	2	4	4	3	1
10	4	3	3	3	4	4	3	2
11	4	2	3	3	4	4	4	1
12	4	4	3	3	4	4	4	2
13	4	3	3	2	4	4	3	1
14	4	2	2	2	4	3	3	1
15	4	2	1	1	4	4	2	1
16	3	2	2	2	4	3	3	1
17	3	2	2	2	4	4	3	2

Note. Crimes: A = arson; B = burglary; C = counterfeiting; F = Forgery; H = Homicide; K = kidnapping; M = mugging; R = receiving stolen goods. Seriousness: 1 = somewhat serious; 2 = moderately serious; 3 = very serious; 4 = extremely serious.

```
3  2  2  2  4  3  3  1
4  3  2  2  4  4  4  3
4  3  2  2  4  4  3  2
4  3  3  2  4  4  3  2
4  1  2  2  4  4  2  1
4  4  2  2  4  4  3  2
3  2  1  2  4  4  3  1
4  3  3  3  4  4  3  2
4  2  3  3  4  4  4  1
4  4  3  3  4  4  4  2
4  3  3  2  4  4  3  1
4  2  2  2  4  3  3  1
4  2  1  1  4  4  2  1
3  2  2  2  4  3  3  1
3  2  2  2  4  4  3  2
```

```
            *** TRANSFORMED DATA MATRIX E ***
 -8    0    6    9   -4   -4   -4    9    3    3  -10
 -8    0    4    8   -4   -4   -4    8    8    2  -10
 -8    0    8    4   -4   -4   -4   10    4    4  -10
-10   -4    2    7   -1   -7   -7    7    7    7   -1
-10   -2    4    8    1   -6   -6    8    8    1   -6
```

-10	-1	4	8	0	0	-7	8	8	0	-7
-6	2	4	8	-9	-2	-2	8	8	-2	-9
-10	-2	2	7	7	-6	-6	7	7	0	-6
-6	0	6	3	-3	-9	-3	9	9	3	-9
-10	-6	4	8	-1	-1	-1	8	8	-1	-8
-8	-4	2	7	-6	-1	-1	7	7	7	-10
-10	-6	0	6	6	-3	-3	6	6	6	-8
-8	-4	4	8	0	0	-6	8	8	0	-10
-8	0	6	9	-4	-4	-4	9	3	3	-10
-4	2	4	8	-1	-8	-8	8	8	-1	-8
-8	0	8	4	-4	-4	-4	10	4	4	-10
-10	0	6	3	-5	-5	-5	9	9	3	-5

*** THE TOTAL VARIANCE TO BE ACCOUNTED FOR = .3820321

SOLUTION 1

SQUARED CORRELATION RATIO = 0.32724
MAXIMUM PRODUCT-MOMENT CORRELATION = 0.57205
PERCENTAGE HOMOGENEITY = 32.72
DELTA (TOTAL VARIANCE ACCOUNTED FOR): PARTIAL = 85.66
CUMULATIVE = 85.66

*** OPTIMAL WEIGHT VECTORS ***

SUBJECTS		CATEGORY BOUNDARIES		STIMULUS VALUES	
1	1.0484	1	-1.4558	1	1.1824
2	1.0696	2	-0.2791	2	-0.3445
3	1.0268	3	0.7734	3	-0.6968
4	0.9312			4	-0.7715
5	1.0477			5	1.4342
6	1.0241			6	1.1781
7	0.9475			7	0.3959
8	0.9237			8	-1.4162
9	1.0169				
10	0.9921				
11	0.9748				
12	0.8747				
13	1.0330				
14	1.0484				
15	0.9812				
16	1.0268				
17	1.0105				

******** END OF ANALYSIS *********

Notice that δ is 85.66%, meaning that the first solution captures most of the information in the data. This aspect of the data structure is also reflected in the weights for subjects which are all close to 1. As you recall, the weights for subjects are all equal to 1 when $\delta = 1$. Let us assume that one solution represents the data adequately, and plot Solution 1 as shown in Figure 12.3.

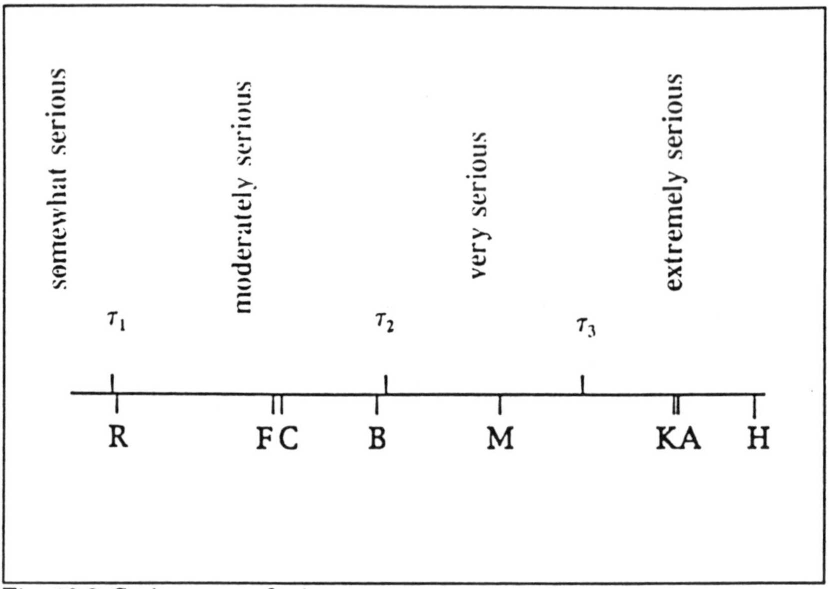

Fig. 12.3. Seriousness of crimes.

12.2.2 Time Series Data

There are occasions in which subjects are all involved in data collection over a period of time. Data sets at time 1, time 2,. . ., at Time T constitute time series data. Pretreatment data and posttreatment data are one example, and tallies of a specified number of goods sold in each of the past 12 months are another example. Let us indicate by $\mathbf{F}_t$ the subjects-by-stimuli table of successive categories data, obtained at Time t, and define the following time series data $\mathbf{F}$, where $\mathbf{F} = [\mathbf{F}_1 \ \mathbf{F}_2 \ldots \mathbf{F}_T]$, that is, the subjects-by-(stimuli times T) matrix. Dual scaling of $\mathbf{F}$ provides a single set of weights for the subjects, a single set of category boundaries, and T sets of stimulus scale values (i.e., n scale values of stimuli at time 1, time 2, . . . , and at Time T). Therefore, one can evaluate how stimulus scale values have changed over T time points in reference to the common set of category boundaries.

In the pretreatment and posttreatment situation, one can evaluate the effects of treatment on different types of behavior. In an applied scaling class at the Ontario Institute for Studies in Education, Melinda Head reported the following results of dual scaling. Successive categories data were collected from a random sample of mothers who participated in a counseling training session for handling behavioral problems of preschool children. The data were collected once prior to the training and once after the training. A number of questions related to problematic types of behavior were rated in terms of their frequencies, and dual scaling of $[\mathbf{F}_1 \;\; \mathbf{F}_2]$ from the two time points was carried out. Since we lack the space to reproduce all of the results from the questionnaire, we present only points of specific interest here.

1. Two parallel scales with the same origin and category boundaries were prepared for the precounseling and the postcounseling results.
2. Two scale values of each type of problematic behavior were plotted on their respective scales, and were connected by a straight line, from which one could tell whether or not the frequency of the behavior had increased (i.e., failure of counseling), or decreased (i.e., success of counseling).
3. The training successfully reduced frequencies of such behavior patterns as "excitable, impulsive"; "easily frustrated in efforts"; "restless in the squirmy sense"; "cries easily or often"; "depends on grown-ups too much"; "pouts and sulks"; "feelings easily hurt"; "disturbs other children"; "fails to finish"; "fights constantly."
4. The training did not influence such bahavior types as "restless, always up and on the go"; "sassy to grown-ups"; "gets into more trouble than others same age."
5. The training led to the increase of such behavior types as "bullies others" and "denies mistakes, or blames others."

This is only an example. Is it not very useful that we can demonstrate the effectiveness of a training program graphically as in this example? Time series data are of course not restricted to successive categories data. We can collect other types of data from the same group of subjects at different time points, and can observe, for example, how people's ranking of politicians has changed over a number of years. If we extract more than one solution, the interpretation of the results will naturally become more complicated, and interesting methodological problems will arise, which are beyond the scope of the current book.

PART IV
Special Topics

We have so far discussed all the basic forms of dual scaling, as applied to incidence data and dominance data. Part 4 contains some advanced topics, special applications of the basic procedure and other problems of importance, especially limitations and "unusual" results that require special attention. The main object of Part 4 is therefore two-fold, to enhance the applicability of dual scaling and to make this book self-contained. Parts 3 and 4 cover developments that have been carried out mainly under the name of dual scaling.

CHAPTER 13
Forced Classification and Focused Analysis

13.1 INTRODUCTION

Forced classification is a procedure of dual scaling, which carries out both discriminant analysis and analysis with the focus on a particular variable or a set of variables. The procedure was developed in 1984 and was generalized in 1986.[1] Its mathematical properties were clarified, and several applications were discussed.[2] In this chapter, we look at forced classification as applied to different types of data. Before discussing its applications, however, we must know the basic idea behind it.

The procedure is based on two principles, the principle of equivalent partitioning (PEP) and the principle of internal consistency (PIC). The PEP was proposed by Nishisato (1984b), and it is closely related to the principle of distributional equivalence (Benzécri et al., 1973). To explain the PEP, let us consider partitioning a data table (matrix) arbitrarily into a table with proportional rows and/or proportional columns, and observe that the squared correlation ratio and distinct optimal weights remain invariant over proportional row, column, or both row and column partitioning (Table 13.1).

Can you see how Example 1 is obtained from the original table? The original row 2 was divided into "1 to 2" proportions, that is, 15 was divided into 5 and 10, 30 into 10 and 20, and 45 into 15 and 30. These two rows in Example 1 have the same optimal weights as that of the original row, that is, row 2. Examples 3 and 4 show partitioning of both rows and columns, yet as you see the distinct weights remain the same, indicating invariance in data structure. That data structure remains invariant over

---

[1] See Nishisato (1984b, 1986b).

[2] See, for example, Nishisato (1987, 1988c, 1988d), Day (1989), Nishisato and Gaul (1990), and Halpine (1991).

TABLE 13.1
Examples of Equivalent Partitioning

(Original Table)	1	2	3	y_i
1	30	10	10	1.78
2	15	30	45	-0.47
3	10	40	20	-0.67
x_j	1.67	-0.67	-0.51	$\eta_1^2 = 0.1859$

(Example 1)	1	2	3	y_i
1	30	10	10	1.78
2	5	10	15	-0.47
2	10	20	30	-0.47
3	10	40	20	-0.67
x_j	1.67	-0.67	-0.51	$\eta_1^2 = 0.1859$

(Example 2)	1	2	3	y_i
1	15	5	5	1.78
1	15	5	5	1.78
2	15	30	45	-0.47
3	5	20	10	-0.67
3	5	20	10	-0.67
x_j	1.67	-0.67	-0.51	$\eta_1^2 = 0.1859$

(Example 3)	1	2	2	3	y_i
1	30	5	5	10	1.78
2	15	15	15	45	-0.47
3	5	10	10	10	-0.67
3	5	10	10	10	-0.67
x_j	1.67	-0.67	-0.67	-0.51	$\eta_1^2 = 0.1859$

(Example 4)	1	1	1	1	1	2	2	3	y_i
1	6	6	6	6	6	5	5	10	1.78
2	1	1	1	1	1	5	5	15	-0.47
2	1	1	1	1	1	5	5	15	-0.47
2	1	1	1	1	1	5	5	15	-0.47
3	1	1	1	1	1	10	10	10	-0.67
3	1	1	1	1	1	10	10	10	-0.67
x_j	1.67	1.67	1.67	1.67	1.67	-0.67	-0.67	-0.51	$\eta_1^2 = 0.1859$

row and column partitioning is the essence of the principle of equivalent partitioning, a special property of dual scaling.

The second principle used in forced classification is that of internal consistency (PIC), due to Guttman (1950). As discussed in chapter 4, dominant patterns of responses, or homogeneous patterns, constitute a major source for determining a solution. Therefore, for a given matrix of response patterns, if you augment it by repeating a response pattern matrix of N subjects to a particular item many times, these repeated elements become dominant in the augmented data matrix. Or, more precisely, the repeated item becomes the determining force of the solution. Since dual scaling quantifies data so as to maximize internal consistency, it follows that as the number of repetitions of the response pattern matrix of a particular item increases, the patterns of this item become the patterns of the solution itself. For example, if the repeated item is on the presence or the absence of abnormal EKG (electrocardiogram), dual scaling of the augmented matrix would yield scores for subjects that maximally distinguish between those with abnormal EKG and those without. Furthermore, as the number of repetitions increases, the correlation between the repeated item (hereafter called the criterion item) and the total score approaches 1. In other words, as the number of repetitions increases, the criterion item becomes more like the total score. Recall that one of the properties of dual scaling is to quantify option weights so as to maximize r_{jt}, the product-moment correlation between each item and the total score. Thus, it follows that as the number of repetitions increases, dual scaling derives optimal weights of items that increase, and eventually maximize, the average correlation between items and the criterion item.

Let us now combine the PEP and the PIC into the procedure of forced classification. The PEP tells us that repeating one item k times is equivalent to multiplying the item by k, thus without expanding the size of the data matrix, one can attain the same results as the case with many repetitions. For instance, the following two examples with the third item being the criterion provide identical results:

(Example A)

$$\mathbf{F} = [\mathbf{F}_1\ \mathbf{F}_2\ \mathbf{F}_3\ \mathbf{F}_3\ \mathbf{F}_3] = \begin{bmatrix} 1 & 0 & 0 & 0 & 0 & 1 & 0 & 1 & 0 & 0 & 1 & 0 & 0 & 1 & 0 \\ 0 & 1 & 0 & 0 & 0 & 1 & 0 & 1 & 0 & 0 & 1 & 0 & 0 & 1 & 0 \\ 0 & 0 & 1 & 1 & 0 & 0 & 1 & 0 & 0 & 1 & 0 & 0 & 1 & 0 & 0 \\ 0 & 0 & 1 & 0 & 1 & 0 & 1 & 0 & 0 & 1 & 0 & 0 & 1 & 0 & 0 \\ 0 & 1 & 0 & 1 & 0 & 0 & 0 & 0 & 1 & 0 & 0 & 1 & 0 & 0 & 1 \end{bmatrix}$$

(Example B)

$$\mathbf{F} = [\mathbf{F}_1\ \mathbf{F}_2\ 3\mathbf{F}_3] = \begin{bmatrix} 1 & 0 & 0 & 0 & 0 & 1 & 0 & 3 & 0 \\ 0 & 1 & 0 & 0 & 0 & 1 & 0 & 3 & 0 \\ 0 & 0 & 1 & 1 & 0 & 0 & 3 & 0 & 0 \\ 0 & 0 & 1 & 0 & 1 & 0 & 3 & 0 & 0 \\ 0 & 1 & 0 & 1 & 0 & 0 & 0 & 0 & 3 \end{bmatrix}$$

Now we are ready to stretch our imagination into forced classification. Suppose we have multiple-choice data from N subjects on n items. Let the $N \times m_j$ data matrix of response patterns for item j be indicated by $\mathbf{F}_j$. If we choose item j to be the criterion, forced classification of data matrix $\mathbf{F}$ means dual scaling of the following modified data with the constant k being sufficiently large to make r_{jt} indistinguishable from 1,

$$\mathbf{F}_{(k)} = [\mathbf{F}_1\ \mathbf{F}_2\ \mathbf{F}_3\ \ldots\ \mathbf{F}_{j-1}\ k\mathbf{F}_j\ \mathbf{F}_{j+1}\ \ldots\ \mathbf{F}_n]\ . \tag{13.1}$$

When k is sufficiently large, the weights of options of the noncriterion items are determined so as to produce m_j distinct scores. Do you know why it is so? Recall that $\mathbf{F}_j$ has m_j options and that item j identifies m_j subgroups in terms of the option each subject has chosen. Thus, the fact that r_{jt} is indistinguishable from 1 (perfect correlation) means m_j distinct scores being generated by dual scaling of the modified data matrix.

Let us briefly look at the choice of k. Up to this point, we have considered only what will happen if we increase k to a large value. In practice, it is important for the investigator to know how to choose the value of k. Let us look at an example (Table 13.2), and at forced classification results on the criterion-total correlation, r_{ct}, by using the criterion item c_1, and item-total correlation, r_{jt}, as a function of k (Table 13.3). Since the last three items identified by c are uncorrelated, they are not included in Table 13.3. In this particular instance, each r_{jt} increases steadily as k increases, and each case shows that an asymptote is almost reached when k is only 5 or 6. Thus, for this particular example, a k value of 10 would be sufficient for forced classification analysis. Such a criterion value, however, depends not only on the number of variables, but also on the data types. The question is if the chosen k value is likely to make the criterion variable dominant in the data set. Generally speaking, the k value must increase as the number of variables becomes larger, and a larger k value is required for dominance data than for incidence data. We do not wish to use unnecessarily large k value because the information contained in the variables other than the criterion may become totally negligible in proportion. The practical guideline is, therefore, to use a k value just large enough to make

TABLE 13.2
Multiple-Choice Data With Criterion Variable

Subject	Item 1	2	3	4	5	c_1	c_2	c_3
1	1	3	2	1	1	1	1	1
2	1	3	2	1	2	1	1	1
3	1	3	2	2	2	1	1	1
4	1	1	1	1	1	2	1	1
5	2	1	1	1	2	2	1	1
6	3	2	3	1	2	2	1	1
7	3	4	3	5	4	1	2	1
8	4	5	3	5	4	1	2	1
9	5	5	4	5	5	1	2	1
10	2	2	3	3	2	2	2	1
11	2	3	3	4	2	2	2	1
12	2	4	4	4	2	2	2	1
13	2	3	2	2	2	1	1	2
14	2	4	2	3	2	1	1	2
15	2	4	2	4	2	1	1	2
16	5	5	4	2	2	2	1	2
17	5	5	4	2	3	2	1	2
18	4	5	4	3	3	2	1	2
19	5	4	4	3	4	1	2	2
20	5	5	5	4	4	1	2	2
21	5	5	5	5	5	1	2	2
22	5	5	5	4	4	2	2	2
23	5	5	5	5	5	2	2	2
24	4	5	4	5	5	2	2	2

TABLE 13.3
Correlations r_{ct} and r_{cj} as a Function of k

k	r_{ct}	Correlation of Items with c_1, Item 1	2	3	4	5
1	0.06	0.15	0.04	0.22	-0.17	-0.05
2	0.13	0.15	0.07	0.25	-0.17	-0.04
3	0.77	0.18	0.42	0.54	-0.16	0.15
4	0.97	0.23	0.57	0.63	0.18	0.38
5	0.99	0.24	0.57	0.63	0.20	0.39
6	0.99	0.25	0.58	0.63	0.20	0.40
7	0.99	0.25	0.58	0.63	0.21	0.40
8	1.00	0.25	0.58	0.63	0.21	0.40

the criterion-total correlation to be close to 0.999. This remains, however, to be an empirical question.

TABLE 13.4
Item-Total Correlation Under Forced Classification with Item c_1 as the Criterion as a Function of k

k	Item 1	2	3	4	5	c_1	c_2	c_3
1	0.93	0.93	0.91	0.84	0.92	0.03	0.69	0.58
2	0.93	0.93	0.91	0.84	0.92	0.05	0.69	0.58
3	0.94	0.92	0.92	0.83	0.91	0.11	0.69	0.58
4	0.50	0.72	0.80	0.12	0.46	0.94	0.19	0.15
5	0.32	0.68	0.73	0.15	0.42	0.98	0.03	0.02
10	0.27	0.63	0.67	0.19	0.41	1.00	0.00	0.00
20	0.26	0.60	0.65	0.20	0.41	1.00	0.00	0.00
P_cF	0.25	0.58	0.63	0.21	0.40	1.00	0.00	0.00

TABLE 13.5
Correlation Between the Criterion Item and Other Items Under Forced Classification with Different Values of k

k	Item 1	2	3	4	5	c_2	c_3	Σr_{cj}^2
1	0.15	0.04	0.27	-0.17	-0.05	0.00	0.00	0.088
2	0.15	0.07	0.25	-0.17	-0.04	0.00	0.00	0.092
3	0.18	0.42	0.54	-0.16	-0.15	0.00	0.00	0.108
4	0.23	0.57	0.63	-0.18	0.38	0.00	0.00	0.789
5	0.24	0.57	0.63	0.20	0.39	0.00	0.00	0.972
10	0.25	0.58	0.63	0.21	0.40	0.00	0.00	1.000
20	0.25	0.58	0.63	0.21	0.40	0.00	0.00	1.000
P_cF	0.25	0.58	0.63	0.21	0.40	0.00	0.00	1.000

To see the effects of k, we can also look at the same phenomenon as the above in terms of item—total correlation, r_{jt} (Table 13.4), and the sum of squares of r_{jc} over items (the last column of Table 13.5). Observe that the criterion is initially not highly correlated with the total score, but that it eventually takes over the quantification space as reflected on the correlation of 1 with the total score. The last column of Table 13.5 is of particular importance, for the entries there can be used to judge if the k value is large enough. Examine Tables 13.3, 13.4, and 13.5 to understand what is going on in this simple procedure of forced classification.

With this much introduction to the idea of forced classification, let us look now at forced classification with specific data types and its generalizations.

13.2 MULTIPLE-CHOICE DATA

The idea of forced classification can be applied most readily to multiple-choice data, as already explained, and also to paired comparision and rank-order data with slight modifications. Let us first look at special problems associated with multiple-choice data.

The basic idea of forced classification was explained in Section 13.1, using multiple-choice data. In practice, we need to define discriminant scores in the following way. When one of the n items is chosen as the criterion and this item's correlation with the total score is increased to 1 by a sufficiently large k value, our specific interest would be in the question of how well the $(n-1)$ noncriterion items can classify subjects into m_j distinct subgroups associated with the criterion item. The scores of subjects calculated from the $(n-1)$ items, using forced classification option-weights, are called *discriminant scores.* Suppose that forced classification was carried out, using expression 13.1. Let us denote by $\mathbf{F}^*$ the data matrix obtained from Formula 13.1 by striking out $k\mathbf{F}_j$, and by $\boldsymbol{x}^*$ the weight vector of the forced classification option-weights, corresponding to the columns of $\mathbf{F}^*$, that is, the vector of option weights without those of the options of the criterion item. Then the vector of discriminant scores of the subjects, $\mathbf{y}^*$, can be defined as

$$\mathbf{y}^* = \frac{c}{n-1}\mathbf{F}^*\mathbf{x}^* \tag{13.2}$$

where c is a scaling constant.[3]

We have so far talked about only a solution from forced classification. The fact is that dual scaling of Formula 13.1 with a sufficiently large k value yields $(m_j - 1)$ solutions, for each of which the correlation between the criterion and the total score is equal to, or nearly equal to, 1. In other

---

[3] This constant may be determined as before in such a way that the sum of squares of the weighted responses is equal to the number of responses in $\mathbf{F}^*$, that is, $(n - 1)N$. Since the row marginals of $\mathbf{F}^*$ are each $(n - 1)$, $\mathbf{y}^{*\prime}\mathbf{D}^*\mathbf{y}^* = (n-1)\mathbf{y}^{*\prime}\mathbf{y}^*$. This is set equal to $(n-1)N$, hence $\mathbf{y}^{*\prime}\mathbf{y}^* = N$. But, $\mathbf{y}^{*\prime}\mathbf{y}^* = c_2\mathbf{x}^{*\prime}\mathbf{F}^{*\prime}\mathbf{F}^*\mathbf{x}^*/(n-1)^2$. Therefore, the desired constant is given by

$$c = (n-1)(N/\mathbf{x}^{*\prime}\mathbf{F}^{*\prime}\mathbf{F}^*\mathbf{x}^*)^{1/2}. \tag{13.3}$$

words, forced classification with item j as the criterion produces $(m_j\ \text{-}1)$ proper solutions. For instance, suppose that the criterion item is, "What is your view about religious education? ( ) for; ( ) against; ( ) indifferent." Since there are three options for this criterion item, the first two solutions of forced classification will be proper, and three options will be maximally separated in the two-dimensional space. Why there are $(m_j - 1)$ solutions will become clear shortly.

Now that we have discussed proper solutions, we may ask the question of what the remaining solutions are after $(m_j - 1)$ forced classification solutions are extracted. To look into this question as well as the questions of why $(m_j - 1)$ proper solutions and how to determine an appropriate value of k, it is convenient to know that there exists an exact solution to the forced classification problem, rather than the asymptotic procedure discussed previously. Consider multiple-choice data $\mathbf{F} = [\mathbf{F}_1\ \ \mathbf{F}_2\ \ \mathbf{F}_3]$, extracted from Example A. Let Item 3 be the criterion. Then there are two ways to obtain exact values of option weights, expected from forced classification with sufficiently large k value for $\mathbf{F}_3$.

(a) Method 1. Dual scaling of $\mathbf{C}_{3,12} = \mathbf{F}_3'[\mathbf{F}_1\ \ \mathbf{F}_2]$, that is, the contingency table of the "options of Item 3"-by-"options of Item 1, options of Item 2." If we use those three submatrices in Example A in Section 13.1, this contingency matrix is given by

$$\mathbf{C}_{3,12} = \mathbf{F}_3'\ [\mathbf{F}_1\ \ \mathbf{F}_2] = \begin{bmatrix} 0 & 0 & 2 & 1 & 1 & 0 \\ 1 & 1 & 0 & 0 & 0 & 2 \\ 0 & 2 & 0 & 2 & 0 & 0 \end{bmatrix}.$$

It is obvious that there are two soltuions from this matrix, namely, $\min(n,m) - 1$.

(b) Method 2. Dual scaling of $\mathbf{P}_3\mathbf{F} = \mathbf{F}_3(\mathbf{F}_3'\mathbf{F}_3)^{-1}\mathbf{F}_3'\mathbf{F}$, that is, matrix $\mathbf{F}$ projected onto the subspace spanned by the columns of $\mathbf{F}_3$. Using the same submatrices in Example A, this matrix is given by

$$\mathbf{P}_3\mathbf{F} = \frac{1}{2}\begin{bmatrix} 1 & 1 & 0 & 0 & 0 & 2 & 0 & 2 & 0 \\ 1 & 1 & 0 & 0 & 0 & 2 & 0 & 2 & 0 \\ 0 & 0 & 2 & 1 & 1 & 0 & 2 & 0 & 0 \\ 0 & 0 & 2 & 1 & 1 & 0 & 2 & 0 & 0 \\ 0 & 2 & 0 & 2 & 0 & 0 & 0 & 0 & 2 \\ 0 & 2 & 0 & 2 & 0 & 0 & 0 & 0 & 2 \end{bmatrix}.$$

Noting the equivalence of the three methods (these two and forced classification), we can now explore the nature of the solutions obtained after proper solutions are extracted. In Method 2, **P** is an orthogonal projection operator generated by the columns (options) of the criterion matrix $\mathbf{F}_c$. Thus, if dual scaling of $\mathbf{P}_c\mathbf{F}$ provides proper solutions, dual scaling of $(\mathbf{I} - \mathbf{P}_c)\mathbf{F}$ yields solutions expected when the effects of the criterion item are partialed out. In other words, dual scaling of $(\mathbf{I} - \mathbf{P}_c)\mathbf{F}$ offers conditional analysis that the solutions are uncorrelated to those of $\mathbf{P}_c\mathbf{F}$. It is known (Nishisato, 1988c) that $(m_j - 1)$ proper solutions correspond to the $(m_j - 1)$ solutions from $\mathbf{P}_c\mathbf{F}$, and that the next $(m_j - 1)$ forced classification solutions correspond to the $(m_j - 1)$ solutions of $(\mathbf{I} - \mathbf{P}_c)\mathbf{F}$. Table 13.6 shows results of forced classification, using Item 4 as the criterion. Since the item has five options, four forced classification solutions are proper. For this comparison, eight solutions were extracted by forced classification, using the weight k of 50, and four solutions from dual scaling of each of $\mathbf{P}_4\mathbf{F}$ and $(\mathbf{I} - \mathbf{P}_4)\mathbf{F}$.

TABLE 13.6
Proper Solutions and Conditional Solution

Solution	Method	Item 1	Item 2	Item 3	Item 4	Item 5	c_1	c_2	c_3
1	FC	0.51	0.79	0.71	1.00	0.81	0.14	0.77	0.37
1	$\mathbf{P}_4\mathbf{F}$	0.69	0.78	0.69	1.001	0.79	0.17	0.76	0.36
2	FC	0.49	0.35	0.45	1.00	0.70	0.08	0.25	0.41
2	$\mathbf{P}_4\mathbf{F}$	0.49	0.35	0.43	1.00	0.69	0.08	0.24	0.48
3	FC	0.35	0.41	0.27	1.00	0.37	0.12	0.25	0.10
3	$\mathbf{P}_4\mathbf{F}$	0.34	0.40	0.27	1.00	0.36	0.12	0.24	0.10
5	FC	0.54	0.69	0.80	0.00	0.44	0.22	0.00	0.25
1	$(\mathbf{I}-\mathbf{P}_4)\mathbf{F}$	0.62	0.68	0.78	0.01	0.48	0.20	0.00	0.28
6	FC	0.46	0.54	0.70	0.00	0.15	0.45	0.02	0.06
2	$(\mathbf{I}-\mathbf{P}_4)\mathbf{F}$	0.54	0.58	0.71	0.01	0.17	0.37	0.04	0.05
7	FC	0.56	0.34	0.43	0.00	0.26	0.08	0.02	0.03
3	$(\mathbf{I}-\mathbf{P}_4)\mathbf{F}$	0.23	0.63	0.71	0.05	0.25	0.15	0.02	0.05
8	FC	0.54	0.27	0.33	0.00	0.26	0.01	0.02	0.01
4	$(\mathbf{I}-\mathbf{P}_4)\mathbf{F}$	0.45	0.13	0.43	0.03	0.48	0.03	0.06	0.00

In Section 13.1, it was mentioned that r_{jt} being indistinguishable from 1 meant m_j distinct scores. This point requires an explanation. Since our interest lies in discriminant scores as defined by Formula 13.2 and not in the total scores which include the criterion item, r_{jt} being 1 does not generally mean that each of the $(m_j - 1)$ proper solutions provides m_j distinct discriminant scores. In fact, if the noncriterion items are all poorly correlated with the criterion item, one may not even be able to see any clusters of distinct scores. This means that all the proper solutions are not always useful in data analysis.

13.2.1 As Compared with An Average Person

Recall the questionnaire "As Compared with an Average Person" in Table 2.4 of chapter 2. An example of a sample analysis was presented in Table 2.5. Thus instead of presenting the data matrix separately, let us look at the output of the DUAL3, which contains the data matrix. Since all the questions are dichotomous (i.e., two options per question), there is only one proper solution. The criterion used in the example of Table 2.5 was "I am more high-handed and opinionated." The analysis captured an interesting set of personality traits. For the present example, let us choose as the criterion something which seems unrelated to any of the questions, namely, "I eat more," and see what personality traits, if any, may be remotely correlated. The following is an abbreviated version of the DUAL3 output:

```
                    DUAL SCALING OF
                  MULTIPLE-CHOICE DATA
                 (FORCED CLASSIFICATION)
            (c) Copyright 1986 by MicroStats

      TITLE OF STUDY:   "I Eat More" As Criterion

            *** INPUT DATA ***
2 1 2 2 2 1 1 2 2 2 2 2 1 2 2 2 1 2 2 2 2 2 2 2 2 2 2 2 2 1
1 1 1 2 2 2 1 2 2 1 2 1 1 2 2 1 1 1 2 1 2 1 2 1 2 2 1 2 1 1
1 2 1 2 2 2 1 2 1 1 1 1 2 1 2 1 1 2 2 1 1 1 1 1 2 2 1 2 1 1
2 1 1 2 1 2 1 1 2 2 1 2 1 1 1 1 2 1 1 2 1 2 1 2 2 2 2 2 1 1
1 1 2 2 1 2 1 2 1 1 2 1 1 2 1 2 2 2 2 1 1 1 2 1 2 2 2 1 1 1
2 1 1 2 1 2 1 1 2 1 2 2 1 1 2 1 1 1 2 1 1 2 1 1 2 2 2 1 1 1
2 1 2 1 1 2 1 1 2 2 1 2 2 2 2 1 2 1 1 1 2 2 1 2 1 1 1 1 1 1
1 2 1 2 1 2 1 1 1 1 1 1 2 1 2 1 2 2 1 1 1 1 1 1 2 1 1 2 1 2
1 2 1 2 1 2 2 1 1 1 1 1 1 1 1 1 2 2 2 1 1 1 1 1 2 1 1 1 2 1
2 1 2 1 1 2 2 1 2 2 2 2 2 1 1 2 1 2 2 1 2 2 2 2 2 2 2 2 2 2
1 1 1 2 1 2 1 2 2 1 1 2 1 1 2 1 2 2 2 1 2 2 1 1 2 2 1 2 2 1
1 2 1 2 1 2 1 1 1 1 2 2 1 1 1 2 2 1 1 1 2 1 1 1 1 2 2 2 1 2
2 2 0 2 1 1 1 2 2 2 1 2 1 1 2 1 2 2 2 1 1 2 2 1 2 2 2 1 2 2
1 1 1 2 2 2 1 2 1 1 1 1 1 1 1 2 2 2 1 1 1 1 1 1 2 2 1 2 1 2
2 1 1 2 2 1 1 1 2 1 1 2 1 2 2 2 1 1 2 1 2 2 1 1 2 2 2 1 1 1
1 1 1 2 1 2 2 2 1 2 2 1 2 1 2 2 2 2 2 1 2 1 2 1 2 2 2 1 1 1
1 2 1 2 1 2 1 2 1 1 2 1 1 2 2 1 2 2 2 1 1 1 1 1 1 2 1 1 2 1
2 1 2 2 2 1 1 2 2 2 2 1 1 1 2 0 1 2 1 1 1 2 2 2 2 2 2 1 1 1
1 2 1 2 1 2 2 1 1 2 1 2 1 1 1 1 2 2 1 1 1 1 2 2 2 2 1 1 1 2
2 2 2 2 1 2 2 2 1 1 2 2 1 2 2 2 2 2 2 1 1 2 1 1 2 2 1 2 1 2
```

```
1 1 1 2 2 2 1 1 2 1 2 1 2 1 2 1 2 2 2 1 1 1 1 1 2 2 2 1 2 2
1 1 1 2 2 2 0 2 2 1 2 2 1 1 2 2 1 2 2 1 2 2 2 1 2 2 1 2 2 1
2 1 1 2 2 2 1 0 2 0 0 0 0 2 0 2 2 1 2 1 2 2 1 1 2 0 0 2 1 2
2 1 1 2 1 2 2 1 2 1 1 1 1 1 1 1 2 2 1 1 1 1 1 1 1 1 1 1 1 2
1 2 1 2 1 2 1 2 2 1 2 1 2 1 2 1 2 2 2 1 1 1 1 1 1 2 1 1 1 2
1 2 1 2 1 1 1 2 1 1 2 1 1 1 2 1 2 2 2 1 1 1 1 1 2 2 1 1 1 1
2 1 1 2 2 2 2 2 1 1 1 1 1 1 2 1 2 2 2 1 2 2 1 1 2 2 1 1 1 2
1 2 1 2 1 2 1 2 2 2 1 2 1 1 2 1 2 2 2 1 2 2 1 1 2 2 1 1 2 1
2 2 1 2 2 2 1 2 2 2 2 2 1 1 2 2 2 2 2 1 2 2 1 1 2 2 2 2 2 1

  *** THE TOTAL VARIANCE TO BE ACCOUNTED FOR = 1.00811  ***

        SOLUTION 1

        SQUARED CORRELATION RATIO = 0.65677
        MAXIMUM PRODUCT-MOMENT CORRELATION = 0.81042
        PERCENTAGE HOMOGENEITY =   65.68

              *** OPTIMAL SCORES AND WEIGHTS ***

        DISCRIMINANT SCORE          OPTION   NORMED ADJUSTED
      1  -0.13     15  -0.08        1  1    -0.07   -0.06
      2  -0.13     16  -0.07        1  2     0.08    0.07
      3  -0.04     17  -0.07        2  1     0.04    0.03
      4   0.24     18   0.04        2  2    -0.07   -0.05
      5  -0.03     19   0.23        3  1    -0.01   -0.01
      6  -0.02     20  -0.05        3  2     0.16    0.13
      7   0.25     21  -0.04        4  1     0.66    0.54
      8   0.17     22  -0.24        4  2    -0.05   -0.04
      9   0.12     23  -0.07        5  1     0.17    0.14
     10   0.07     24   0.27        5  2    -0.29   -0.24
     11  -0.10     25   0.02        6  1    -0.23    0.19
     12   0.14     26  -0.05        6  2     0.05    0.04
     13  -0.08     27  -0.03        7  1     0.02    0.01
     14   0.08     28  -0.09        7  2     0.06    0.05
                   29  -0.16        8  1     0.77    0.62
                                    8  2    -0.45   -0.37
                                    9  1     0.16    0.13
                                    9  2    -0.12   -0.10
                                   10  1    -0.15   -0.12
                                   10  2     0.34    0.27
                                   11  1     0.52    0.42
                                   11  2    -0.40   -0.33
                                   12  1     0.03    0.02
                                   12  2     0.02    0.02
                                   13  1     0.02    0.02
                                   13  2     0.05    0.04
                                   14  1     0.16    0.13
                                   14  2    -0.43   -0.35
                                   15  1     0.98    0.80
```

15	2	-0.36	-0.29
16	1	0.06	0.05
16	2	-0.28	-0.23
17	1	-0.43	-0.35
17	2	0.16	0.13
18	1	0.41	0.34
18	2	-0.14	-0.11
19	1	1.99	1.62
19	2	-0.76	-0.62
20	1	-0.05	-0.04
20	2	0.61	0.50
21	1	0.28	0.23
21	2	-0.35	-0.29
22	1	0.23	0.19
22	2	-0.22	-0.18
23	1	0.07	0.06
23	2	-0.16	-0.13
24	1	-0.29	-0.23
24	2	1.08	0.88
25	1	0.91	0.74
25	2	-0.19	-0.16
26	1	1.35	1.09
26	2	-0.19	-0.16
27	1	0.10	0.08
27	2	-0.08	-0.06
28	1	-0.06	-0.05
28	2	0.07	0.06
29	1	0.40	0.32
29	2	-0.77	-0.62
30	1	-0.28	-0.23
30	2	0.39	0.32

*** ITEM STATISTICS ***

ITEM(j)	SS(j)	R^2(jt)	R(jt)
1	0.18	0.004	0.064
2	0.08	0.002	0.043
3	0.14	0.003	0.056
4	0.95	0.022	0.147
5	1.47	0.033	0.183
6	0.32	0.007	0.085
7	0.01	0.000	0.015
8	9.91	0.225	0.474
9	0.57	0.013	0.114
10	1.51	0.034	0.185
11	5.95	0.135	0.367
12	0.00	0.000	0.005
13	0.00	0.000	0.010
14	1.99	0.045	0.213
15	10.28	0.233	0.483
16	0.79	0.018	0.134
17	2.04	0.046	0.215
18	1.60	0.036	0.190

19	44.04	0.997	0.999
20	0.82	0.019	0.136
21	2.89	0.065	0.256
22	1.49	0.034	0.184
23	0.33	0.008	0.087
24	8.91	0.202	0.449
25	5.05	0.114	0.338
26	8.11	0.184	0.429
27	0.23	0.005	0.071
28	0.12	0.003	0.052
29	8.88	0.201	0.448
30	3.18	0.072	0.268

********* END OF ANALYSIS *********

From this output, let us choose arbitrarily those questions with r_{jt} [R(jt) in the computer output] greater than 0.300. Since the criterion-total correlation is 0.999, the item-total correlation is in essence the same as the criterion-item correlation. Therefore, we are indeed looking at the criterion-item correlation of higher than 0.300. From the item statistics in the previous output, the questions of our interest are questions 15 (0.483), 8 (0.474), 24 (0.449), 29 (0.448), 26 (0.429), 11 (0.367), and 25 (0.338).

Now we must look at the option weights of these items as well as those of the criterion item in the output. The option weights of the criterion item are 1.99 for "I eat more," and -0.76 for "I eat less." Recall that in dual scaling the sum of weighted responses to each item is zero. Therefore, if we indicate by v the number of subjects who said "eat more," then the number of subjects who said "eat less" is $(29 - v)$. Thus, we can calculate v, hence $(29 - v)$, from the constraint used in dual scaling: $1.99v + (-0.76)(29 - v) = 0$, that is, $v = 8$. There are 8 subjects who said "eat more" out of 29 subjects. Let us examine the option—weights of the other items on the above list. Table 13.7 shows the summary of the relations.

Is this not interesting? Who could imagine that this group of 29 subjects revealed such connections of personality to eating? We can consider a number of applications of forced classification with multiple-choice data, to name a few, identify biographical correlates of the propensity to drop out of school, explore what attributes of cosmetics make a particular brand a smashing hit, find out personality traits that distinguish successful executives from unsuccessful ones, and investigate characteristics of cars, if any, that are related to the increased sales, as opposed to the decrease.

TABLE 13.7
If I Eat More Than an Average Person. . .

If I eat more, then I am	r_{cj}	If I eat less, then I am
More presumptuous	0.483	Less presumptuous
More high-handed and more opinionated	0.474	Less high-handed and less opinionated
Less pleasant in dealing with others	0.449	More pleasant in dealing with others
More fortunate and lucky	0.448	Less fortunate and lucky
More of a spendthrift	0.429	Less of a spendthrift
Less patient	0.367	More patient
More flamboyant	0.338	Less flamboyant

13.3 SORTING DATA

As we have already discussed, sorting data can be handled in the same way as multiple-choice data, except that the roles of rows and columns of the data matrix are interchanged. Therefore, the procedure of forced classification for sorting data employs a subject as the criterion, rather than an item. This means that if the criterion subject has used the geographical proximity as the criterion for similarity among nations, then the analysis will reveal which other subjects may have used the same criterion through intersubject correlation. As far as the method is concerned, there is nothing new to be added to the discussion in the previous section. Let us therefore look at an example of its application.

13.3.1 Sorting Countries by Language

In chapter 9, we looked at the data set in which 39 countries were sorted into piles by 21 subjects. Most subjects appeared to have used geographical proximities, or related concepts, in sorting countries. Can we consider a criterion that seems unrelated to geographical proximity? Recall that Solution 5 appeared related to language. thus, let us use the same data set, and augment it by adding one "fictitious" subject who classifies the countries into those of English-speaking and the others. In the data set, countries where English is a major language are Britain, Canada, New

Zealand, South Africa, and the USA. These 5 countries are coded as 1 and the remaining 34 countries as 2 in the condensed format. This substantial imbalance in group sized (i.e., 5 vs. 34) is not desirable, and reduces somewhat its value as a good example. But, not knowing what else may be more interesting than the languages as a criterion, we will look at the results of this forced classification. Notice that the five countries as a whole are by no means neighbors geographically.

Because most people created many piles, we needed the weight of 200 to make the criterion-total correlation close to 1.000. Table 13.8 shows the scores of countries arranged in descending order.

TABLE 13.8
Forced Classification of Countries by Language

1.91 Canada	-0.11 Tanzania	-0.37 Lebanon
1.90 USA	-0.12 Spain	-0.38 Hungary
1.89 New Zealand	-0.18 Finland	-0.40 Libya
1.59 Britain	-0.19 Greece	-0.41 Cuba
1.42 South Africa	-0.20 Norway	-0.42 Iran
0.38 France	-0.22 Nigeria	-0.44 China
0.28 W. Germany	-0.24 Singapore	-0.44 Pakistan
0.02 Switzerland	-0.26 Egypt	-0.44 USSR
-0.01 Denmark	-0.30 Ethiopia	-0.47 Peru
-0.05 Netherlands	-0.30 Thailand	-0.51 Yugoslavia
-0.07 Italy	-0.35 India	-0.52 Romania
-0.08 Sweden	-0.35 Poland	-0.52 Guatemala
-0.11 Japan	-0.36 Venezuela	-0.54 Brazil

Since the fictitious subject, our criterion subject, classified 39 countries into only two groups, the very small group of English-speaking countries is expected to be cleanly identified as is the case. The remaining 34 countries were placed into a single pile by the criterion subject, and the weights (scores) of those countries are bound to vary more than those of the smaller group by the judgment of the remaining subjects. This point will become clear shortly when we look at inter-subject correlation and the original sorting data. Table 13.9 lists correlation coefficients between the criterion subjects and each of the remaining subjects.

If you examine these coefficients in comparison with the sorting data, you can see that Subjects 16 and 20 both classified the five English-speaking countries into a single pile. But they did neither classify the remaining countries into a single pile, nor sort them into the same number of piles. The discrepancies between the two subjects over the classification of the 34 countries do not matter very much because the forced classification pro-

TABLE 13.9
Correlation Between Criterion and Each Subject

Subject i	r_{ci}	Subject *i*	r_{ci}	Subject *i*	r_{ci}
1	0.460	8	0.920	15	0.700
2	0.374	9	0.636	16	1.000
3	0.282	10	0.347	17	0.575
4	0.796	11	0.444	18	0.516
5	0.910	12	0.656	19	0.474
6	0.650	13	0.904	20	1.000
7	0.255	14	0.379	21	0.674

cedure determine the weights for those 39 countries to be as similar as possible.

From Table 13.9, we can tell varying degrees of subject contributions to the criterion classification. Subjects 8, 5, and 13 also show high values of correlation. Let us look at the five highest correlation coefficients, and how those subjects classified the five English-speaking countries:

Subject 16 (Britain, Canada, New Zealand, South Africa, USA).
Subject 20 ((Britain, Canada, New Zealand, South Africa, USA).
Subject 8 (Britain, Canada, New Zealand), (South Africa), (USA, China, USSR).
Subject 5 (Britain, Canada, USA), (New Zealand, South Africa).
Subject 13 (Britain, Canada, New Zealand, USA), (South Africa, Egypt, Ethiopia, Nigeria, Tanzania).

Those parentheses indicate piles. These comparisons are by no means sufficient to understand the intersubject correlation. We need to look into the correlation between, for instance, China and the other countries. Then, we may be able to tell what effect of the combination of USA and China (see Subject 8) had on the correlation r_{c8}. We should be careful about accepting the correlation coefficient at its face value. In other words, a comparatively high value of correlation should not be immediately used to conclude that the subject classified the countries into English-speaking and non-English—speaking countries. Subjects 13 and 8 provide examples of this danger: Subject 13 must have used two different criteria for the classification, Anglo-Saxon and geographical, and Subject 8 shows a different scheme of classification, isolating South Africa from the rest. In order to draw an interpretation of correlation more directly, the investigator should consider a more finely divided criterion with a comparatively even distribution of objects over piles. For instance, one can introduce a fictitious subject as the criterion who classifies countries into North America, Central America, South America, Asia, Europe, Africa, and Australia. As the number of piles in the criterion increases, however, one must look into more solutions to explain the results. More specifically, one must look at $(g - 1)$ solutions, where g is the number of piles in the criterion. Comparatively speaking, sorting data are difficult to analyze, and much more theoretical work on interpretation is needed.

13.4 PAIRED COMPARISON DATA

Incidence data, such as contingency tables and multiple-choice data, are absolute measurements in the sense that each datum is meaningful by itself and can be quantified. In contrast, dominance data, such as paired comparison, rank-order, and successive categories data, are comparative measurements such that each datum requires others to be meaningful. This comparative aspect of dominance data has an important implication for forced classification: dominance data require two variables to form the criterion. Let us look at a small example to introduce the procedure of forced classification for paired comparison data.

Consider two subjects participated in paired comparisons of four stimuli with the results as shown in Table 13.10, where responses are coded as 1, 0, and -1, respectively, for the preference of stimulus p to stimulus q, a tied judgement, and the preference of stimulus q to p.

TABLE 13.10
Paired Comparison Data

Subject	(X_1,X_2)	(X_1,X_3)	(X_1,X_4)	(X_2,X_3)	(X_2,X_4)	(X_3,X_4)
1	1	1	-1	1	-1	-1
2	1	-1	-1	-1	-1	-1

In the standard analysis, the dominance matrix is given by

$$\mathbf{E} = \begin{bmatrix} 1 & -1 & -3 & 3 \\ -1 & -3 & 1 & 3 \end{bmatrix}.$$

This matrix can be obtained by the multiplication of the data matrix **F** and the paired comparison design matrix **A**,[4] that is, **E** = **FA**. In the present example, we have

$$\mathbf{E} = \mathbf{FA} = \begin{bmatrix} 1 & 1 & -1 & 1 & -1 & -1 \\ 1 & -1 & -1 & -1 & -1 & -1 \end{bmatrix} \begin{bmatrix} 1 & -1 & 0 & 0 \\ 1 & 0 & -1 & 0 \\ 1 & 0 & 0 & -1 \\ 0 & 1 & -1 & 0 \\ 0 & 1 & 0 & -1 \\ 0 & 0 & 1 & -1 \end{bmatrix}.$$

Suppose that the pair (X_1, X_3) was chosen as the criterion pair with the weight of 3, that is, k=3. This is equivalent to repeating the corresponding column of the data matrix **F** three times, and the new augmented matrix is subjected to dual scaling. In this example, the augmented matrix $\mathbf{F}_1$ is given as

$$\mathbf{F}_1 = \begin{bmatrix} 1 & 1 & 1 & 1 & -1 & 1 & -1 & -1 \\ 1 & -1 & -1 & -1 & -1 & -1 & -1 & -1 \end{bmatrix}.$$

Is this clear? Note that the second column of the original **F** is repeated three times. Thus, if we multiply the second column by 3 instead of repeating it three times, the equivalent matrix for forced classification is given by

$$\mathbf{F}_2 = \begin{bmatrix} 1 & 3 & -1 & 1 & -1 & -1 \\ 1 & -3 & -1 & -1 & -1 & -1 \end{bmatrix}.$$

The dominance matrix **E** for this forced classification can be obtained by $\mathbf{F}_2\mathbf{A}$, or $\mathbf{F}_1\mathbf{A}^*$, where

$$\mathbf{A}^* = \begin{bmatrix} 3 & -3 & 0 & 0 \\ 1 & 0 & -1 & 0 \\ 3 & 0 & 0 & -3 \\ 0 & 1 & -1 & 0 \\ 0 & 1 & 0 & -1 \\ 0 & 0 & 1 & -1 \end{bmatrix}.$$

This matrix is obtained from the original design matrix **A** by multiplying row 1 and row 3 (i.e., the criterion pair) by 3. Thus, in general, if the pair

[4] See Bock and Jones (1968).

(X_p, X_q) is chosen as the criterion with the weight k, then the dominance matrix **E** can be obtained in two ways: one is to multiply the criterion column of the data matrix **F** by k and postmultiply it by the ordinary paired comparison design matrix **A**, and the other is to premultiply the data matrix **F** by the modified design matrix **A***, where **A*** is obtained from **A** by multiplying row p and row q by the weight k. In the current example, the dominance matrix **E** is given by the two methods as follows:

$$\mathbf{E} = \begin{bmatrix} 1 & 3 & -1 & 1 & -1 & -1 \\ 1 & -3 & -1 & -1 & -1 & -1 \end{bmatrix} \begin{bmatrix} 1 & -1 & 0 & 0 \\ 1 & 0 & -1 & 0 \\ 1 & 0 & 0 & -1 \\ 0 & 1 & -1 & 0 \\ 0 & 1 & 0 & -1 \\ 0 & 0 & 1 & -1 \end{bmatrix} = \begin{bmatrix} 3 & -1 & -5 & 3 \\ -3 & -3 & 3 & 3 \end{bmatrix}.$$

$$\mathbf{E} = \begin{bmatrix} 1 & 1 & -1 & 1 & -1 & -1 \\ 1 & -1 & -1 & -1 & -1 & -1 \end{bmatrix} \begin{bmatrix} 3 & -3 & 0 & 0 \\ 1 & 0 & -1 & 0 \\ 3 & 0 & 0 & -3 \\ 0 & 1 & -1 & 0 \\ 0 & 1 & 0 & -1 \\ 0 & 0 & 1 & -1 \end{bmatrix} = \begin{bmatrix} 3 & -1 & -5 & 3 \\ -3 & -3 & 3 & 3 \end{bmatrix}.$$

The total number of responses for each row, that is, each of the diagonal elements of the matrix $\mathbf{D}_r$, is constant and equal to $n(n - 1)+2(k - 1)$. The number of responses for each column, however, is not constant: it is given by $N(n - 1)$ if the stimulus is not a criterion variable, and it is $N(n + k - 2)$ if the stimulus is a criterion variable. Thus, diagonal matrices $\mathbf{D}_r$ and $\mathbf{D}_c$ must be modified. In this example, $N = 2$, $n = 4$, and $k = 3$. Thus,

$$\mathbf{D}_r = \begin{bmatrix} 16 & 0 \\ 0 & 16 \end{bmatrix}, \quad \mathbf{D}_c = \begin{bmatrix} 10 & 0 & 0 & 0 \\ 0 & 6 & 0 & 0 \\ 0 & 0 & 10 & 0 \\ 0 & 0 & 0 & 6 \end{bmatrix}.$$

Since the paired comparision data (chapter 10) and the rank-order data (chapter 11) provide similar results, we do not look at an example of forced classification analysis here, but see the results obtained from the rank-order data in the next section.

13.5 RANK-ORDER DATA

It was earlier mentioned that paired comparison data and rank-order data can be handled in the same way, except that the computation of the dominance matrix **E** is easier for rank-order data than for paired comparison data. In forced classification, rank-order data are computationally more complicated than paired comparison data. Let us look at a simple example to derive a procedure of forced classification for rank-order data. Suppose that three subjects ranked five stimuli and provided responses as shown in Table 13.11.

9

TABLE 13.11
Rank-Order Data

Subject	X_1	X_2	X_3	X_4	X_5
1	1	2	3	4	5
2	2	3	5	1	4
3	3	1	4	2	5

Suppose that the investigator chose Stimuli 2 and 4 as the criterion pair, and decided to use the weight of 3, that is, k=3. Then, if we repeat the responses to these two stimuli three times, the augmented data matrix **F** is given as

$$\mathbf{F} = \begin{bmatrix} 1 & 2 & 2 & 2 & 3 & 4 & 4 & 4 & 5 \\ 2 & 3 & 3 & 3 & 5 & 1 & 1 & 1 & 4 \\ 3 & 1 & 1 & 1 & 4 & 2 & 2 & 2 & 5 \end{bmatrix}.$$

We now have nine columns, and we must adjust these rank numbers for the range from 1 to 9. The adjusted rank-order matrix **F*** is given as follows,

$$\mathbf{F}* = \begin{bmatrix} 1 & 3 & 3 & 3 & 5 & 7 & 7 & 7 & 9 \\ 4 & 6 & 6 & 6 & 9 & 2 & 2 & 2 & 8 \\ 7 & 2 & 2 & 2 & 8 & 5 & 5 & 5 & 9 \end{bmatrix}.$$

Is this conversion clear? Those repeated rank numbers are averages of the rank numbers that they would have taken. For instance, response "3" of (3,3,3) of Subject 1 to Stimulus 2 is the average of three rank numbers 2,

3 and 4. Similarly, "7" of (7,7,7) of the samo oubject is the average of 6, 7 and 8. Using the formula to convert rank numbers to dominance numbers, we obtain the following dominance matrix **E**.

$$\mathbf{E} = \begin{bmatrix} 8 & 4 & 4 & 4 & 0 & -4 & -4 & -4 & -8 \\ 2 & -2 & -2 & -2 & 8 & 6 & 6 & 6 & -6 \\ -4 & 6 & 6 & 6 & -6 & 0 & 0 & 0 & -8 \end{bmatrix}.$$

Using the principle of equivalent partitioning, we can combine those repeated columns to obtain the dominance matrix **E** for forced classification. The matrix with combined columns is given by

$$\mathbf{E} = \begin{bmatrix} 8 & 4\times3 & 0 & -4\times3 & -8 \\ 2 & -2\times3 & 8 & 6\times3 & -6 \\ -4 & 6\times3 & -6 & 0\times3 & -8 \end{bmatrix} = \begin{bmatrix} 8 & 12 & 0 & -12 & -8 \\ 2 & -6 & 8 & 18 & -6 \\ -4 & 18 & -6 & 0 & -8 \end{bmatrix}.$$

The generation of this final 3x5 matrix from the original 3x5 rank-order matrix is not very simple. We must generate the matrix **E** for any number of subjects, that of stimuli and the weight k. We must obtain the general expression for the elements of **E**.[5] Let K_{ip} and K_{iq} be the ranks of the two criterion stimuli given by Subject i, and consider comparing K_{ip}, K_{iq}, and K_{ij}, where K_{ij} is the rank number given to a noncriterion stimulus by Subject i. Then the element e_{ij} of the dominance matrix **E** is given by

$$e_{ij} = n+2k-1-2K_{ij}, \text{ if } K_{ij}<K_{ip}, \ K_{ij}<K_{iq} .$$

$$e_{ij} = n+1-2K_{ij}, \text{ if } K_{ip}<K_{ij}<K_{iq}, \text{ or if } K_{iq}<K_{ij}<K_{ip} .$$

$$e_{ij} = n+3-2k-2K_{ij}, \text{ if } K_{ij}>K_{ip}, \text{ and if } K_{ij}>K_{iq} .$$

$$e_{it} = (n+k-2K_{ij})k, \text{ where } t = \min(K_{ip},K_{iq}) .$$

$$e_{it} = (n+2-k-2K_{ij})k, \text{ where } t = \max(K_{ip},K_{iq}) .$$

The number of responses involved in each row is the same as that of the corresponding element of paired comparison data. The number of

[5] See Nishisato (1984b).

responses of the column of the criterion variable is $Nk(n+2k-3)$, and that of a noncriterion column is $N(n+2k-3)$. Once these quantities are specified together with the value of k, the standard procedure of dual scaling will yield forced classification results. Let us look at an example.

13.5.1 Extracting the Cost Dimension of Christmas Party Plans

In chapters 10 and 11, people's preferences for eight Christmas party plans were analyzed, using paired comparison data and rank-order data, respectively. Those eight plans are:

1. A pot-luck at someone's home in the evening.
2. A pot-luck in the group room.
3. A pub/restaurant crawl after work.
4. A reasonably priced lunch in an area restaurant.
5. Keep to one's self.
6. An evening banquet at a restaurant.
7. A pot-luck at someone's home after work.
8. A ritzy lunch at a good restaurant (tablecloths).

Both analyses revealed three interpretable solutions, conviviality dimension, cost dimension, and time dimension. Let us use the data in Table 11.14, and choose party plan 7 ("a pot-luck at someone's home after work"—an inexpensive plan) and party plan 8 ("a ritzy lunch at a good restaurant [tablecloths]"—an expensive plan) as the criterion pair, representing the cost dimension. To see the effect of the k value, forced classification was carried out with different k values. Results of the analyses are as shown in Table 13.12. Notice that in this table stimuli and subjects are listed in rows and k values in columns, unlike those previous tables in this chapter.

When k is 1, the scale values of the eight plans show the conviviality dimension, with only two subjects showing negative weights. Plans 8 and 9, used as the criterion pair, are located both on the positive side in close proximity. Once k is doubled, the criterion stimuli swiftly move toward both extreme ends, and the scale shows definitely the cost dimension, with expensive plans on the positive side (Plans 8, 6, and 4), and with inexpensive plans on the negative side (Plans 7, 2, 5, and 1). If we look at the weights of subjects, we can clearly see that as the k value increases the weights change toward either -1 or 1 (see when k = 50). One interesting point associated with this change is that the weights of the criterion pair also move toward the same extreme values, that is, -1 and 1. We can carry out forced classification of the paired comparison data, presented in chapter

TABLE 13.12
Weights and Scores as a Function of k

	k					
Stimulus	1	2	3	4	5	50
1	0.35	-0.64	-0.67	-0.68	-0.68	-0.71
2	-1.26	-0.83	-0.66	-0.59	-0.56	-0.44
3	0.87	0.35	0.24	0.20	0.17	0.14
4	0.56	-0.18	-0.20	-0.21	-0.22	-0.27
5	-2.00	-0.68	-0.62	-0.64	-0.70	-0.70
6	0.04	0.67	0.56	0.50	0.45	0.30
7(c)	0.59	-1.04	-1.07	-1.06	-1.05	-1.01
8(c)	0.93	1.69	1.52	1.41	1.34	1.04
Subject						
1	-1.08	-1.19	-1.09	-1.06	-1.04	-1.00
2	1.40	0.92	0.86	0.86	0.87	0.98
3	1.31	1.02	0.93	0.91	0.91	0.98
4	1.31	-0.49	-0.66	-0.73	-0.77	-0.97
5	0.34	-1.32	-1.32	-1.29	-1.27	-1.05
6	0.80	-0.95	-0.99	-1.00	-1.00	-1.00
7	1.22	-0.30	-0.48	-0.58	-0.64	-0.95
8	1.09	1.14	1.10	1.09	1.07	1.01
9	0.55	0.74	0.85	0.91	0.93	0.99
10	0.05	-1.09	-1.10	-1.09	-1.08	-1.01
11	1.11	0.99	1.02	1.04	1.04	1.01
12	0.72	-0.94	-1.04	-1.07	-1.07	-1.02
13	-0.46	-1.07	-1.05	-1.04	-1.03	-1.00
14	1.32	1.31	1.19	1.14	1.12	1.01

10, using the same two plans as the criterion pair. The results are remarkably similar to those presented here, hence are not shown.

13.6 GENERALIZED FORCED CLASSIFICATION

The discussion in the previous sections is on the original formulation of forced classification as applied to different types of data. The idea was then extended to a more general case in two respects: the weight k can take not only a large value for analysis, but also a fraction if the effect of a certain variable needs to be supressed, and the criterion for forced classifcation may be a set of variables or sets of variables from the rows and the columns of a two-way table. This extension is referred to as generalized forced classification. Although there are yet many unknown characteristics of this generalization, Day (1989) has successfully applied it to the classification of a large number of subjects into 16 personality types, using the Myers-Briggs Type Indicator. In analyzing responses to this personality inventory, Day augmented the data matrix with "ideal" subjects and their responses. In other words, theory-driven responses were added to the data matrix, and those ideal subjects were used as the criterion set for forced classification. The matrix of intersubject correlations then revealed very clear patterns, every subject showing an exceptionally high correlation coefficient to only one of the ideal subjects, indicating which personality type the person belongs to.

Since generalized forced classification covers a wide variety of applications, and its full potential cannot be adequately discussed here. Interested readers are referred to Nishisato (1986b), and Gibson 1993). Let us move to another equally important topic of graphical display.

CHAPTER 14
Graphical Display

14.1 SPECIAL PROBLEMS

French correspondence analysis is often described as a graphical technique for data analysis.[1] In a review of Nishisato's 1980 book, Tenenhaus (1982) was critical about the paucity of graphical analysis in the book, stating that "Thus the simultaneous graphical representation of rows and columns of a contingency table and indices, which characterize the quality of the representation, are standard products of correspondence analysis. Is it possible to imagine a multidimensional scaling without a mapping?" (p. 120). Granting that graphical display is almost indispensable for interpretation of the quantification results, one should nevertheless be aware of some potential pitfalls and logical difficulties associated with graphical display.[2] During the past several years, a number of relevant papers were published, in particular, a proposal known as CGS scaling by Carroll, Green, and Schaffer (1986, 1987, 1989), strong criticism of CGS scaling by Greenacre (1989), a proposal of measures of "badness of graphical display" by Nishisato (1988a), and a proposal of another method of graphical display (Nishisato, 1990). In the 1988 article, Nishisato also stated that CGS scaling does not offer a solution to the problem.

What appears to be the major problem is the fact that the fundamental problems inherent in joint graphical display are neither clearly delineated

---

[1] For example, Tenenhaus (1982), Greenacre (1984), and Hoffman and Franke (1986).

[2] For example, "it is legitimate to interpret distances among elements of one set of points. It is also ligitimate to interpret the relative positions of one point of one set with respect to *all the points* of the other set. Except in special cases, it is extremely dangerous to interpret the proximity of *two* points corresponding to different sets of points" (Lebart, Morineau, and Warwick, 1984, p. 46). In this quotation, two sets mean row weights and column weights.

nor well understood in general. It would, therefore, be instructive to look at the important problems associated with graphical presentation of quantified results.[3]

14.1.1 Incidence Data and Similarity Between Row and Column Spaces

Suppose that the data in Table 14.1 can be perfectly accommodated in two dimensions. Suppose further that those data can be plotted as shown in Figure 14.1. Subject 1, indicated by y_1 , is located at the centroid of the three options that he or she has chosen, and Option 1, x_1, is positioned at the centroid of the three subjects who have chosen it, that is, y_1, y_3, and y_4.

TABLE 14.1
Hypothetical Two-Dimensional Data

	x_1	x_2	x_3	x_4	x_5	x_6	x_7	x_8	x_9
y_1	1	0	0	1	0	0	1	0	0
y_2	0	0	1	0	1	0	0	0	1
y_3	1	0	0	0	1	0	0	1	0
y_4	1	0	0	0	0	1	0	0	1

Let y_{ik} be the score of Subject i on dimension k, and x_{pk} the weight of Option p on dimension k. Then, their relation can be expressed as

(14.1) $$y_{ik} = \sum_{p=1}^{m} \frac{f_{ip}x_{pk}}{f_{i\cdot}}, \qquad x_{pk} = \sum_{i=1}^{N} \frac{f_{ip}y_{ik}}{f_{\cdot p}}.$$

This is a *perfect reciprocal averaging* relationship between row weights, y_i, and column weights, x_p. In other words, y_i is the average of x_p, and x_p is the average of y_i on each dimension. In this case, one can state that y_i and x_p belong to the same space.

Such a relation between y_i and x_p as above, however, does not generally hold. In fact, dual scaling is a technique to approximate a perfect reciprocal averaging relationship as much as possible. Equivalently to this approximation, dual scaling can be stated as a method to determine those option-weights x_1, x_4, and x_7 as close to y_1 as possible, and scores y_1, y_3, and y_4 as close as possible to x_1 in Figure 14.1. However, such a configuration

[3] There are relevant and important studies on Gabriel's biplot. Interested readers are referred to Gabriel (1971) and Bradu and Gabriel (1978).

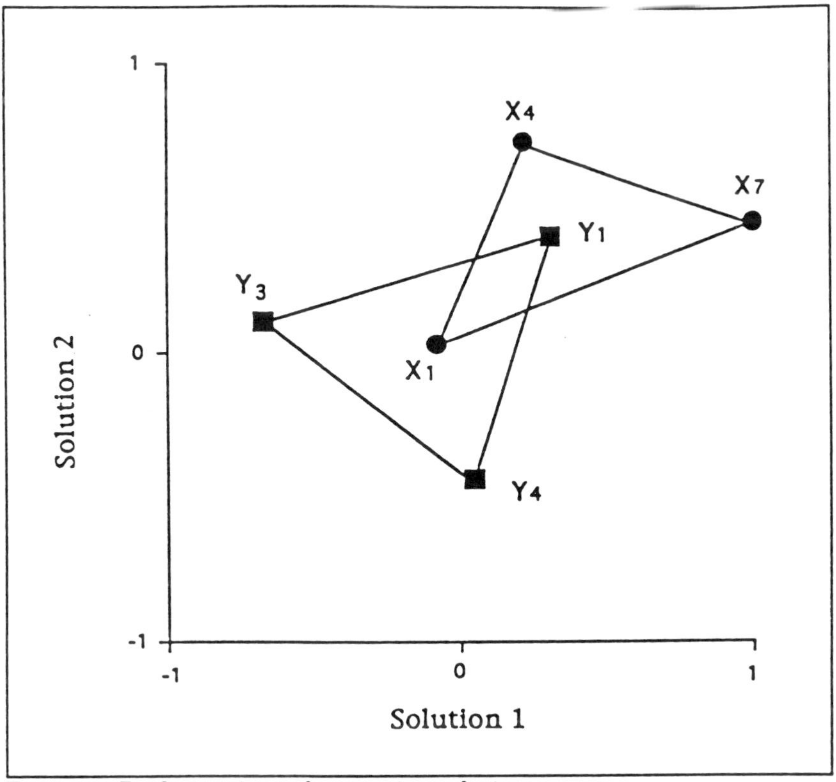

Fig. 14.1. Perfect reciprocal averaging relations.

as Figure 14.1 is not likely to materialize, for the perfect reciprocal averaging relation between y_i and x_p implies that y_1 must coincide with x_1, x_4, and x_7, and x_1 with y_1, y_3, and y_4. In other words, it is attained only when ρ = 1. This perfect correlation is attained only when y_1, y_3, y_4, x_1, x_4, and x_7 all occupy the same point, a case which would not require graphical display for interpretation of the results.

Table 14.2 shows a more general case of a perfect reciprocal averaging relation in the condensed format. Let us change the notation for option weight x_p from x_p to x_{jp} to indicate the weight for Option *p* of Item *j*, and also denote by f_{ijp} the response of Subject *i* to Option *p* of Item *j*. As shown in Figure 14.2, all the *related* points occupy a single point, a *sufficient*, but *not necessary*, condition for ρ to be 1. In this example, two solutions account for 100% of the total association ($\delta_1 = \delta_2 = 50\%$), and $\eta_1^2 = \eta_2^2 = 1$.

This example shows a perfect triangle, which is due not to perfect internal consistency, but to the fact that different response patterns appear with the same frequency. Thus, it may be instructive to look at another example of perfect internal consistency in two-dimensional space, in which

TABLE 14.2
Multiple-Choice Data of Perfect Internal Consistency in Condensed Format

Subject	Item 1	Item 2	Item 3
1	1	2	3
2	1	2	3
3	1	2	3
4	2	3	1
5	2	3	1
6	2	3	1
7	3	1	2
8	3	1	2
9	3	1	2

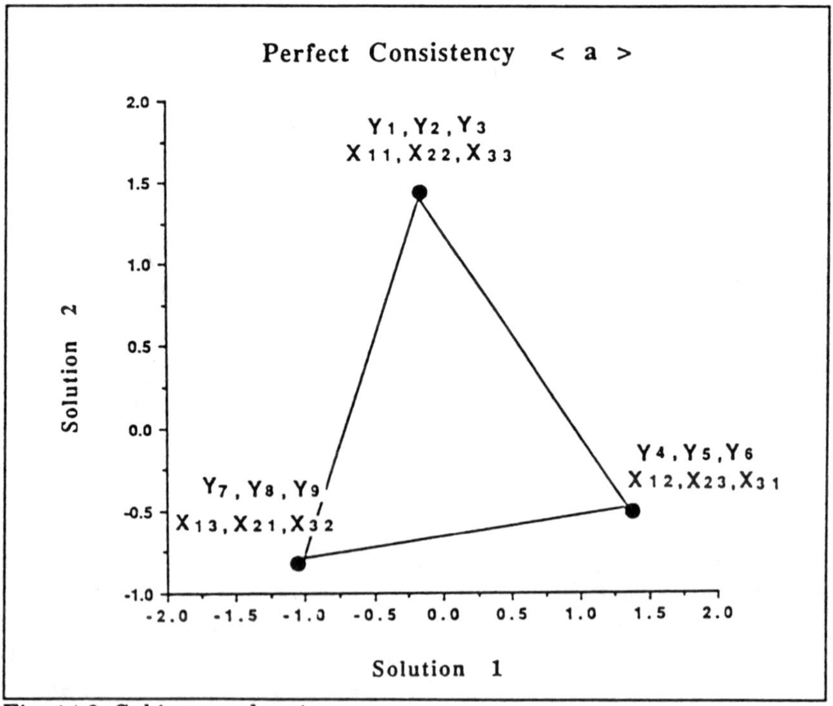

Fig. 14.2. Subjects and options.

different patterns appear with different frequencies. See Table 14.3 and Figure 14.3. You can see the effect of pattern frequencies on the locations of the points. This is an important point for understanding what dual

TABLE 14.3
Another Example of Perfect Internal Consistency in Two Dimensions

	Item		
Subject	1	2	3
1	1	2	3
2	2	3	1
3	2	3	1
4	2	3	1
5	2	3	1
6	2	3	1
7	3	1	2
8	3	1	2
9	3	1	2

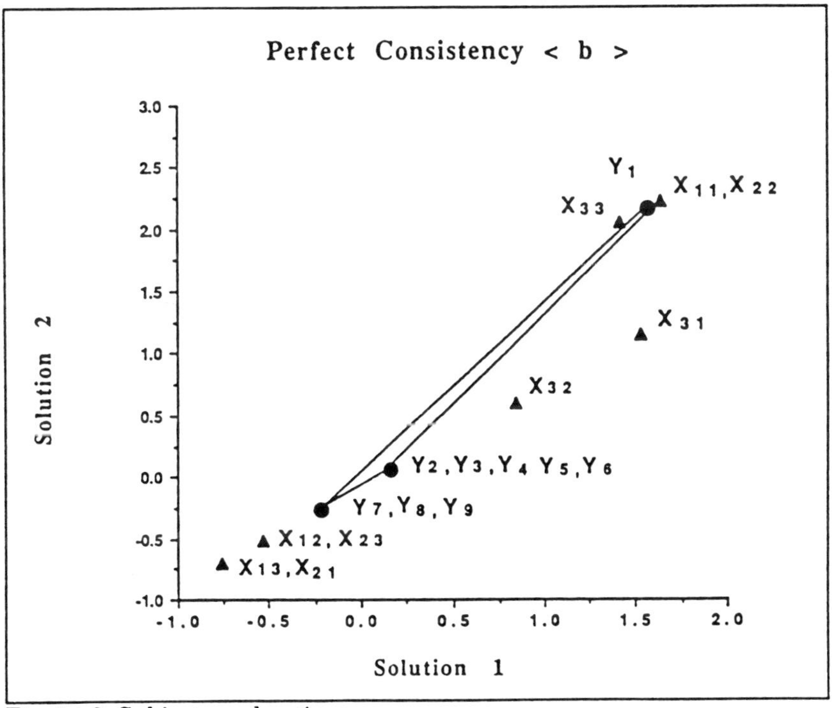

Fig. 14.3. Subjects and options.

scaling does with incidence data, and it would be useful to look at it right now.

Incidence data are frequency data even if they are coded only 1 or 0, and are influenced by frequencies in the process of quantification. In contrast, dominance data are based on an equal number of responses in each cell of a data matrix, that is, $(n - 1)$ responses as you would recall, and hence are not subject to the effect of response frequencies. Look at Figure 14.3: y_1 is at the centroid (average) of x_{11}, x_{21}, and x_{33}, each of y_2, y_3, y_4, y_5, and y_6 is at the centroid of x_{12}, x_{23}, and x_{31}, and each of y_7, y_8, and y_9 is at the centroid of x_{13}, x_{21}, and x_{32}. Thus, in spite of the fact that $\rho = 1$ these points do not occupy the same place. Figure 14.3 suggests that a single solution can capture the locations of subjects and options. Yet dual scaling tells us otherwise: $\rho_1 = 1$, $\delta_1 = 50\%$, $\rho_2 = 1$, and $\delta_2 = 50\%$. This peculiar outcome seems to be related to the fact that δ_1 is not 100% when $\rho_1 = 1$. Another strange point about this particular case is that the options of Item 3 have a smaller total sum of squares over the two solutions than those of other items. This is unexpected, and these two points appear to be causing the unexpected outcome. The clarification of the second point has yet to be carried out.

Now let us get back to reality. We cannot expect our real data to show perfect internal consistency. What we will see is more like a data set as shown in Table 14.4 and a two-dimensional graph, based on the first two solutions (Figure 14.4).

TABLE 14.4
Sample Data for Graphical Display

	Item		
Subject	1	2	3
1	1	2	2
2	2	1	1
3	2	2	1
4	1	2	2
5	2	1	2
6	1	2	2

Joint display of y_i and x_{jp} is the traditional procedure of French correspondence analysis, and it is indeed the most popular way of graphical display. However, the fact is that the space for y_i is different from the space for x_{jp}. For example, observe that y_3 is not at the centroid of x_{12}, x_{22}, and x_{31}, and that x_{31} is not in the middle of y_2 and y_3 who chose it. Figure 14.4 is an overlay of two different spaces. Suppose that, as some people say, we should not calculate the distances between y_i and x_p in the graph. Can we then infer from their closeness in the graph accurate relations between y_i and x_{jp}? Can we draw a correct conclusion on their relation on

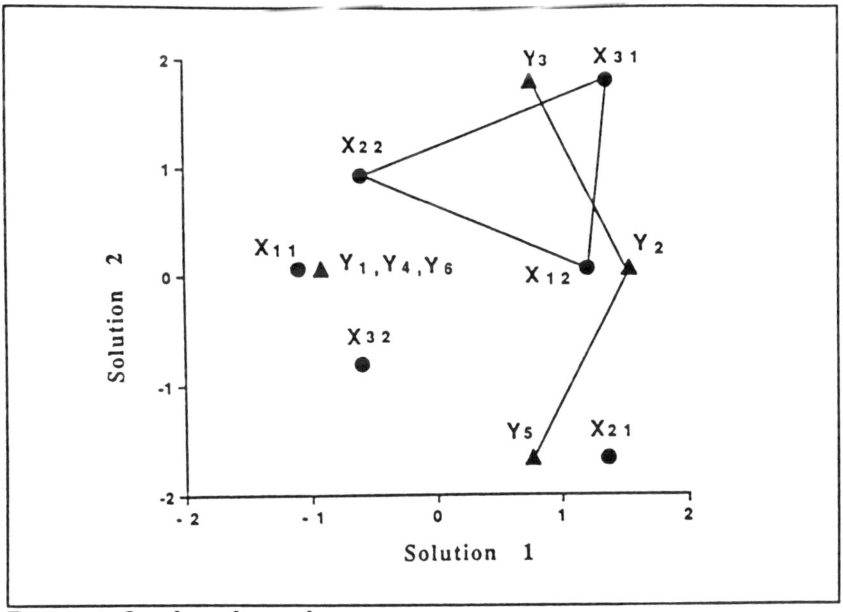

Fig. 14.4. Overlay of two plots.

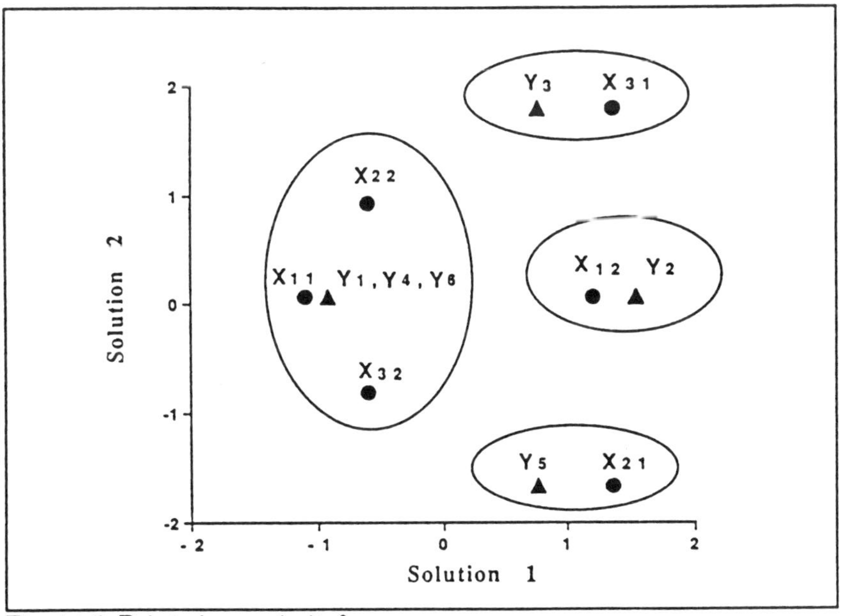

Fig. 14.5. Points in proximity?

the basis of the joint display of y_i and x_{jp}? Let us look at the same configuration as Figure 14.4, and enclose those closely located data points as in Figure 14.5. Who in the world could guess from this graph that Subject 2, for example, chose not only x_{12} but also x_{21} and x_{31}? Similarly, it is very unlikely that one can relate x_{12} to y_2, y_3, and y_5, isn't it?

The perfect reciprocal averaging relation between y_i and x_{jp} does notin general hold for real data. In other words, the formulas of 14.1 are two unrelated equations. In contrast, simultaneously derived equations from a single data set are *dual relations*, as mentioned earlier, that is,

$$(14.2) \qquad y_i = \frac{1}{\rho}\sum_{j=1}^{n}\sum_{p=1}^{m_j}\frac{f_{ijp}x_{jp}}{f_{i\cdot}}, \qquad x_{jp} = \frac{1}{\rho}\sum_{i=1}^{N}\frac{f_{ijp}y_i}{f_{\cdot jp}}.$$

These equations hold for any data matrix $\mathbf{F} = (f_{ip})$. Alternative expressions of dual relations are

$$(14.3) \qquad \rho y_i = \sum_{j=1}^{n}\sum_{p=1}^{m_j}\frac{f_{ijp}x_{jp}}{f_{i\cdot}}, \qquad \rho x_{jp} = \sum_{i=1}^{N}\frac{f_{ijp}y_i}{f_{\cdot jp}}$$

or, in matrix notation,

$$\rho\mathbf{y} = \mathbf{D}_r^{-1}\mathbf{F}\mathbf{x}\text{ , and } \rho\mathbf{x} = \mathbf{D}_c^{-1}\mathbf{F}'\mathbf{y}\ .$$

There are two important points to note about dual relations. First, the perfect reciprocal averaging relation is attained when $\rho = 1$, an improbable case and an uninteresting situation from the visual-presentation point of view. Secondly, $\rho\mathbf{y}$ is a projection of $\mathbf{y}$ onto the subspace spanned by $\mathbf{x}$, and $\rho\mathbf{x}$ that of $\mathbf{x}$ onto the subspace of $\mathbf{y}$. In other words, $\rho\mathbf{y}$ and $\mathbf{x}$ span the same space, and $\mathbf{y}$ and $\rho\mathbf{x}$ belong to the same space. Figure 14.6 shows plots of ($\rho\mathbf{y}$, $\mathbf{x}$) and ($\mathbf{y}$, $\rho\mathbf{x}$). It is clear that in the plot of ($\rho\mathbf{y}$, $\mathbf{x}$), $\rho\mathbf{y}$ is at the centroid of $\mathbf{x}$, and that in the plot of ($\mathbf{y}$, $\rho\mathbf{x}$), $\rho\mathbf{x}$ is at the centroid of $\mathbf{y}$, showing a common space in each case. Thus, Figure 14.6 poses no problem as far as the common space is concerned. But, upon close scrutiny, it is clear that the projected points lie closer to the origin of the graph than the nonprojected weights. This is due to the well-known fact that the norm of a projection is generally smaller than that of a nonprojected variable in Euclidean space. This difference in the norm between the two sets of weights, therefore, leaves the joint plot of ($\rho\mathbf{y}$, $\mathbf{x}$) or that of ($\mathbf{y}$, $\rho\mathbf{x}$) unsatisfactory from the interpretation point of view.

In conclusion, the joint plot of y and x, though commonly used, is theoretically unsound and practically misleading. The graph often fails even to relate the plot of variables to the input data, as demonstrated earlier in the inference of which options a subject may have chosen. The joint plot of ($\rho\mathbf{y}$,$\mathbf{x}$), or ($\mathbf{y}$, $\rho\mathbf{x}$) poses no problems of disparate spaces, but it

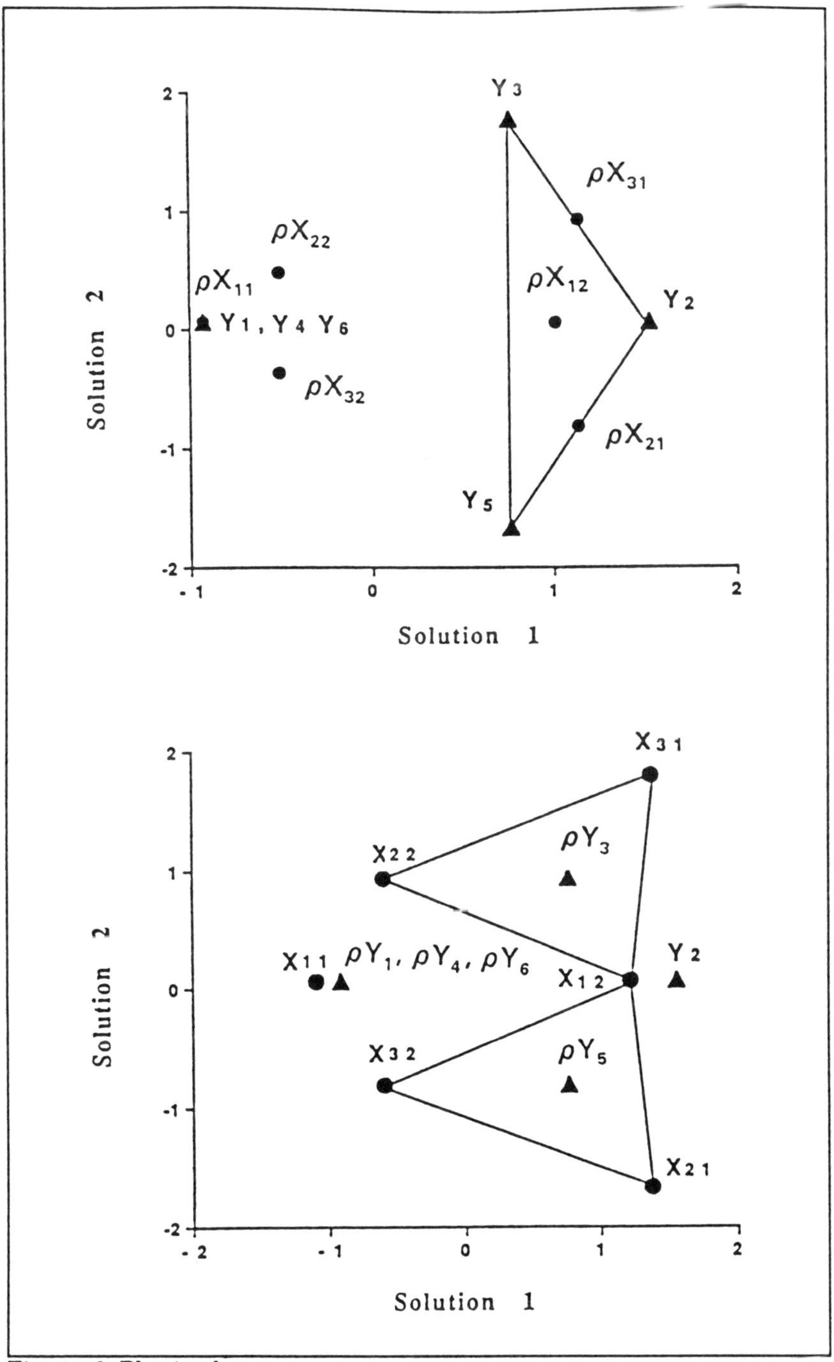

Fig. 14.6. Plot in the same space.

is practically unsatisfactory since the two sets of weights are not directly comparable due to the discrepancy in their norms. Thus, whether you plot **y** and **x**, or (ρ**y**, **x**), or (**y**, ρ**x**), joint graphical display of rows and columns is unsatisfactory. What alternatives do we then have? This question is discussed in Section 14.3.

14.1.2 Joint Plot of Dominance Data

Scale values for the columns (stimuli) of a dominance matrix are adjusted to the zero origin and a constant unit, while weights for the rows (subjects) have an arbitrary origin for each dimension. Thus, the weights for the columns are similar to those for rows and columns of an incidence matrix, and weights for the rows are nothing but quantities which indicate the relative importance of subjects' response patterns in determining maximally discriminative values for the columns (stimuli). Thus, if all the subjects agree more or less about the ranking of crimes in terms of seriousness, for instance, all the weights for the subjects are positive and close to 1. This is definitely not the case with weights for the columns of a dominance matrix, which are summed to zero. Values given to the rows are "weights" and those to the columns are estimates of "location parameters." Since these two sets of scale values are generated from the original ordinal measurements, our question is whether or not the joint graph still retains ordinal information contained in the input data. Let us examine this question, using small examples. Consider the following subject-by-stimulus rank-order data:

$$\mathbf{F}_1 = \begin{bmatrix} 1 & 2 & 3 & 4 & 5 \\ 1 & 2 & 3 & 4 & 5 \\ 1 & 2 & 3 & 4 & 5 \\ 1 & 2 & 3 & 4 & 5 \end{bmatrix}.$$

This can be exhaustively explained by one solution, as you can see from the identical ranking of stimuli by all the subjects. Thus, $\mathbf{F}_1$ offers a good starting point for discussion. Results of dual scaling of $\mathbf{F}_1$ are:

$\rho_1 = 0.5000$, $\delta_1 = 100\%$ (i.e., one-dimensional data)
$y_1 = y_2 = y_3 = y_4 = 1.0000$ (ρy_i are all equal to 0.7071)
$x_1 = 1.4142$, $x_2 = 0.7071$, $x_3 = 0.0000$, $x_4 = -0.7071$, $x_5 = -1.4142$
$\rho x_1 = 1.0000$, $\rho x_2 = 0.5000$, $\rho x_3 = 0.0000$, $\rho x_4 = -0.5000$, $\rho x_5 = -1.0000$.

The results show two important points. First, while the stimulus values are widely spaced with zero origin, the normed weights for the subjects are all equal to 1, that is, the origin is not zero. Second, the original ranking of stimuli by Subject i can be generated from the rank order of the proximity of the subject's position, y_i, to individual stimuli projected onto the subspace of y, ρx, that is, the absolute differences $|\rho x_j - y_i|$. In the current example,

$$|\rho x_1 - y_1| = 0.00\ ,\ |\rho x_2 - y_1| = 0.50\ ,\ |\rho x_3 - y_1| = 1.00\ ,$$
$$|\rho x_4 - y_1| = 1.50\ ,\ |\rho x_5 - y_1| = 2.00, \text{ so that}$$
$$k_{11} = 1,\ k_{12} = 2,\ k_{13} = 3,\ k_{14} = 4, \text{ and } k_{15} = 5.$$

Thus, if we consider ρx_j as the stimulus scale value and y_i the location of Subject i (i.e., "ideal point i" in Coombs' unfolding technique; Coombs, 1964), unidimensional rank order data can be interpreted in the same way as Coombs' unfolding technique for rank order data. Indeed, dual scaling provides a practical method to solve Coombs' problem, namely, dual scaling determines ρ, y_i, and x_j in such a way that

$$(14.5) \qquad |\rho x_j - y_i| \prec |\rho x_r - y_i| \text{ whenever } k_{ik} \prec k_{ir}\ .$$

This is very encouraging, but what will happen when δ_1 is not 100%?

We can speculate on this question from two different points of view. First, unlike incidence data, dominance data are collected in such a way that each entry of a data matrix is based on the same number of responses (see chapters 10, 11, and 12), provided that there are no missing responses. Thus, dominance data do not have to suffer such consequences of uneven response frequencies as we have observed in the previous sections. This is a positive aspect for the question. Second, however, as we have witnessed (and see again in chapter 17), the value of ρ for dominance data can be considerably lower than its absolute maximum of 1 even when δ_1 is 100%. The maximum of 1 is attained when the number of stimuli is two, and the value declines as the number of stimuli increases. As we will see later in this chapter, the discrepancy between row space and column space widens as the value of ρ becomes smaller, and common sense tells us that expected distance relations between row i and column j of dominance data will become increasingly more difficult to obtain as the number of stimuli increases—this would be so even if we could maintain the same value of ρ. Thus, this second view is a negative aspect for the question.

As may be obvious, it should be remembered that Formula 14.5 as applied to multidimensional analysis is expected to hold true, not with respect to each solution, but to the distance relations in the multidimensional space. Let us look at two-dimensional cases, $\mathbf{F}_2$ and $\mathbf{F}_3$:

$$\mathbf{F}_2 = \begin{bmatrix} 1 & 2 & 3 & 4 & 5 & 6 & 7 & 8 & 9 \\ 1 & 2 & 3 & 4 & 5 & 6 & 7 & 8 & 9 \\ 1 & 2 & 3 & 6 & 5 & 4 & 7 & 8 & 9 \\ 1 & 2 & 3 & 6 & 5 & 4 & 7 & 8 & 9 \end{bmatrix}, \quad \mathbf{F}_3 = \begin{bmatrix} 1 & 2 & 3 & 4 & 5 & 6 & 7 & 8 \\ 1 & 2 & 3 & 4 & 5 & 6 & 7 & 8 \\ 8 & 1 & 7 & 2 & 6 & 3 & 5 & 4 \\ 8 & 1 & 7 & 2 & 6 & 3 & 5 & 4 \end{bmatrix}.$$

To restrict the dimensionality to two, two sets of identical response patterns are used in both cases. One is a case in which the first solution is dominant ($\delta_1 = 95.24\%$, $\delta_2 = 4.76\%$), and the second example is a case in which both solutions have similar contributions ($\delta_1 = 59.52\%$, $\delta_2 = 40.48\%$). For $\mathbf{F}_2$, $\mathbf{y}$ and $\rho\mathbf{x}$ for the two solutions are as follows:

$$(\mathbf{y}_1 \quad \mathbf{y}_2) = \begin{bmatrix} 1 & 1 \\ 1 & 1 \\ 1 & -1 \\ 1 & -1 \end{bmatrix}, \quad (\rho_1\mathbf{x}_1 \quad \rho_2\mathbf{x}_2) = \begin{bmatrix} 1.0000 & 0.0000 \\ 0.7143 & 0.0000 \\ 0.4286 & 0.0000 \\ -0.1429 & 0.2857 \\ -0.1429 & 0.0000 \\ -0.1429 & -0.2857 \\ -0.1429 & 0.0000 \\ -0.7143 & 0.0000 \\ -1.0000 & 0.0000 \end{bmatrix}.$$

You can see that the difference between two distinct rankings is looked after by Solution 2 by relocating Subjects 3 and 4 as well as Stimuli 4 and 6. For $\mathbf{F}_3$, $\mathbf{y}$ and $\rho\mathbf{x}$ for the two solutions are as follows:

$$(\mathbf{y}_1 \quad \mathbf{y}_2) = \begin{bmatrix} 1 & 1 \\ 1 & 1 \\ -1 & 1 \\ -1 & 1 \end{bmatrix}, \quad (\rho_1\mathbf{x}_1 \quad \rho_2\mathbf{x}_2) = \begin{bmatrix} 1.0000 & 0.0000 \\ -0.1428 & 0.8572 \\ 0.5714 & -0.1429 \\ -0.2857 & 0.4286 \\ 0.1428 & -0.2857 \\ -0.4286 & 0.0000 \\ -0.2858 & -0.4285 \\ -0.5717 & -0.4285 \end{bmatrix}.$$

This case looks complicated, and it would be best to look at it in graph form. Figure 14.7 shows what happens to the unidimensional model in two dimensions. It is remarkable that the stimulus ranking by each subject can be correctly reproduced from Figure 14.7, even in the case of $\mathbf{F}_3$. It is

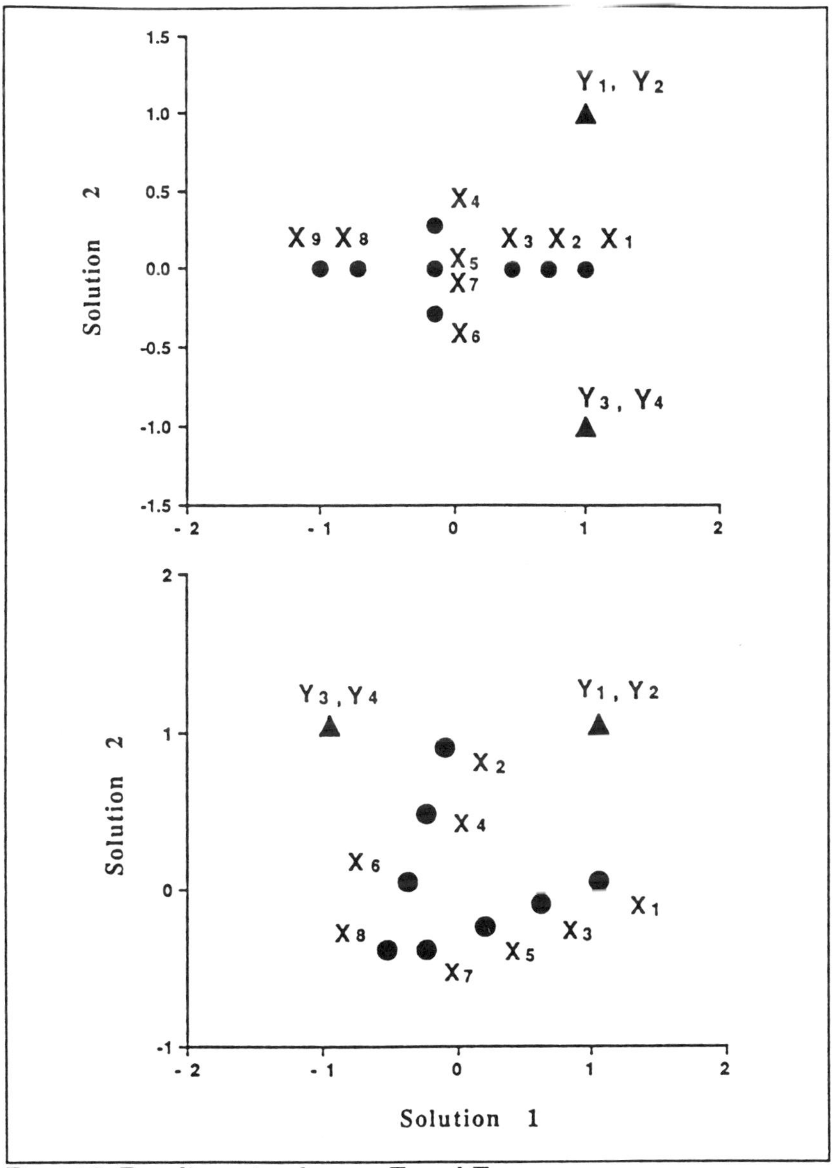

Fig. 14.7. Two-dimensional cases, $\mathbf{F}_2$ and $\mathbf{F}_3$

important to note, however, that an almost negligible solution such as the second solution of $\mathbf{F}_2$ is necessary to reproduce the stimulus ranking correctly. It is also important to note that the plot of $\mathbf{y}$ and $\mathbf{x}$, rather than that of $\mathbf{y}$ and $\rho\mathbf{x}$, would not reproduce the correct rank-order of stimuli.

It seems that paired comparison and rank-order data are two types of dominance data, for which dual scaling can be considered as a multidimensional extension of Coombs' unfolding technique. When we look at successive categories data, however, even if all the possible solutions were extracted it would generally not be possible to reproduce input data exactly. The fact that a subject places, for example, three stimuli in one category means that the stimuli will be given the same rank in our formulation. However, it is typically the case that those stimuli are given different locations in multidimensional space because other subjects would not necessarily put them into a single category. In this regard, we must be satisfied with the idea that the current formulation provides a least-square type approximation to the correct ranking of each individual subject. Further clarification of this problem, and possibly a new formulation of dual scaling for successive categories data are left for future work.

14.1.3 Shape of the Total Space

Dual scaling is also called principal component analysis of categorical data (PCAC) (Torgerson, 1958), as opposed to PCA, which is formulated for continuous variables (Hotelling, 1933). It seems instructive to look at PCA of standardized continuous variables, to find out the shape of the total space. PCA is to find a set of linear combinations of variables that provide the maximal variance, and it amounts to solving an eigenequation of the correlation matrix, **R**. If the rank of **R** is 2, all the variables can be fully represented in two-dimensional space, with each variable being located somewhere on the circle with the diameter of 1. In other words, the variance of 1 of each variable is decomposed into two orthogonal components as shown in Figure 14.8

Similarly, if the rank of **R** is three, all the variables are located on the surface of a sphere at a distance of 1 from the origin. Thus, no matter what the dimensionality of the total space may be , each variable in the principal hyperspace is located at a distance of 1 from the origin. This kind of homogeneity of the total space can be observed in the representation of dominance data, namely, the contribution of each stimulus to the total space is constant and equal to n - 1. This equality of contributions of the variables to the total space cannot in general be observed for incidence data (Nishisato, 1991). The jaggedness of the column space for multiple-choice data can be quite extensive, and deserves some scrutiny. To look into the matter, let us consider multiple-choice data as partitioned into items, that is, $\mathbf{F} = [\mathbf{F}_1, \mathbf{F}_2, ..., \mathbf{F}_n]$. As discussed in chapter 6, the total contribution of Option p of Item j increases as the number of subjects who choose the option decreases. Intuitively, this may not sound right, but it is a direct

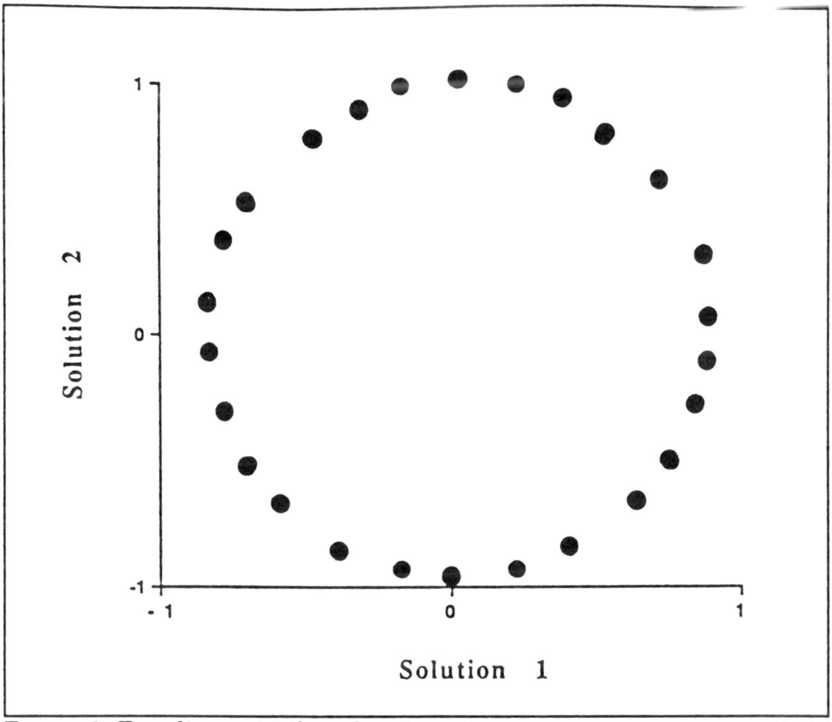

Fig. 14.8. Two-dimensional configuration.

outcome of removing the second trivial solution (see chapter 6 for the mathematical derivation of dual scaling). How drastically the jaggedness of the space shows up in data analysis can be illustrated by a numerical example as shown in Table 14.7. This is an example based on Tung's questionnaire (Table 14.5) and his data (Table 14.6). As is obvious, those options chosen by one subject or two have exceptionally large sums of squares, meaning that these options are likely to play greater roles than the other options in explaining data structure. Table 14.7 also shows the contibutions of items to the total sum of squares. It can clearly be seen that the contribution of an item is proportional to the number of options of the item: the greater the number of options of an item, the greater the total sum of squares of the item. Generally speaking, therefore, it is not solely the problem of discrepant row and column spaces, but also that of the jaggedness of the space that make row-column joint graphical display of multiple-choice data unfit for interpretation.

TABLE 14.5
Tung's Eleven-Item Questionnaire

(1) Your sex: 1 = male; 2 = female.

(2) Your school: 1 = MSS; 2 = YLP; 3 = SS; 4 = SKW.

(3) Your class: 1 = grade 10; 2 = grade 12.

(4) After graduation from high school, will you use English to talk to others?
1 = most definitely; 2 = very likely; 3 = likely; 4 = unlikely; 5 = don't know.

(5) After graduation from your school, will you use English to talk to others at home?
1 = yes; 2 = no; 3 = don't know.

(6) After graduation from your school, will you use English to talk with others at school gathering?
1 = yes; 2 = no; 3 = don't know.

(7) After graduation from your school, will you use English to talk with others at your job?
1 = yes; 2 = no; 3 = don't know.

(8) After graduation from your school, will you use English to talk with others in pursuing your future studies?
1 = yes; 2 = no; 3 = don't know.

(9) Please give a rough estimate of the total family income (i.e., that earned by people who support and live with you).
1 = $10,000 or more; 2 = between $5,000 and $9,999; 3 = between $3,000 and $4,999; 4 = between $1,500 and $2,999; 5 = below $1,500.

(10) How many people have you counted in Question 9 in calculating the total monthly income of your family?
1 = one person; 2 = two persons; 3 = three persons; 4 = four or more persons.

(11) Would you say the main wage earner of your family is
1 = white-collar, or 2 = blue-collar worker?

TABLE 14.6
Tung's Questionnaire and Data

	Question (Item)						11 Questions				
Ss	1	2	3	4	5	6	7	8	9	10	11
1	2	1	1	1	2	1	1	1	2	2	1
2	2	1	1	1	1	1	1	1	1	1	1
3	2	1	1	1	2	1	1	1	1	1	1
4	2	1	1	2	2	2	1	1	2	2	1
5	2	1	1	1	2	1	1	1	1	3	1
6	2	1	2	2	2	3	1	1	2	2	1
7	2	1	2	1	2	1	1	1	2	1	1
8	2	1	2	1	3	3	1	1	2	2	1
9	2	1	2	2	2	1	1	1	2	1	1
10	2	1	2	2	2	1	1	1	2	2	1
11	1	2	2	1	2	3	1	1	4	2	2
12	1	2	2	2	1	1	1	1	2	1	1
13	1	2	2	2	3	1	1	1	5	1	2
14	1	2	2	3	2	3	1	1	4	2	2
15	1	2	2	2	2	3	3	1	1	3	2
16	1	2	1	1	2	3	1	1	5	1	2
17	1	2	1	3	1	3	3	2	5	2	2
18	1	2	1	1	2	1	1	1	2	1	1
19	2	2	1	1	2	1	1	1	2	1	1
20	2	2	1	4	2	2	2	1	2	4	1
21	2	3	1	3	2	1	1	1	4	3	2
22	2	3	1	3	2	1	1	1	3	2	2
23	2	3	1	5	2	1	1	2	3	2	2
24	2	3	1	2	2	1	1	1	2	2	1
25	2	3	1	2	1	1	1	1	2	3	2
26	2	3	2	1	3	1	1	1	3	4	1
27	2	3	2	2	2	1	1	1	3	3	2
28	2	3	2	1	1	1	1	1	4	1	2
29	2	3	2	1	3	1	1	1	2	4	1
30	2	3	2	5	2	3	3	1	2	4	1
31	1	4	1	1	1	1	1	1	2	3	1
32	1	4	1	2	3	3	1	1	3	2	1
33	1	4	1	1	3	1	1	1	5	1	1
34	2	4	1	1	3	1	1	1	3	2	1
35	2	4	1	1	1	1	1	1	3	2	1
36	2	4	2	1	2	1	1	1	3	2	1
37	2	4	2	1	1	1	1	1	3	1	1
38	2	4	2	1	3	1	1	1	2	2	1
39	1	4	2	1	3	1	1	1	3	1	2
40	1	4	2	2	2	1	1	1	1	2	1
41	2	5	2	3	2	1	1	1	3	4	2
42	1	5	2	2	2	1	3	1	4	2	2
43	2	5	2	3	2	1	1	1	4	3	1
44	1	5	2	2	2	1	1	1	5	1	2
45	2	5	2	1	1	1	3	1	3	3	1
46	2	5	1	1	2	1	1	1	2	3	1
47	2	5	1	1	1	1	1	1	2	3	1
48	2	5	1	1	3	1	1	1	4	3	2
49	2	5	1	3	2	1	1	1	4	3	2
50	2	5	1	2	2	1	1	1	5	1	2

Table 14.7
Jagged Total Non-Trivial Space of Twenty-Five Dimensions
Example: Data from Peter Tung (N=50, m=36, n=11)

Item	Option	f_{jp}	$\Sigma f_{jp} x^2_{jp}$	Σx^2_{jp}	x_1	SS(item)
8	2	2	528	264.0	-2.08	550
	1	48	22	0.5	.09	
1	1	15	385	25.7	-1.21	550
	2	35	165	4.7	.52	
11	2	19	341	17.9	-1.34	550
	1	31	209	6.7	.92	
3	1	25	275	11.0	.31	550
	2	25	275	11.0	-.31	
7	2	1	539	539.0	10.05	1,100
	3	5	495	99.0	-1.38	
	1	44	66	1.5	-.07	
6	2	2	528	264.0	6.65	1,100
	3	9	451	50.1	-.93	
	1	39	121	3.1	-.13	
5	1	10	440	44.0	-.39	1,100
	3	10	440	44.0	-.09	
	2	30	220	7.3	.16	
10	4	5	495	99.0	2.64	1,650
	3	12	418	34.8	-.52	
	1	15	385	25.7	-.29	
	2	18	352	19.6	-.14	
4	4	1	539	539.0	10.05	2,200
	5	2	528	264.0	-.07	
	3	7	473	67.6	-1.41	
	2	15	385	25.7	-.25	
	1	25	275	11.0	.15	
9	1	5	495	99.0	.03	2,200
	5	6	484	89.7	-1.65	
	4	8	462	57.8	-1.42	
	3	12	418	34.8	-.28	
	2	19	341	17.9	1.29	
2	1	10	440	44.0	1.11	2,200
	2	10	440	44.0	-.06	
	3	10	440	44.0	-.06	
	4	10	440	44.0	-.13	
	5	10	440	44.0	-.86	

14.2 BADNESS OF JOINT GRAPHICAL DISPLAY

If we were to overlay two different graphs, it would be of interest to see the extent to which the resultant graph fails to fulfil our optimism. In Section 14.1.1, we saw displacement effects of ρ on the positions of between-set points. We also noted that the triangle in the example used becomes smaller as the value of ρ increases. In Chapter 8, we observed that ρ^2 is the average of squared item-total correlations r_{jt}^2, and that r_{jt}^2 is proportional to the sum of the squares of the weighted responses of item j. Considering all these relations, it is not difficult to derive a few measures that indicate how poorly the joint graph represents configurations of variables lying in two disjointed spaces. Let us consider the following measures of the badness of joint display of multiple-choice data.[4]

14.2.1 Discrepancy Between Symmetric and Nonsymmetric Scaling

By symmetric scaling, we refer to **y** and **x** that are normed to the same constant: in this book, $\mathbf{x}'\mathbf{D}_c\mathbf{x} = \mathbf{y}'\mathbf{D}_r\mathbf{y} = f_t$. In contrast, under nonsymmetric scaling, we calculate **y** and ρ**x**, or ρ**y** and **x**, in which the two vectors in each pair have different norms. In the previous section, ρ**y** was shown to be at the centroid of **x**, and ρ**x** at the centroid of **y**. This ideal relation of reciprocal means, however, does not hold between **x** and **y** unless ρ is 1, and as the value of ρ decreases, the discrepancy between the space for **x** and that for **y** widens. The first measure of badness of joint display is based on this discrepancy:

$$\delta_{1ik} = (1-\rho_k)^2 y_{ik}^2 \tag{14.6}$$

$$\delta_{1pk} = (1-\rho_k)^2 x_{pk}^2 \;, \tag{14.7}$$

where y_{ip}, x_{pk}, and ρ_k are respectively the ith and the pth elements of **y** and **x** of Solution k, and ρ_k the square root of the kth value of ρ^2. This measure δ_1 is a straightforward statistic of the discrepancy between the complete reciprocal averaging relations and dual relations. The French school uses the concept of inertia to indicate the contribution of a variate to the total variance, which can be expressed as $g_i y_{ik}^2$, and $f_p x_{pk}^2$, where g_i and f_p are marginals of row i and column p, respectively. Since "badness" can be

[4] These measures were proposed by Nishisato (1988a).

considered opposite to inertia, we may incorporate the concept of inertia in measures of badness of joint display. Consider multiple-choice data, and let N be the number of subjects, n the number of items, f_{jp} the frequency of Option p of Item j, and m_j the number of options of Item j. Consider a weighting function for δ_{1pk} as follows: the weight for Option p of Item j, w_{jp}, is 1 if the frequency of Option p of Item j is equal to the expected value, that is, N/m_j; if the frequency is greater than N/m_j, the weight is less than 1; and if the frequency is smaller than N/m_j, the weight is greater than 1. Hence,

(14.8)
$$\delta_{2pk} = [1 - \frac{(f_{jp} - \frac{N}{m_j})}{N}]\delta_{1pk} = [1 + \frac{1}{m_j} - \frac{f_{jp}}{n}]\delta_{1pk} = w_{jp}\delta_{1pk}$$

As for row i, $g_i = n$, assuming no missing responses. Therefore, $w_i = 1$, and $\delta_{1ik} = \delta_{2ik}$. The measure of Formula 14.8 can be simplified by removing the effect of m_j from the measure, yielding

(14.9)
$$\delta_{3pk} = (1 - \frac{f_{jp}}{N})\delta_{1pk} \; .$$

14.2.2 Lack of Between-Set Representation

If y_1, for example, is a linear combination of x_a, x_b, and x_c, we would ideally like to see that these three values of x are the three closest points to y_1, out of all elements of $\mathbf{x}$. Similarly, if x_1 is chosen by Subjects 2, 5, and 9, then y_2, y_5, and y_9 should be the three closest points to x_1 out of all the elements of $\mathbf{y}$. To arrive at a measure of badness of joint display that incorporates this notion, we need to calculate the distance between y_i and x_p in K-dimensional space, weighted by ρ_k, by

(14.10)
$$d_{ip} = \sqrt{\sum_{k=1}^{K} \rho_k (y_{ik} - x_{pk})^2}$$

Since there are N subjects, there are N interpoint distances d_{ip} for each p. Arrange d_{ip} according to the order of their values. Out of f_{jp} subjects who chose Option p of Item j, it is likely that some of them are not among the f_{jp} closest to the location of Option p of Item j. Thus, in order to capture this unexpected outcome, let us indicate by n_{pj} the number of subjects out of f_{jp} who are among the f_{jp} closest persons to Option p of Item j. Then a measure of badness of option p of item j is given by

$$\delta_{4pj} = \frac{f_{jp} - n_{jp}}{f_{jp}} . \quad (14.11)$$

Those f_{jp} closest subjects to Option p of Item j are called the *satellite elements* of Option p, $p = 1, 2, \ldots, m$. Assuming no missing responses, the idea of δ_{4pj} can also be defined for Subject i as the proportion of n weights of those options chosen by Subject i which are not in the set of the n smallest distances out of m d_{ip},

$$\delta_{4i} = \frac{(n - n_i)}{n}, \quad (14.12)$$

where n_i is the number of those x's chosen by Subject i that belong to the set of the n smallest d_{ip}, $p = 1,2, \ldots, m$. When $\delta_4 = 0$, each point is generated by those which are located closest to the point, and when $\delta_4 = 1$, those close to a point are not the ones that define it. In practice, the distance may be calculated for dimension k, and we can define the two measures for each dimension, indicating them respectively by δ_{4pjk} and δ_{4ik}.

We can also define δ_4 in terms of interpoint distances as

$$\delta_{5q} = \frac{(d_q - d_{\min})}{(d_{\max} - d_{\min})}, \quad q = i, p, \quad (14.13)$$

where d_q is the average distance between point q and the other n_q points involved in calculations of point q, d_{min} is the average of the n_q smallest distances between point q and the other points, and d_{max} is the average of the n_q largest distances between point q and the others. The measure $\delta_5 = 0$ when n_q points describing point q are the n_q closest points; $\delta_5 = 1$ when n_q points describing point q are located among the farthest from point q. If δ_5 is defined for dimension k, it can be indicated by δ_{5pk} and δ_{5ik}.

14.2.3 Influence of Poor Items

The total inertia is the sum of ρ_k^2, and each ρ_k^2 can be expressed as

$$\rho_k^2 = \frac{\sum_{j=1}^{n} r_{jt}^2}{n} \quad (14.14)$$

Measures reflecting weak item-total correlation for Item j and Subject i, given dimension k, may be given respectively by

$$\delta_{6jk} = (1 - r_{jt}^2) x_{jk}^2 , \quad (14.15)$$

(14.16)
$$\delta_{6ik} = \frac{\sum_{j=1}^{n}\sum_{p=1}^{m_j}(1-r_{jt}^2)e_{ip}x_{jk}^2}{n},$$

where e_{ip} is the element of the ith row and the pth column of the response-pattern matrix, and is either 1 or 0. Since δ_6 resembles δ_1, we may consider another measure derived from both of them,

(14.17)
$$\delta_{7jk} = (1 + \frac{1}{m_j} - \frac{f_j}{N})\delta_{6jk},$$

or, alternatively,

(14.18)
$$\delta_{8jk} = (1 - \frac{f_j}{N})\delta_{6jk}.$$

14.2.4 Influence of Outliers

In chapter 8, we noted that an option with a smaller number of responses contributes more to r_{jt}^2. Thus, the instability of quantification may be captured by $(N - f_{jp})/N$, which, however, is also directly influenced by the number of options of Item j, m_j. To free it from the effect of m_j, we may propose the following:

(14.19)
$$\delta_{9p} = (\frac{N}{m_j} - f_{jp})\frac{m_j}{N} = 1 - \frac{m_j f_{jp}}{N}.$$

When responses are evenly distributed over options of an item, δ_9 is zero for all the options; when the number of responses in an option is greater than the average, δ_9 is negative. Note that unlike other measures δ_9 is defined for a given data set, and not for each item, option, or subject. The corresponding measure for the group of subjects is

(14.20)
$$\delta_{9i} = \frac{\sum_{j=1}^{n}\sum_{p=1}^{m_j} x_{pk}^2 e_{ij}(1 - \frac{m_j f_{jp}}{N})}{n}.$$

14.2.5 Average Instability

Since dual scaling minimizes the within-subject sum of squares and the within-option sum of squares, stability of each data point may be assessed by the following two measures:

(14.21)
$$\delta_{10ik} = \frac{x_k'(H_i - \frac{H_i 11' H_i}{n})x_k}{n},$$

(14.22)
$$\delta_{10pk} = \frac{y_k'(H_p - \frac{H_p 11' H_p}{f_{jp}})x_k}{f_{jp}},$$

where $\mathbf{H}_i$ and $\mathbf{H}_p$ are diagonal matrices of response-patterns of row i and column p, respectively. We can also assess tightness of the defining configuration of each point in k-dimensional space by

(14.23)
$$\delta_{11q} = d_q, \quad q = i, p,$$

where d is, as defined earlier, the average distance of satellite elements from Subject i or Option p. If we define average satellite distances for each dimension, we obtain δ_{11ik} and δ_{11pk}. Stability can also be expressed in terms of the variance of d as follows:

(14.24)
$$\delta_{12i} = \frac{\sum_{j=1}^{m}(d_{ip} - d_i)^2}{n_i},$$

(14.25)
$$\delta_{12p} = \frac{\sum_{i=1}^{N}(d_{ip} - d_p)^2}{f_{jp}},$$

where d_i and d_p are average distances. If these measures are defined for each dimension, then they can be denoted δ_{12ik} and δ_{12pk}.

All these measures are related to the badness or inappropriateness of visually represented data points. Some of them are correlated in various degrees with those other measures previously discussed. What seems certain is that these measures cannot be combined into a single measure of badness, for principal component analysis of those measurements from artificial data has shown five or six dominant components. Thus, it is recommended that these measures be used to alert the investigators about data elements that may require scrutiny before accepting the outcome at face value. These measures are presented here only tentatively, and further work on them is needed in order to put them into practical use. To give an idea of what these measures capture, Tables 14.8 and 14.9 show

TABLE 14.8
Sample Values of Measures of Badness for Subjects

Subj	δ_1	δ_4	δ_5	δ_6	δ_9	δ_{10}	δ_{11}	δ_{12}
1	2	18	0	7	-4	7	28	1
2	0	45	3	5	0	6	22	1
3	1	27	1	5	-1	5	22	1
4	19	36	13	58	98	112	91	40
5	1	27	1	5	0	6	22	1
6	0	54	7	8	-3	13	31	3
7	1	36	2	8	-5	9	29	1
8	1	54	6	8	-3	12	30	3
9	1	54	4	8	-5	9	29	1
10	1	54	3	7	-4	9	29	1
11	4	54	7	11	4	11	35	2
12	0	54	9	8	-2	12	22	6
13	4	63	7	11	5	10	37	1
14	7	45	7	12	6	12	36	6
15	5	45	6	10	7	10	34	3
16	4	63	8	12	6	14	38	2
17	18	27	6	27	22	16	46	13
18	0	45	7	7	-3	11	23	6
19	1	27	1	5	-5	6	23	1
20	182	18	4	249	604	468	237	48
21	2	81	10	9	3	15	38	2
22	0	72	10	6	2	11	29	2
23	1	72	11	15	11	16	36	4
24	1	36	2	5	-4	6	23	1
25	0	54	10	7	-3	12	23	6
26	1	45	7	16	10	20	36	7
27	0	54	5	4	0	6	20	2
28	1	63	7	8	1	10	30	2
29	4	36	6	18	6	21	42	6
30	1	45	13	21	10	34	45	13
31	0	54	10	8	-2	12	25	5
32	0	54	8	5	2	8	22	3
33	0	63	11	9	3	13	30	4
34	0	45	2	2	0	3	15	0
35	0	54	4	2	0	3	15	1
36	0	54	3	2	-1	3	14	1
37	0	54	5	2	0	4	15	1
38	0	54	4	4	-4	7	24	1
39	2	54	5	7	2	7	28	0
40	0	45	6	4	0	6	16	3
41	0	72	17	21	13	34	37	19
42	8	45	7	13	7	11	36	6
43	1	63	10	9	1	15	33	4
44	5	54	7	13	5	11	36	3
45	0	54	8	6	2	11	25	4
46	0	45	5	7	-4	10	24	4
47	0	54	8	7	-4	11	25	3
48	1	72	9	9	1	12	33	2
49	3	72	8	11	3	15	38	3
50	2	63	9	10	3	13	35	2

TABLE 14.9
Sample Values of Measures of Badness for Item Options

Op	δ_1	δ_2	δ_3	δ_4	δ_5	δ_6	δ_7	δ_8	δ_9	δ_{10}	δ_{11}	δ_{12}
1-1	9	10	6	40	11	35	42	24	40	8	33	5
1-2	1	1	0	17	30	6	5	1	-40	31	33	22
2-1	7	7	6	40	3	32	41	25	0	5	34	0
2-2	0	0	0	80	77	0	0	0	0	121	66	72
2-3	0	0	0	90	24	0	0	0	0	8	25	1
2-4	0	0	0	40	8	0	0	0	0	2	13	0
3-1	4	4	3	70	17	19	24	15	0	6	25	4
3-2	0	0	0	48	56	2	2	1	0	48	40	32
4-1	0	0	0	60	35	2	2	1	0	11	29	2
4-2	0	0	0	36	17	0	0	0	-150	6	20	2
4-3	0	0	0	66	36	0	0	0	-50	17	31	7
4-4	12	13	10	71	13	18	25	16	30	7	40	3
4-5	620	731	607	0	0	937	1387	918	90	868	249	620
5-1	0	0	0	100	13	0	0	0	80	14	26	0
5-2	0	1	0	80	25	4	5	3	40	10	25	4
5-3	0	0	0	46	72	0	0	0	-80	43	41	26
6-1	0	0	0	90	26	0	0	0	40	9	27	1
6-2	0	0	0	12	13	0	0	0	-134	7	24	1
6-2	271	351	260	0	0	560	817	537	88	199	164	199
7-1	5	6	4	55	25	11	14	9	46	17	35	9
7-2	0	0	0	6	17	0	0	0	-164	10	26	3
7-3	620	814	607	0	0	1037	1535	1016	94	868	249	620
8-1	11	14	10	40	14	19	27	17	70	20	41	13
8-2	0	0	0	2	66	0	0	0	-92	29	32	18
8-3	26	39	25	50	6	121	177	116	92	20	51	20
9-1	0	0	0	100	18	0	0	0	50	10	25	2
9-2	10	8	6	26	18	30	33	18	-90	43	58	18
9-3	0	0	0	50	10	1	1	1	-20	3	15	1
9-4	12	13	10	62	9	36	49	30	20	1	35	1
9-5	16	18	14	50	8	49	68	43	40	6	42	4
10-1	0	0	0	66	29	1	2	1	-20	9	27	1
102	0	0	0	61	45	0	0	0	-44	19	34	6
103	1	1	1	58	19	6	7	4	4	5	20	2
104	42	48	37	60	28	154	216	139	60	152	126	6
111	4	3	1	9	15	13	11	5	-24	30	39	18
112	11	12	6	10	2	35	39	22	24	4	35	2

their applications to the Tung data, discussed in the previous section. Notice that these measures can definitely warn investigators that they must pay attention to Items 13, 19, 22, and Subject 22. See in chapter 15 that responses to these three items of Subject 22 are judged to be outliers in this data set.

14.3 GRAPHICAL DISPLAY OF RESPONSE PATTERNS

From the previous sections, it is obvious that joint graphical display of row weights and column weights is at best problematic. It is also clear that it is difficult to find a graphical method which is free from the problems mentioned here. What then would be our desiderata for a practical graphical method? It seems that such a method should be at least logically sound, and that a graph should be a concise description of the input data. No matter how appealing a graph may be, it is worthless if it neither resembles nor accounts for the input data. This section contains a description of a graphical method that aims to avoid some of the aforementioned difficulties. This method was proposed for multiple-choice data by Nishisato (1990).

Bahadur (1961) presented a model for joint responses to n dichotomous items. His model describes such a response pattern as the sum of the patterns predicted by item independence, two-item association, three-item association, and up to n-item association. This mode of decomposition tells us that response patterns constitute the entire information in the data. If we go one step further, we can say that dual scaling solutions indeed capture different distributions of response patterns. This assertion can be easily verified by the numerical example discussed in chapter 8. Recall the Singapore data and the analysis, in which we observed that when the input data are rearranged in order of scores of subjects, determined by each solution, the response patterns revealed very systematic changes. We have also observed that plotting subjects and labeling them by their response patterns indicated easily interpretable representations of solutions. This is the essence of what Nishisato proposed in 1990.

Since there is no logical difficulty in calculating intersubject distances from the scaling outcome, this method is based on a sound rationale. It is easy to see clusters of response patterns, as we saw earlier in chapter 8, and clusters of response patterns yield easy and immediately discernable interpretations. As we discussed, it would be necessary to cluster subjects by ignoring some items that are not shared by those subjects in a cluster. Interpretation of a cluster also becomes easier by looking at only its common response patterns. For a given cluster of subjects, those items whose options the subject did not share can be indicated by an asterisk (*) so that we may characterize the cluster in terms of a smaller number of item options. In this representation, we can indeed relate row weights and column weights without involving such logical difficulty as due to two disparate spaces. When the number of items increases, this method may encounter some difficulty because the response pattern will eventually become too long. This may be true so far as the graphical display is concerned, but for the purpose of interpretation we can always introduce more asterisks by reducing the size of each cluster. Apart from this kind of possible practical difficulty, this method appears to be free from logical problems, which have plagued the traditional method of joint graphical display.

From the application point of view, we can further investigate the use of traditional cluster analysis in identifying clusters of subjects and clusters of contributing items.[5] At this point, it seems opportune to discuss a fundamental problem of multidimensional interpretation of data. This problem pertains to the basic difference between cluster analysis and a group of projection methods (e.g., dual scaling, multidimensional scaling, principal component analysis, factor analysis). The latter methods decompose data in such a way that projections of data points onto orthogonal subspaces are calculated or estimated. This is fundamentally different from partitioning data in multidimensional space. In other words, there exists a clear distinction between finding clusters in multidimensional space and decomposing data into orthogonal components by projection, that is, mutually exclusive clusters (cluster analysis by partitioning) versus non-mutually exclusive clusters (e.g., dual scaling). In the latter case, a variable can be a main contributor to more than one dimension, while in the former (i.e., cluster analysis[6]), one variable belongs to only one cluster. How to arrive at the interpretation of results from these two distinct approaches is difficult to explain, but it is important to be aware of the

---

[5] The use of cluster analysis to reorganize quantification outputs has been tried by a number of people. See, for example, Mucha (1992).

[6] Nonhierarchical clustering is equivalent to partitioning, and provides mutually exclusive clusters (i.e., one variable belongs to only one cluster). Hierarchical clustering is typically represented in the form of a dendrogram, or inverted tree structure, and is a method to generate nested clusters. Yet, once the number of clusters is decided, the inverted tree is cut horizontally by a line, and the resultant clusters are mutually exclusive. In other words, in this nested clustering, too, one variable in a cluster does not belong to another for the given number of clusters.

differences between them. Suppose that a tight cluster of data points can be found in the three-dimensional space. In partitioning, it will be captured as a single cluster, which seems to be a natural way. In contrast, the projection method will describe the same cluster as consisting of three components, that is, three projections of this cloud of data points onto three orthogonal axes. Consider another, but different, case in which four tight clusters are located at the corners of the quadrants in two-dimensional space. The partitioning method will identify them as four clusters, while the projection approach will capture them as two bipolar components, that is, each component consisting of highly positive and highly negative weights for the variables.

Remember that we are discussing the results of dual scaling, which is a projection method. Yet, in interpreting the results, it is still possible to incorporate the idea of partitioning. In the proposed method of graphical display, the two approaches are used in a complementary way. Three variations are considered.

1. Plot subjects in the traditional way, that is, in two-dimensional space for a pair of solutions. Then, partition subjects into groups, and then identify those options shared by subjects in each group. This is the method used in chapter 8.
2. The same as Variation 1, except that for each group of subjects response options are subjected to cluster analysis to identify key options for characterizing each group.
3. Extract as many solutions as appropriate. For instance, the criterion of η^2 greater than or equal to $1/n$ (i.e., $\alpha \geq 0$) can be used to determine the number of solutions. Then, using the subject-by-solution matrix of scores of "appropriate" solutions, calculate the matrix of intersubject distances, which is then subjected to cluster analysis. Once clusters of subjects are obtained, interpret them in terms of key options.

These are some alternatives within the confines of the response pattern approach. Nishisato and Kolić (1992) compared the outcome of Procedure 1 with that of Ward's hierarchical clustering based on the Hamming distances of response options calculated from response patterns. While Procedure 1 was clear as to how clusters were formed (i.e., shared options), cluster analysis yielded clusters which could not be related to clusters of options. In some cases, two clusters from cluster analysis consisted of the same set of shared options, leaving the impression that the output from cluster analysis requires careful inspection. This comparison seems to suggest that partitioning subjects in terms of subsets of shared response patterns may be more meaningful than rather mechanical cluster analysis of clouds of data points in terms of distances—and note, too, that there are different measures of distance and a number of ways to define the distance between two clusters.

Within a cluster of subjects, all the subjects are likely to share a subset of response options. This entire group can be partitioned into m_j subgroups in terms of their responses to an item, of which the options were not shared by all the subjects, thus reducing the number of asterisks in the label by one. These subgroups are further divided in terms of another item. This leads to a tree structure, an outcome of hierarchical clustering. Unlike the case of cluster analysis, however, a single data set can be partitioned in a number of ways by this procedure, for one can choose any one of the items with asterisks to partition the current cluster. Research on this approach is currently under way.

14.4 RESPONSE PATTERNS AND ITEM STATISTICS

Plotting subjects and labeling them by their response patterns can reveal interesting and useful information about items. Recall Figure 8.3 (a), (b), (c), and (d), in which distributions of subjects who chose different options were shown item by item. It was mentioned in chapters 6 and 8 that the maximized squared correlation ratio η^2 is equal to the average of the squared item-total correlation $\Sigma r_{jt}^2/n$. Then, how and how much does r_{jt} influence graphical display of items? Figure 8.3 is based on the first two solutions, for which the interitem correlation coefficients of the individual items are:

$$r_{1t} = (0.931,\ 0.832),\ r_{2t} = (0.827,\ 0.552),\ r_{3t} = (0.492,\ 0.151),$$
$$r_{4t} = (0.793,\ 0.876).$$

The first figure in each set of parentheses is for Solution 1, and the second for Solution 2 (see these coefficients in the computer output in chapter 8). As you can see, the higher the correlation, the better the separation of three groups of subjects who chose different options of an item. From the graphical point of view, we would be interested in determining how high r_{jt} should be for the item to partition the space cleanly by its options, hence judged to be a "key item" for clustering. We can use linear boundaries as typically used in partitioning, or more leniently, any monotone nonlinear boundaries that minimize the number of misclassifications. For a given definition of boundaries, we can then propose a measure of classifiability of Item j by

$$w_j = \frac{\text{Number of correctly classified subjects}}{\text{Total number of subjects}} \tag{14.26}$$

For a given data set in which all the items have the equal number of options, it is almost certain that w_j is monotonically related to r_{jt}. However, such a monotone relation between w_j and r_{jt} will deteriorate as (a) the number of dimensions used decreases, and possibly as (b) the number of options increases. Point (a) is relevant to our earlier discussion of the two different approaches to multidimensional analysis, that is, clustering in multidimensional space and projections onto orthogonal spaces. What parameters would influence w_j and the relation between w_j and r_{jt}? Can we find a practical way to determine the magnitude of r_{jt} which yields clean separations of subjects who chose different options? Call such a critical value of r_{jt} "graphically significant" or "practically significant."

Let us not be swayed again only toward the "significance" of a statistic. From the data analytic point of view, those items which are not graphically significant can be as important as those which are. To make this point, let us look at a data set reported by Komazawa (1978). Data were collected as answers to the following questions:

1. Age: 1 = 49—50; 2 = 55—59.
2. Intraocular pressure: 1 = normal; 2 = abnormal.
3. Electrocardiogram (EKG): 1 = normal; 2 = abnormal.
4. Maximal blood pressure: 1 = below 139; 2 = 140—170; 3 = over 170.
5. Minimal blood pressure: 1 = below 84; 2 = 85—95; 3 = over 95.
6. Pulse rate: 1 = below 7.4; 2 = 7.5—8.5; 3 = over 8.5.
7. Cholesterol level: 1 = below 169; 2 = 170—210; 3 = over 210.
8. Four illnesses: 1 = cerebral hemorrhage; 2 = brain infarction; 3 = miocardial infarction; 4 = angina pectoris.

Data were obtained from 52 subjects, and are given in Komazawa (1978, Pp.171—172). Using the variable "four illnesses" as the criterion, forced classification was carried out. Since the criterion variable has four options, there were three forced classification solutions, and four illnesses were maximally separated in three-dimensional space. To highlight the point we want to make, let us use only two solutions, and construct the match-mismatch table of forced classification analysis (Table 14.10).

There are 38 patients correctly classified, out of 50. Thus, the percentage of the correct classification is 76%, which is very good. With this in mind, look at the variable "cholesterol level." The analysis has shown the correlation between this variable and the criterion variable "illnesses" on three appropriate solutions, as follows:

TABLE 14.10
Match-Mismatch Table of Four Illnesses

		Predicted			
		Angina	Brain Infarction	Hemor-rhage	Miocardial
	Angina	8	1	1	2
Observed	Brain Infarction	1	12	1	1
	Hemorrhage	0	2	10	1
	Miocardial	0	1	1	8

$$r_{\text{cholesterol,illness}} = (0.236, 0.075, 0.115).$$

Are these not unexpectedly low values of correlation for cholesterol and illnesses? What has happened to our expectation that cholesterol level is highly correlated with illnesses? If you plot those patients with a high cholesterol level, using the first two solutions, you will see that the patients are scattered all over the two-dimensional space which relatively cleanly separates four regions of the illnesses. This is the reason why cholesterol level is so poorly correlated with illnesses. It affects all the illnesses under investigation!

From the data analysis point of view, therefore, a low value of correlation is as important as a high value of correlation. To satisfy your curiosity, you can create a variable that affects only one kind of illnesses, another that affects two, another that affects three, and so on, to see how those variables would correlate with the variable "illnesses." You will discover that different values of correlation tell different stories, and that all the stories are equally important for some people. Thus, the classifiability of a variable is one thing, and the information content is another. To understand this kind of information supplied by data analysis, graphical display is essential. As the foregoing discussion would suggest, there are still many problems that need to be clarified. This is especially true with the quantification of categorical data.

CHAPTER 15
Outliers and Missing Responses in Multiple-Choice Data

An outlier response in dual scaling is defined as a response that exerts an excessive influence on the outcome of quantification. This excessive influence shows up in the form of spurious results on weights or other statistics. In dual scaling, such outlier responses are observed almost exclusively in multiple-choice data. Therefore, our discussion of outliers is restricted to this data type. In the first section, we see a case in which a fraction of the data set exerts such a strong effect that it determines the characteristics of the first solution. In other words, the first solution is determined by only a few outliers in a large data set.

The second topic of this chapter is the problem of missing responses in multiple-choice data. Although there is no direct causal relationship between outliers and missing responses, it is interesting to know that missing responses can be instrumental to outlier responses. In a broader context, however, the handling of missing responses is a serious problem not only in dual scaling, but also in data analysis in general. Irrespective of your area of interest, missing responses have constantly plagued data analysts. Ask someone who is involved in marketing or political surveys. The percentage of missing responses in survey data is astonishingly high, particularly in the case of a disappointingly low return rate when a study is conducted by mail. Not knowing why there are so many missing responses, what would you do to analyze data? Would you discard all the respondents who missed even a few questions? Would you ignore the existence of a substantial number of missing responses and go ahead with analysis? In this chapter, we will look at these ubiquitous but practical problems, and see if we can find appropriate ways to handle them.

15.1 PROCEDURES FOR OUTLIERS

To understand the problem of outlier responses in multiple-choice data, it is important to remember a few facts about dual scaling. The first and probably most important fact is that the total information of Option p of Item j is inversely proportional to the number of subjects who choose that option.[1] The sum of squared option weights over all possible solutions is given by

$$\sum_{k=1}^{t} f_{pj} x_{pjk}^2 = \frac{N - f_{pj}}{nN} \tag{15.1}$$

where f_{pj} is the number of subjects who have chosen Option p of Item j, x_{pjk} is the weight of Option p of Item j in Solution k, and t = the total number of options (m) minus the number of items (n). Formula 15.1 means that the fewer the subjects who choose the option, the greater the total contribution of the option to explaining the data. On average, therefore, an option chosen by a small number of subjects is given a weight of comparatively large absolute value. Consider an extreme case of the Tung data (chapter 14) in which Option p of Item j is chosen by only one subject out of 500. Then, this option is likely to be the main factor that determines Solution 1. In other words, Item j and other items that are comparatively highly correlated with Item j are likely to be the main contributors to the solution.

The second fact is that the correlation of Item j with the total score, r_{jt}, is strictly proportional to the sum of squares of responses weighted by $\mathbf{x}_j$,

$$r_{jt} \propto \mathbf{x}_j' \mathbf{D}_{cj} \mathbf{x}_j \tag{15.2}$$

where $\mathbf{x}_j$ is the vector of weights for options of Item j, and $\mathbf{D}_{cj}$ is the diagonal matrix of the frequencies of the m_j options of Item j. This relation suggests that a single extreme value in $\mathbf{x}_j$ will increase r_{jt} spuriously.

The third fact is that the total contribution of item j is proportional to the number of options of that item, that is, m_j,

$$\mathbf{x}_j' \mathbf{D}_{cj} \mathbf{x}_j = m_j - 1 \ . \tag{15.3}$$

This relation[2] suggests the possibility that an item with a large number of options may produce option weights of large absolute values, hence a large value of the item-total correlation r_{jt}.

To demonstrate these points, let us look again at the first solution from Tung's data. In chapter 14, we observed that three response options had extreme weights: two of them were chosen by only one subject, and the third one by two. Three of these four responses came from one subject. To

---

[1] See Nishisato (1991, 1993), Yamada and Nishisato (1993).

[2] See Nishisato (1991, 1993).

see the effects of the three responses from this one subject on dual scaling results, let us carry out two analyses, one using the complete data, consisting of 550 responses (i.e., 50 subjects times 11 responses), and the other using the data without these three responses, that is, the data with 547 responses. Tables 15.1, 15.2, and 15.3 are, respectively, the matrix of interitem correlation based on the first solution of the complete data, **R**, that of the reduced data, $\mathbf{R}_{red}$, and $\mathbf{R}$-$\mathbf{R}_{red}$. Three items that involve the three responses, eliminated in calculating $\mathbf{R}_{red}$, are Items 4, 6 and 7.

TABLE 15.1
Inter-Item Correlation Matrix Based on 550 Responses, **R**

Item	1	2	3	4	5	6	7	8	9	10	11
1	100										
2	20	100									
3	13	00	100								
4	11	07	14	100							
5	05	12	-06	02	100						
6	21	13	23	65	12	100					
7	15	04	19	92	12	69	100				
8	09	02	-20	10	11	08	10	100			
9	35	45	02	28	-02	25	19	17	100		
10	19	-06	-22	38	12	26	38	03	20	100	
11	30	40	04	26	-12	19	15	26	66	14	100

TABLE 15.2
Inter-Item Correlation Based on 447 Responses, $\mathbf{R}_{res}$

Item	1	2	3	4	5	6	7	8	9	10	11
1	100										
2	36	100									
3	13	00	100								
4	03	32	03	100							
5	-11	15	-06	38	100						
6	38	22	20	17	-02	100					
7	18	34	15	20	12	31	100				
8	09	16	-20	35	03	17	25	100			
9	38	44	01	31	-03	19	06	15	100		
10	-09	-03	-22	15	12	03	04	06	03	100	
11	30	51	04	45	12	20	11	26	66	12	100

Notice that the three responses have exerted strong effects on the correlational structure. It is nearly unbelievable that 3 responses out of

TABLE 15.3
Difference Matrix, **R** minus $\mathbf{R}_{res}$

Item	1	2	3	4	5	6	7	8	9	10	11
1	00										
2	-16	00									
3	00	00	00								
4	08	-25	11	00							
5	-16	-03	00	-36	00						
6	-12	-09	03	48	14	00					
7	-03	-30	04	72	00	38	00				
8	00	-14	00	-25	08	-09	-15	00			
9	-03	01	01	-03	01	06	13	02	00		
10	28	-03	00	23	00	23	34	-03	17	00	
11	00	-11	00	-19	-24	-01	04	00	00	02	00

TABLE 15.4
Optimal Scores under Different Transformations

Subject	S	S*	0	I	R
1	37	-42	-70	-50	-59
2	20	-62	-75	-45	-61
3	25	-52	-69	-46	-57
4	95	06	-52	-67	-66
5	22	-47	-68	-48	-52
6	21	-08	-26	-18	-24
7	30	-48	-71	-44	-58
8	22	-35	-49	-26	-44
9	27	-36	-56	-37	-47
10	28	-30	-51	-37	-45
11	-46	60	61	76	76
12	-05	00	-09	15	03
13	-48	32	43	69	57
14	-61	89	86	88	106
15	-52	79	88	92	93

550 could have such decisive influences on correlation which is calculated not from a few pairs of numbers but from 50 pairs! In addition, we saw briefly in chapter 14 that the three responses are the main factor determining Solution 1! Indeed, Solution 2 for the complete data of 550 responses is very similar to Solution 1 for the reduced data set of 447 responses. See Table 15.4, which shows sample scores of the first 15 subjects, obtained from Solution 1 of the reduced data [*R*], Solution 1 of the complete data [*C*1], and Solution 2 of the complete data [*C*2]. The column [*S*] list scores of Solution 1 from the complete data with option standardization, to be

discussed in 15.1.6. The similarity between [*R*] and [*C*2] lends strong support to the statement that the three responses have determined Solution 1. Incredible as it may sound, this is an aspect of dual scaling that *capitalizes* on an idiosyncratic pattern of responses.

Because they reveal unusual circumstances, outlier responses are themselves an interesting topic for investigation. In this sense, scrutinizing outlier responses is a legitimate part of our investigation. At the same time, it could be argued that quantification should not be influenced by a few aberrant or unusual responses. From this point of view, let us look at some methods that are resistant to outlier responses.[3]

15.1.1 Methods of Reciprocal Medians

Recall that dual relations are basically the formulas for the method of reciprocal averages (MRA). It is well known that the average is affected by an extreme value, and that in the presence of an extreme value the central tendency of the data set is better approximated by the median. Thus, the first approach is to substitute the median for the average in the reciprocal computation. Unlike in the case of the reciprocal averaging, there is no guarantee that this method of reciprocal medians (MRM) will ever converge to a stable point. Therefore, we will introduce a small value of discrepancy in the computation, and define the process of reciprocal medians by

$$y_i = \mathrm{Mdn}_j\,[f_{ij}x_j] + e_i$$

$$x_j = \mathrm{Mdn}_i\,[f_{ij}y_i] + e_j\ ,$$

where e_i and e_j are as negligible as possible, and Mdn is the median operator over chosen responses. For instance, $y_i = \mathrm{Mdn}_j[f_{ij}x_j] + e_i = \mathrm{Mdn}_j[0x_1, 0x_2, 1x_3, 0x_4, 1x_5, 0x_6, 1x_7, 0x_8, 0x_9] + e_i = \mathrm{Mdn}_j[x_3, x_5, x_7] + e_i = \mathrm{Mdn}_j[0.89, 0.34, -1.97] + e_i = 0.34 + e_i$. The residual terms e_i and e_j were added to the set of reciprocal median equations because an empirical study had shown occasional failures for the process to converge.

15.1.2 Alternating Reciprocal Averages and Reciprocal Medians

This is a combination of a mathematically convergent method of reciprocal averages and a resistant process of reciprocal medians. Empirical work has shown that the process converges to two stationary points, corresponding to the two phases, reciprocal averaging and reciprocal medians.

---

[3]See Nishisato (1984a, 1984b, 1987, 1991).

15.1.3 Method of Trimmed Reciprocal Averages

This method is a generalized scheme that accommodates MRA and MRM as special cases. In the method of trimmed reciprocal averages, the q largest and the q smallest values in each set of $[f_{ij}y_i]$ and $[f_{ij}x_j]$ are omitted from the averaging. The value of q may change from 0 (i.e., MRA) to half the number of elements in $[f_{ij}y_i]$ and $[f_{ij}x_j]$ (i.e., MRM). Empirically, q=1 is large enough to attain convergence of the process most of the time. To be on the safe side, we recommend that q = 2 or 3. It should be noted, however, that the value of q must be determined in relation to the data size and distributions of responses over different options. A systematic study on the appropriate value of q has yet to be done.

15.1.4 Method of Generalized Forced Classification

When dual scaling yields option weights with excessively large absolute values, we can multiply those options in the data matrix by k, which is approximately 0.5, and subject the modified (weighted) data to dual scaling. This is an example of generalized forced classification. Under this scheme, k can also be set equal to 0, which amounts to ignoring those options. When $0 < k < 1$, we can expect some suppressing effects on the quantification of those options. Empirical research (Nishisato, 1987; Gibson, 1993) has so far shown that k of 0.5 leads to a reasonable outcome. This, too, requires further work.

15.1.5 Projection Method

Data are first subjected to dual scaling to identify outlier options. Then, those outlier options are deleted from the data matrix $\mathbf{F}$ to form the reduced matrix $\mathbf{F}_r$. If there are three outlier options, then $\mathbf{F}_r$ has three fewer columns than $\mathbf{F}$. Now, construct the projection operator $\mathbf{P}_r$ for a subspace spanned by the rows of $\mathbf{F}_r$ by

$$\mathbf{P}_r = \mathbf{F}_r(\mathbf{F}_r'\mathbf{F}_r)^{-1}\mathbf{F}_r' \ .$$

Then, subject the matrix $\mathbf{P}_r\mathbf{F}$ to dual scaling. In this case, the weights for those outlier options are determined in relation to their contributions to the rest of the data. Although the idea of this approach is interesting, this is not an attractive alternative, at least from the computational point of view. Who would want to calculate the inverse of $\mathbf{F}_r'\mathbf{F}_r$?

15.1.6 Standardization

In chapter 14, we noted that the total sum of squares of a response option is inversely related to the frequency of the option. This means that those options chosen by only a few subjects have a greater impact on the analysis—the essence of the problem of outliers. It is empirically known that standardizing the data in terms of option sums of squares practically eliminates the contributions of outlier responses. See [*S*] in Table 15.4. This approach is therefore good for reducing the influence of outliers, but at the same time it will no longer optimize our objective function, that is, the squared correlation ratio.

Let us look at numerical examples of some of these approaches, using Tung's data. Table 15.5 summarizes the weights for three outlier options of Items 4, 7 and 6, obtained by the standard procedure of dual scaling (**F**), dual scaling of the reduced matrix ($\mathbf{F}_r$), the method of trimmed reciprocal averages with q being 1 (Q1) and 2 (Q2), and the generalized forced classification with weights of 0.5 for the three options (GF0.5). In this table, the outlier options are indicated by the enclosed parentheses.

TABLE 15.5
Comparison of Five Approaches to Outliers in Terms of Optimal Option Weights

		Method				
Item	Option	**F**	$\mathbf{F}_r$	Q1	Q2	GF0.5
4	1	.15	.94	.87	.99	.92
	2	-.24	-.27	-.87	-.55	-.33
	3	-1.14	-2.41	-1.97	1.40	-2.40
	(4)	10.05	0	3.33	3.29	1.92
7	1	-.07	.30	.22	.41	.27
	(2)	10.05	0	3.43	3.02	1.92
	3	-1.38	-2.75	-2.66	-3.00	-2.72
6	1	-.13	.37	.38	.64	.35
	(2)	6.65	2.40	2.46	1.34	1.96
	3	-.93	-1.92	-2.18	-2.47	-1.92

We can see that the generalized forced classification weights are very similar to those obtained from $\mathbf{F}_r$, except that $\mathbf{F}_r$ assigns the weight of 0 to each of the discarded options. Furthermore, generalized forced classfication is computationally fast. In this comparison, the method of option standardization is not included because the scaling unit is different from the other cases, making the comparison difficult. Another possibility for handling outlier responses is that if the outlier responses are the main contributors to the first solution, we should extract two solutions, and look at the second solution, as demonstrated by [$C2$] in Table 15.4. In this procedure, however, there remains the issue that the extraction of the first solution may dilute the contributions of nonoutlier responses to the remaining solutions.

These are some possible alternatives that we can consider in dealing with outlier responses. As mentioned earlier, however, eliminating the contributions of outliers may not always be the best strategy in interpreting the data. In this context, the methods presented here should be considered not as better than the standard procedure of dual scaling, but as outlier-resistant or outlier-robust procedures. The final decision of whether we should use such a robust method must be left to our best judgment.

15.2 PROCEDURE FOR MISSING RESPONSES

How to handle missing responses is a universal problem in data analysis. When we see a number of missing responses in a data set, it is difficult to identify why and how those missing responses have occurred. There are a number of reasonable explanations for missing responses, such as recording error, equipment failure, the subject's lack of interest, the reluctance of the subjects in answering specific questions, irrelevance of the questions (i.e., "not applicable"), no opinion of the subject, and partial loss of response sheets. Such missing responses may or may not occur randomly. Even in the situation in which a strictly controlled experimental design is used, it is difficult to introduce an appropriate model for missing responses. In the case of dual scaling, this difficulty is compounded by the fact that it deals with quantification of categorical data. Can we quantify missing responses or should we ignore them completely? Whatever we do, it will affect the outcome of scaling.

A problem peculiar to dual scaling is the fact that when a subject misses a multiple-choice question, for example, the task is not to estimate the missing value, but to identify which one of the multiple options the subject would have chosen. How can we identify the option? This task is not easy because there is neither a statistical model for dual scaling, nor a widely acceptable criterion for correct identification. As for missing responses in paired comparision, rank-order, and successive categories data, the respective chapters have already discussed how to handle them.

15.2.1 Four Approaches

We can identify at least four distinct approaches to the problem of missing responses: (a) deletion of rows and columns of a data matrix that contain missing responses, referred to as the method of listwise deletions, (b) analyzing only observed responses, thus ignoring missing responses, (c) accommodating missing responses as part of the data by creating an extra category for each multiple-choice item that has missing responses, and (d) filling in missing cells with reasonable numbers, called the imputation approach.

The method of listwise deletions is widely used, but it often wastes too much information. Suppose that you collect data by asking people 100 questions on the nation's economic prospects. Would you discard a subject from the data analysis because of one or two missing responses? If so, you would do it very reluctantly. What if everyone misses at least one response? Thus, this method is useful only when the data set contains a very small number of missing responses.

The second procedure of analyzing only observed responses is probably the best overall approach to missing responses so long as they occur infrequently and randomly. This is the method adopted in the DUAL3 package.

The third procedure creates an extra category for each item involving at least one missing response, and each missing response is entered in that category. This arrangement changes imcomplete data to complete data, and gives a chance for missing responses to contribute to data analysis and interpretation. The procedure is very appealing especially when missing responses occur in a nonrandom fashion, a situation which can happen more often than you may think. For example, high income earners may not wish to disclose their financial resources, and thus may not answer some of the related questions. Subjects with certain socioeconomic backgrounds may find some questions too objectionable to answer. Systems analysts may find some technical questions too simple or too boring to respond to. In these circumstances, refusals to respond reflect some tractable attributes or traits that the subjects possess, and missing responses may offer very interesting information about them. When missing responses occur in a random fashion, we would expect that the extra category for missing responses will be weighted so little as to exert practically no influence on the results of the data analysis. This assertion is correct, except that an extra category for each of many items may act as a source of outliers, as discussed in the previous section of this chapter.

The fourth procedure is the imputation approach. There are many ways to impute data for missing cells, and the rest of this section is devoted to this topic. Since this is an important approach to the problem of missing responses, we discuss it in detail.

15.2.2 Imputation Methods

In imputing values for missing responses, there are at least three different principles which can be considered: the principle of maximal internal consistency, the hot-deck principle, and the principle of complete ignorance. Let us look at each one of them.

(a) Principle of Maximal Internal Consistency (MIC). When this principle is used, the task is to identify which option of a missed item should get a response in order to maximize the squared correlation ratio of the quantified data, and to impute a response for that option. For instance, look at the small example in Table 15.6, in which the first of 10 subjects missed Item 3. Since Item 3 has four options, there are four possible imputations, as indicated by (1), (2), (3), and (4) in the table. Out of these four, the imputation (4) yields the largest squared correlation ratio. Thus, (4) is the choice by MIC. Since dual scaling quantifies data so as to maximize η^2, this principle may appear reasonable. With complete data, this maximization is carried out through weighting observed responses, and thus there is no distortion or fabrication of information involved in the maximization. Once the same principle is applied to identifying options for missing responses, however, the implication of using the principle is totally different. New information can be fabricated independent of observed responses. In other words, identification of response options by MIC will produce a higher η^2 as the number of missing responses increases, and if 99% of responses are missing MIC will produce an almost perfect case, that is, η^2 will be almost 1. At this extreme end, a fundamental difference is evident in the role of maximizing η^2 between observed data and missing responses, or rather between legitimacy of the criterion in dual scaling and absurdity of applying it to missing responses. A real danger of MIC is the case in which there are relatively few missing responses because the outcome would look very good, giving the impression that MIC must have recovered "lost" information. This impression is utterly wrong. MIC by itself does not solve the problem of missing responses.

(b) Hot-Deck Principle (HD). In the hot-deck principle, each missing response is allocated to the option chosen by those whose response patterns are most similar to that of the person who missed the item. Thus, the hot-deck principle would somewhat mitigate the problem of the artificial or fictitious structure induced by MIC. Common sense would also tell us that if two subjects answered most items in a similar manner they would answer another item in a similar way as well. In Table 15.6, the response

TABLE 15.6
Data and Ten Different Imputation Results

	Data									
Subject	Item 1			Item 2			Item 3			
1	1	0	0	0	0	1	*	*	*	*
2	0	1	0	0	1	0	0	0	1	0
3	0	1	0	1	0	0	0	0	0	1
4	0	0	1	0	1	0	0	0	1	0
5	1	0	0	0	1	0	1	0	0	0
6	0	0	1	1	0	0	1	0	0	0
7	1	0	0	0	0	1	0	0	0	1
8	0	1	0	0	1	0	0	1	0	0
9	0	0	1	0	1	0	0	0	1	0
10	0	1	0	1	0	0	0	1	0	0

	Imputation				η^2	
(1)	1	0	0	0	.7650	
(2)	0	1	0	0	.7379	
(3)	0	0	1	0	.6847	η^2_{min}
(4)	0	0	0	1	.8101	η^2_{max}
(5)	0	0	0	0	.8048	
(6)	.25	.25	.25	.25	.7060	
(7)	.25	.25	0	.50	.7553	
(8)	.25	0	0	.75	.7884	
(9)	.02	0	0	.98	.8079	
(10)	Subject 1 deleted				.7415	

pattern of Subject 1 to the first two items is (1,0,0,0,0,1). If we look at response patterns of the other subjects to the first two items, we note that Subject 7 has the same pattern as subject 1. Since Subject 1 missed Item 3 and the response pattern of Subject 7 to Item 3 is (0,0,0,1), this is the pattern we should impute for the missing response of Subject 1. In this particular instance, this imputation is the same as the one by MIC. When there are more missing responses involved, we will have to decide how to assess the similarity between the two response patterns. Apart from this extra task, HD appears to be a reasonable approach. But is it really reasonable? In chapter 13, we saw the basic idea of forced classification, and observed a numerical demonstration that a few repetitions of a data segment can have a decisive impact on determining the orientation of the first solution. In other words, by creating a few identical columns in a data set, we can lead the analysis in such a way that the repeated column becomes the determining structure of Solution 1. In this way, we can look at HD as having the same effect on the outcome. HD imputations can shape the nature of the first dual scaling solution, or more blatantly stated, missing responses can determine the first solution. This is not an overstatement of HD's possible drawbacks, for we can easily imagine what HD will do as the number of missing responses increases. Thus, HD alone is not as reasonable as our first impressions might suggest.

(c) Principle of Complete Ignorance (CI). Under this principle, we include any approach that is based on the assumption of no knowlege about missing responses or data structure. We can identify a number of such approaches. Look at (5) in Table 15.6, where pattern (0,0,0,0) was inserted. This approach ignores missing responses completely —and this works reasonably well in most situations.[4] Another approach is shown as (6) in Table 15.6, where the pattern (0.25,0.25,0.25, 0.25) is used for the missing item. This approach includes all the options with equal weights. It is also reasonable because it does not bias the quantification. Like approach (5), η^2 will become smaller as the number of missing responses increases because the insertion of 0's or equal weights for each missing pattern will reduce the discriminability in the data. Another approach based on ignorance is random imputation. For each of the missed items, one of the possible patterns is chosen randomly for imputation. For instance, if the missed item has four options, we choose randomly one of the patterns (1,0,0,0), (0,1,0,0), (0,0,1,0), and (0,0,0,1). How would this method work? If the data set is relatively small, the effect of ignorance may be accentuated on the outcome of quantification. If the data set is large, the method may behave well. Recall that dual scaling, like principal component analysis, is sometimes influenced rather dramatically by only a few responses. Another approach is listwise deletion, as has already been discussed. See (10) of Table 15.6, which is an example of this approach.

These are some of the approaches to the problem of missing responses in dual scaling. Table 15.6 has a few more entries than those discussed here; see (7), (8), and (9). If we use the observed responses to evaluate the likelihood of choosing options, we can better impute these patterns, instead of merely (0,0,0,0) or (0.25,0.25,0.25,0.25). Entries (7), (8), and (9) are examples of fuzzy coding as opposed to crisp coding. The question is how to come up with such patterns as those three. There is an excellent book on this topic by van Rijckevorsel (1987).

15.2.3 How Many is Too Many?

From a practical point of view, we would like to know which method is the "best" in some sense. At the present moment, there does not seem to be a

---

[4] This method is adopted by Nishisato and Nishisato (1984) as the standard procedure for handling missing responses in multiple-choice data.

universally acceptable answer to this question. Nishisato and Levine (1975; see also Nishisato, 1980a) examined the problem of missing responses from a different point of view, and their idea has been pursued further by Ahn (1991; see also Nishisato and Ahn, 1993). The basic idea is to compare the imputation that maximizes η^2 with the imputation that minimizes η^2. This difference indicates the range of η^2 over all possible imputations for a given number of missing responses. Therefore, when the difference becomes statistically significant, we should be concerned with the possibility that different imputations may lead to significantly different scaling results. The difference between the maximal η^2 and the minimal η^2 is a monotonically increasing function of the number of missing responses. Thus, the Nishisato-Levine proposal was to give up data analysis when the number of missing responses is greater than the critical point of statistical significance. Although the proposal does not give us any clue as to how missing responses should be handled, it presents an important safeguard against wasteful attempts to analyze data with many missing responses. Following their idea, Ahn (1991) investigated three questions:

1. What is the maximal number of missing responses that a data set can contain before it is judged unworthy of analysis?
2. What is the maximal number of missing responses that a single item (question) can tolerate?
3. What is the maximal number of missing responses that a single subject is allowed to make while maintaining the validity of the analysis?

The single underlying notion is that as the number of missing responses increases, different imputations will eventually provide significantly different results, at which point the subject should be discarded, the item eliminated, or the entire data analysis be aborted. One difficult problem in carrying out Monte Carlo studies on dual scaling problems is caused by its lack of a statistical model to generate categorical data. Thus, in addition to the problem of generalizability of Monte Carlo results, we have this problem of what data to use. Ahn (1991) generated multivariate normal variates, categorized them into a specific number, and then discarded a fixed number of data elements in a random fashion for two types of imputations, maximal η^2 and minimal η^2. Under different conditions (e.g., the number of response options, data sizes, interitem correlations), a large number of categorized data sets was generated; the maximal and the minimal squared correlation ratios were calculated for each set; and the 95% confidence interval for each of the maximal and the minimal squared correlation ratio was constructed *empirically* for each condition. The extrapolated critical percentages for Question 1 ranged from 11.5% to 25.4%. To be on the conservative side, therefore, we may say that if your data set is comprised of more than 11% missing responses you should be concerned with the possibility that different methods of imputation may lead to significantly different results. As for Question 2 and Question 3, different statistics were used: the difference between the maximal and the minimal item-total correlation coefficients for Question 2 and the difference between the maximal and the minimal scores of the subject for Question 3. Again, conservative estimates are that we should consider discarding an item if it has more than 5% missing responses, and that a subject with more than 11% missing responses may have to be eliminated from our analysis. These results are not conclusive because of the stated reasons, but interested readers are referred to Ahn (1991) and Nishisato and Ahn (1993) for more detailed results.

van Buuren and van Rijckevorsel (1992) have written an interesting article on handling missing responses. From a practical point of view, however, the problem still requires further work. Considering that different causes may exist for missing responses, it is unlikely that a single method for handling all kinds of them will ever be proposed. To start with, we will need models for the different types of missing responses. But so far, such models are nowhere to be found.

CHAPTER 16
Analysis of Multiway Data

Most of the data discussed so far are called two-way data matrices, with the object of quantification being to determine optimal weights for the rows and columns. In multiway data, we introduce a further structure, or a design, for rows and/or columns. For instance, subjects in the rows of a data matrix might be further divided in terms of gender and age, while items in the columns might be categorized into several groups of aptitude questions. With this increased complexity of the data composition, the object of dual scaling changes from that of maximizing between-row discrimination and maximizing between-column discrimination to the maximization of, in this example, gender differences, age differences, gender-age interaction (i.e., combination effects), or discrimination between different aptitudes.

The book *Multiway Data Analysis*, edited by R. Coppi and S. Bolasco (1989), is an excellent compendium of different approaches to the analysis of multiway data, and is recommended to those who wish to know what other methods are available, and what kinds of problems can be investigated with multiway data matrices. When the focus is restricted to the quantification of data, using the framework of dual scaling, the book contains a summary article of several approaches by Nishisato and Lawrence.

It is convenient to classify multiway tables into three types: orthgonal, proportional, and nonorthogonal. Let us look at Table 16.1, in which 12 subjects are classified in terms of two factors, A and B. This table has only a row design, consisting of two classification variables, A and B, each having two levels. Variable A might be, for instance, "gender," with A_1 and A_2 being female and male, respectively. Similarly, the variable "age" may divide subjects into those 20 years of age or younger (B_1) and those over twenty years (B_2). Let us tabulate the numbers of subjects in the four combination cells (Table 16.2). This is an example of *orthogonal design*, which is characterized by *equal cell frequencies*, as seen in Table 16.2.

As you see, there are three subjects in *every* combination of levels of the two factors A and B. You can consider a "12×12" piece of graph paper with *three* sets of 12 boxes in *every* one of the four cells. A slightly different table of cell frequencies (Table 16.3) would be an example of *proportional design*.

TABLE 16.1
Multiple-Choice Data with Row Structure

		Multiple-Choice Item									
		1			2			3			
Groups		1	2	3	1	2	3	1	2	3	4
		1	0	0	1	0	0	1	0	0	0
	B_1	1	0	0	0	0	1	0	1	0	0
		0	1	0	0	1	0	0	0	0	1
		0	0	1	0	0	1	0	0	0	1
A_1	B_2	0	0	1	0	0	1	0	0	1	0
		0	1	0	0	1	0	1	0	0	0
		0	1	0	0	1	0	0	1	0	0
	B_1	0	1	0	0	0	1	0	1	0	0
		0	1	0	0	1	0	0	0	1	0
		0	0	1	0	1	0	0	0	0	1
A_2	B_2	0	0	1	0	1	0	0	1	0	0
		0	0	1	0	0	1	0	1	0	0

TABLE 16.2
Cell Frequencies of An Orthogonal Design

		Factor B		
		B_1	B_2	Total
Factor A	A_1	3	3	6
	A_2	3	3	6
	Total	6	6	12

TABLE 16.3
Cell Frequencies of A Proportional Design

		Factor B		
		B_1	B_2	Total
Factor A	A_1	2	3	5
	A_2	4	6	10
	Total	6	9	15

Consider a 15×15 piece of graph paper, and partition its columns into two with a 6 to 9 ratio (i.e., in proportion to the column totals, 6 and 9), and its rows into two with a 5 to 10 ratio. If you find that the number of 1×1 squares in each cell is equal to 15 times the frequency of the cell in the above table, then the design is *proportional*. Let us consider another table with different cell frequencies (Table 16.4).

TABLE 16.4
Cell Frequencies of A Non-Orthogonal Design

		Factor B		
		B_1	B_2	Total
Factor A	A_1	4	1	5
	A_2	2	8	10
	Total	6	9	15

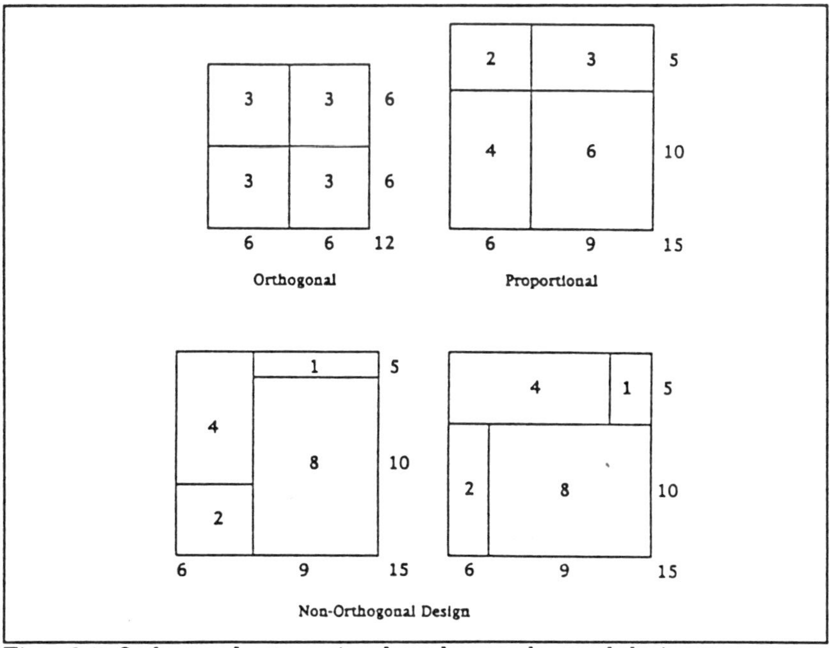

Fig. 16.1. Orthogonal, proportional, and nonorthogonal design.

This is an example of a *nonorthogonal design*. Consider a 15×15 piece of graph paper, and partition its columns into two with a 6 to 9 ratio, and its rows into two with a 5 to 10 ratio. The number of boxes in a cell of this partitioned graph is no longer equal to 15 times the frequency of the cell. In summary, both orthogonal design and proportional design have cell frequencies which are equal to those expected when rows and columns of the respective frequency tables are statistically independent, that is, the frequency (the number of subjects) in cell (A_i, B_j), f_{ij}, is equal to $f_{i.}f_{.j}/f_t$, where $f_{i.}$, $f_{.j}$, and f_t are respectively the marginal frequency of row 1, that of column j, and the total number of subjects. This is not the case with the nonorthogonal design. Look at Figure 16.1 for geometrical representations of these designs. Can you see how these designs are different? Can you stretch your imagination, and speculate why the effects of Factors A and B in a nonorthogonal design might be confounded, and why it would be statistically difficult to isolate their unique effects cleanly? Do you see why the effects of factors in a nonorthogonal design can be decomposed in more than one way?

Why are we discussing these different designs? It is because they have a strong bearing on the way in which multiway data matrices are quantified. Let us start with a general framework for dual scaling of multiway data matrices, assuming that the classification design is orthogonal.

16.1 ORTHOGONAL CLASSIFICATION

Nishisato (1976) proposed a general framework for dual scaling of multiway data matrices. Given the data matrix **F**, dual scaling is carried out with respect to its transform **E**, where

$$\mathbf{E} = \mathbf{M}_r'\mathbf{F}\mathbf{M}_c \,, \tag{16.1}$$

and $\mathbf{M}_r$ and $\mathbf{M}_c$ are matrices to capture respectively the multiway structure of the rows of **F** and that of the columns. This is a very general framework to capture multiway designs of a data matrix. As we will see shortly, $\mathbf{M}_r$ and $\mathbf{M}_c$ can be so-called design matrices, basis matrices, or projection operators. Let us consider the example of Table 16.1, which has only a row design.

In ANOVA, we capture this structure by the design or model matrix, **M**. In the current example, **M** is given as follows:

$$\mathbf{M} = [\mathbf{M}_m\ \mathbf{M}_a\ \mathbf{M}_b\ \mathbf{M}_{ab}] = \begin{bmatrix} 1 & 1 & 0 & 1 & 0 & 1 & 0 & 0 & 0 \\ 1 & 1 & 0 & 1 & 0 & 1 & 0 & 0 & 0 \\ 1 & 1 & 0 & 1 & 0 & 1 & 0 & 0 & 0 \\ 1 & 1 & 0 & 0 & 1 & 0 & 1 & 0 & 0 \\ 1 & 1 & 0 & 0 & 1 & 0 & 1 & 0 & 0 \\ 1 & 1 & 0 & 0 & 1 & 0 & 1 & 0 & 0 \\ 1 & 0 & 1 & 1 & 0 & 0 & 0 & 1 & 0 \\ 1 & 0 & 1 & 1 & 0 & 0 & 0 & 1 & 0 \\ 1 & 0 & 1 & 1 & 0 & 0 & 0 & 1 & 0 \\ 1 & 0 & 1 & 0 & 1 & 0 & 0 & 0 & 1 \\ 1 & 0 & 1 & 0 & 1 & 0 & 0 & 0 & 1 \\ 1 & 0 & 1 & 0 & 1 & 0 & 0 & 0 & 1 \end{bmatrix}.$$

The first column, indicated by $\mathbf{M}_m$, is to capture the mean of scores of all the subjects, the next two columns, $\mathbf{M}_a$, are for Factor A to capture the difference between A_1 and A_2, the next two columns, $\mathbf{M}_b$, are for Factor B to see the difference between B_1 and B_2, and the last four columns are for the interaction AB, a special combination effect of A_i and B_j. As we discussed in chapter 5, we first define hat matrices $\mathbf{H}_m$, $\mathbf{H}_a$, $\mathbf{H}_b$, and $\mathbf{H}_{ab}$ for model matrices $\mathbf{M}_m$, $\mathbf{M}_a$, $\mathbf{M}_b$, and $\mathbf{M}_{ab}$, respectively, as follows,

$$\mathbf{H}_m = \mathbf{1}(\mathbf{1}'\mathbf{1})^{-1}\mathbf{1}',\ \mathbf{H}_a = \mathbf{M}_a(\mathbf{M}_a'\mathbf{M}_a)^{-1}\mathbf{M}_a',\ \mathbf{H}_b = \mathbf{M}_b(\mathbf{M}_b'\mathbf{M}_b)^{-1}\mathbf{M}_b',$$
$$\text{and } \mathbf{H}_{ab} = \mathbf{M}_{ab}(\mathbf{M}_{ab}'\mathbf{M}_{ab})^{-1}\mathbf{M}_{ab}'.$$

Then, orthogonal projection operators are constructed as follows:

$$\mathbf{P}_m = \mathbf{H}_m,\ \mathbf{P}_a = \mathbf{H}_a - \mathbf{H}_m,\ \mathbf{P}_b = \mathbf{H}_b - \mathbf{H}_m,\ \mathbf{P}_{ab} = \mathbf{H}_{ab} - \mathbf{H}_a - \mathbf{H}_b - \mathbf{H}_m.$$

These matrices transform the observation vector, $\mathbf{y}$, to projections onto orthogonal subspaces. In other words, any resultant vectors are mutually orthogonal. For instance, consider two projections, $\mathbf{P}_a\mathbf{y}$ and $\mathbf{P}_b\mathbf{y}$, which have the property that $\mathbf{y}'\mathbf{P}_a\mathbf{P}_b\mathbf{y} = 0$. This follows from the relation that the product of any two of these projectors is a *null* matrix, $\mathbf{O}$, that is, $\mathbf{P}_m\mathbf{P}_a = \mathbf{P}_m\mathbf{P}_b = \ldots = \mathbf{P}_a\mathbf{P}_b = \ldots = \mathbf{P}_b\mathbf{P}_{ab} = \mathbf{O}$.

Suppose that we subject this 12×10 response-pattern matrix (Table 16.1) to dual scaling. We will obtain 10 weights for the options that maximize the between-subject sum of squares. This optimization is carried out without any reference to the design on the subjects. Do we not wish to determine option weights in such a way that the design is taken into consideration? We may wish to maximize, for instance, the gender difference (A_1 - A_2), the age difference (B_1 - B_2), or even the special combination effect (A_1 - A_2) - (B_1 - B_2), called the gender-age interaction. Such a question as this was raised by Fisher in 1948 as a problem of the one-way analysis of variance (ANOVA) of nonnumerical data. During the development of dual scaling, Fisher's problem was extended to the general multi-way design.[1]

Suppose that the 12 subjects in this example had quantitative scores. Let us use subscripts i, j, and k for the scores to indicate the following: j = 1 if Subject i belongs to A_1, and j = 2, if Subject i belongs to A_2; similarly, k = 1, 2 for B_1 and B_2, respectively; and i = 1, 2, 3 to indicate the ith subject for the given (A_j, B_k) combination. Then, the score y_{ijk} can be expressed as follows:

(16.2) $$y_{ijk} = \bar{y} + (\bar{y}_j - \bar{y}) + (\bar{y}_k - \bar{y}) + (\bar{y}_{jk} - \bar{y}_j - \bar{y}_k + \bar{y}) + e_{ijk} ,$$

where

$$\bar{y} = \frac{\sum_{i=1}^{I}\sum_{j=1}^{J}\sum_{k=1}^{K} y_{ijk}}{12} , \quad \bar{y}_j = \frac{\sum_{i=1}^{I}\sum_{k=1}^{K} y_{ijk}}{6} , \quad \bar{y}_k = \frac{\sum_{i=1}^{I}\sum_{j=1}^{J} y_{ijk}}{6} , \quad \bar{y}_{jk} = \frac{\sum_{i=1}^{I} y_{ijk}}{3} .$$

These four quantities are, respectively, the grand mean, the mean of A_j, the mean of B_k, and the mean of (A_j, B_k) cells. The term e_{ijk} is (y_{ijk} - $\bar{y}_{jk}$), that is, the within-cell variation. In matrix notation, the 12 scores are indicated by the vector **y**, and the decomposition can be expressed as

(16.3) $$\mathbf{y} = \mathbf{P}_m\mathbf{y} + \mathbf{P}_a\mathbf{y} + \mathbf{P}_b\mathbf{y} + \mathbf{P}_{ab}\mathbf{y} + \mathbf{P}_e\mathbf{y} ,$$

where the projectors are as defined before, except for $\mathbf{P}_e$, which is given by $\mathbf{I} - \mathbf{P}_m - \mathbf{P}_a - \mathbf{P}_b - \mathbf{P}_{ab} = \mathbf{I} - \mathbf{H}_{ab}$. Thus, each factor (e.g., gender, age, interaction) can be assessed independently. Keep in mind that this decomposition does not hold under nonorthogonal designs, because, as will be discussed later, the factors are then correlated and cannot be isolated individually. Once **y** is given, the decomposition is unique.

It is important to note that in dual scaling the scores of subjects are not given, but must be determined through manipulation of option weights. Thus, in generating scores, the gender effect, for instance, can be maximized or minimized. Although we have discussed the exact expression of **y** in terms of **x**, let us simply indicate the score vector **y** by **Fx**. As you recall, **x** is the vector of option weights that need to be determined, and **F**

---

[1] See Nishisato (1971, 1972, 1976, 1980a, 1984b), Poon (1977), Nishisato and Lawrence (1981, 1982, 1984, 1989), and Lawrence (1985).

is the response-pattern matrix. Then the decomposition (16.3) can be expressed as

(16.4) $$\mathbf{Fx} = \mathbf{P}_m\mathbf{Fx} + \mathbf{P}_a\mathbf{Fx} + \mathbf{P}_b\mathbf{Fx} + \mathbf{P}_{ab}\mathbf{Fx} + \mathbf{P}_e\mathbf{Fx} \,.$$

It is now clear that each factor is influenced by the option weights, and we can manipulate them in any way we like. For Equation 16.4 to hold, we must have $\Sigma\mathbf{P}_k = \mathbf{I}$, that is, the sum of all projection operators is the identity matrix, so that

(16.5) $$\mathbf{Fx} = (\Sigma P_k)\mathbf{Fx} = \mathbf{IFx} = \mathbf{Fx}$$

At this stage, we must make a decision on an important problem of choosing one of the two ways of quantification. For the orthogonal decomposition of $\mathbf{Fx}$ to hold, we must have a single vector of weights, $\mathbf{x}$, to assess the contribution of each ANOVA factor. This is one way. The other way is to discard this restraint, and carry out maximization of each ANOVA factor independently of the other factors. For instance, find $\mathbf{x}_a$ and $\mathbf{x}_b$ that maximize $\mathbf{x}_a\mathbf{P}'_a\mathbf{x}_a$ and $\mathbf{x}_b\mathbf{P}'_b\mathbf{x}_b$, respectively, subject to the usual constraints on each weight vector. In this case, $\mathbf{x}_a$ and $\mathbf{x}_b$ are in general not equal. Considering that we are typically interested in maximizing the effects of a particular factor, and that we are not so concerned with a unique decomposition of the scores into contributions of all the factors and their interactions, we will adopt the position that each factor may be maximized independently of others. That we are not so concerned with a unique decomposition of the scores is due to the fact that unlike ANOVA, which analyzes a given vector of observations, the main task of dual scaling is to generate a vector of scores, that is, there is no fixed vector of scores to start with.

With this choice, if we subject $\mathbf{P}_a\mathbf{F}$ to dual scaling, the vector of option weights of Solution 1, indicated by $\mathbf{x}_a$, generates $\mathbf{y}$ by $\mathbf{Fx}_a$ such that the difference between A_1 and A_2 is a maximum. Similarly, we can subject $\mathbf{P}_b\mathbf{F}$ and $\mathbf{P}_{ab}\mathbf{F}$ to dual scaling to obtain score vectors $\mathbf{Fx}_b$ and $\mathbf{Fx}_{ab}$ that maximize the age effect and the gender-age interaction, respectively. Or, you may wish to subject $(\mathbf{P}_a + \mathbf{P}_b + \mathbf{P}_{ab})\mathbf{F}$ to dual scaling so as to maximize the contribution of the ANOVA model to the explanation of the data, and minimize the residual (i.e., $\mathbf{P}_e\mathbf{F}$), that is, the variation not explained by the model.

In this case and others as well, you may need more than one solution to optimize the effects of your chosen factor(s). The exact number of solutions you will need is equal to the rank of the projection operator(s), which is given by the trace (the sum of the diagonal elements) of the projector(s).

Do you notice the similarity between this approach and forced classification? In chapter 13, we briefly described the relation between forced classification and dual scaling of a projected matrix, $\mathbf{PF}$. Indeed, if we are interested in maximizing the effects of A, B, or AB, we can use forced classification in the following way: Augment the data matrix **F** in Table 16.1 by the following *modified* "design" or "model" matrix **M:**

$$\mathbf{M} = [\mathbf{M}_a \ \mathbf{M}_b \ \mathbf{M}_{ab}] = \begin{bmatrix} 1 & 0 & 1 & 0 & 1 & 0 \\ 1 & 0 & 1 & 0 & 1 & 0 \\ 1 & 0 & 1 & 0 & 1 & 0 \\ 1 & 0 & 0 & 1 & 0 & 1 \\ 1 & 0 & 0 & 1 & 0 & 1 \\ 1 & 0 & 0 & 1 & 0 & 1 \\ 0 & 1 & 1 & 0 & 0 & 1 \\ 0 & 1 & 1 & 0 & 0 & 1 \\ 0 & 1 & 1 & 0 & 0 & 1 \\ 0 & 1 & 0 & 1 & 1 & 0 \\ 0 & 1 & 0 & 1 & 1 & 0 \\ 0 & 1 & 0 & 1 & 1 & 0 \end{bmatrix}$$

The first two columns constitute $\mathbf{M}_a$ for factor A. Similarly, the next two columns belong to $\mathbf{M}_b$ and the last two to $\mathbf{M}_{ab}$. Notice that $\mathbf{M}_m$ is missing, and that $\mathbf{M}_{ab}$ is different from the usual specification of the interaction. In the forced classification approach to multiway data, augmented matrices $[\mathbf{F} \ \ \mathbf{M}_a]$, $[\mathbf{F} \ \ \mathbf{M}_b]$, and $[\mathbf{F} \ \ \mathbf{M}_{ab}]$ are subjected to forced classification, with the variable that is captured by **M** as the criterion. Let us apply this approach to Table 16.1, using the weight of 20. The results of these separate analyses and expected signs of weights are as shown in Table 16.5. You can see the effects of scaling very clearly. These results are almost indistinguishable from those obtained from dual scaling of $\mathbf{P}_a\mathbf{F}$, $\mathbf{P}_b\mathbf{F}$ and $\mathbf{P}_{ab}\mathbf{F}$. We can also carry out forced classification with the matrix $[\mathbf{F} \ \ \mathbf{M}_a \ \ \mathbf{M}_b \ \ \mathbf{M}_{ab}]$, instead of $[\mathbf{F} \ \ \mathbf{M}_a]$, $[\mathbf{F} \ \ \mathbf{M}_b]$, and $[\mathbf{F} \ \ \mathbf{M}_{ab}]$, using one of **M** variables as the criterion. In this case, the presence of two other **M** matrices has some effects on the scores of the subjects. But, on average, such effects may be ignored because all the pairwise correlations of A, B, and AB are 0, this being due to the orthgonal design used.

Let us now get back to Equation 16.1. The matrix $\mathbf{M}_r$ is a matrix that captures the design of rows of the data matrix **F**. In the previous numerical example, $\mathbf{M}_r$ can be $\mathbf{P}_a$, $\mathbf{P}_b$, or $\mathbf{P}_{ab}$. The matrix $\mathbf{M}_c$, on the other hand, is used to look after the design on the columns of the matrix **F**. For example, those multiple-choice questions may be divided into personality questions and aptitude questions. Furthermore, those personality questions may be divided into social introversion items and thinking introversion items, and those aptitude questions into perceptual dimensions and psychomotor

TABLE 16.5
Scores of Subjects Obtained by Forced Classification and Expected Patterns

Subject	A		B		C	
	Score	Pattern	Score	Pattern	Score	Pattern
1	2.42	+	1.43	+	1.48	+
2	0.56	+	0.53	+	0.94	+
3	-0.18	+	0.16	+	-0.21	+
4	0.22	+	-1.05	-	0.23	-
5	-0.28	+	-0.59	-	-0.71	-
6	0.37	+	0.44	-	-0.41	-
7	-0.79	-	0.39	+	-0.08	-
8	-0.28	-	0.48	+	-1.28	-
9	-0.68	-	0.62	+	-1.15	-
10	-0.18	-	-0.91	-	0.36	+
11	-0.79	-	-0.68	-	0.48	+
12	-0.39	-	-0.82	-	0.36	+

dimensions. Then, we may wish to manipulate the scores of subjects in such a way that the effects of the column design may be most cleanly captured. Equation 16.1 shows that both row and column designs can be handled by dual scaling of the transformed matrix **E**. Takane and Shibayama (1991) considered the decomposition:

$$\mathbf{F} = [\mathbf{P}+(\mathbf{I}-\mathbf{P})]\mathbf{F}[\mathbf{Q}+(\mathbf{I}-\mathbf{Q})]$$
$$= \mathbf{PFQ} + \mathbf{PF}(\mathbf{I}-\mathbf{Q}) + (\mathbf{I}-P)\mathbf{FQ} + (\mathbf{I}-\mathbf{P})\mathbf{F}(\mathbf{I}-\mathbf{Q}),$$

where **P** and **Q** are projection matrices. Dual scaling of **PFQ** maximizes the effects of the row and the column designs, and dual scaling of **PF(I - Q)** optimizes the effect of the row design and the information which is not accounted for by the column design. Nishisato and Lawrence (1989) presented a more general expression:

$$\mathbf{F} = (\sum_{i=1}^{n}\mathbf{P}_i)F(\sum_{j=1}^{m}\mathbf{Q}_j) = (\mathbf{P}_1+\mathbf{P}_2+ \ . \ . \ . \ +\mathbf{P}_n)\mathbf{F}(\mathbf{Q}_1+\mathbf{Q}_2+ \ . \ . \ . \ +\mathbf{Q}_m),$$

where $\Sigma\mathbf{P}_i = \mathbf{I}$ and $\Sigma\mathbf{Q}_j = \mathbf{I}$. Dual scaling of $\mathbf{P}_i\mathbf{F}\mathbf{Q}_j$ maximizes the effects of row design i and column design j.

When we wish to maximize the design effect, the use of a projection operator is not always the simplest way to do so. For instance, suppose that we wish to quantify the data so as to maximize the effect of A, that is, $A_1 - A_2$. The projection operation $\mathbf{P}_a$ is given by

$$\mathbf{P}_a = \mathbf{H}_a(\mathbf{H}_a'\mathbf{H}_a)^{-1}\mathbf{H}_a' - \mathbf{1}(\mathbf{1}'\mathbf{1})^{-1}\mathbf{1}' \ .$$

But because we set the sum of weighted responses equal to 0, we do not need to adjust the origin by $\mathbf{1}(\mathbf{1}'\mathbf{1})^{-1}\mathbf{1}'$. Thus, $\mathbf{P}_a = \mathbf{M}_a(\mathbf{M}_a'\mathbf{M}_a)^{-1}\mathbf{M}_a'$. $\mathbf{P}_a$ and $\mathbf{P}_a\mathbf{F}$ are given as follows:

$$\mathbf{P}_a = \mathbf{M}_a(\mathbf{M}_a'\mathbf{M}_a)^{-1}\mathbf{M}_a' = \frac{1}{6}\begin{bmatrix} 1&1&1&1&1&1&0&0&0&0&0&0 \\ 1&1&1&1&1&1&0&0&0&0&0&0 \\ 1&1&1&1&1&1&0&0&0&0&0&0 \\ 1&1&1&1&1&1&0&0&0&0&0&0 \\ 1&1&1&1&1&1&0&0&0&0&0&0 \\ 1&1&1&1&1&1&0&0&0&0&0&0 \\ 0&0&0&0&0&0&1&1&1&1&1&1 \\ 0&0&0&0&0&0&1&1&1&1&1&1 \\ 0&0&0&0&0&0&1&1&1&1&1&1 \\ 0&0&0&0&0&0&1&1&1&1&1&1 \\ 0&0&0&0&0&0&1&1&1&1&1&1 \\ 0&0&0&0&0&0&1&1&1&1&1&1 \end{bmatrix}$$

$$\mathbf{P}_a\mathbf{F} = \begin{bmatrix} 2&2&2&1&2&3&2&1&1&2 \\ 2&2&2&1&2&3&2&1&1&2 \\ 2&2&2&1&2&3&2&1&1&2 \\ 2&2&2&1&2&3&2&1&1&2 \\ 2&2&2&1&2&3&2&1&1&2 \\ 2&2&2&1&2&3&2&1&1&2 \\ 0&3&3&0&4&2&0&3&2&1 \\ 0&3&3&0&4&2&0&3&2&1 \\ 0&3&3&0&4&2&0&3&2&1 \\ 0&3&3&0&4&2&0&3&2&1 \\ 0&3&3&0&4&2&0&3&2&1 \\ 0&3&3&0&4&2&0&3&2&1 \end{bmatrix}$$

Each of the first six rows of $\mathbf{P}_a\mathbf{F}$ is the sum of the first six rows, and each of the last six rows the sum of those rows. It is interesting to know that dual scaling of $\mathbf{P}_a\mathbf{F}$ provides the identical option weights as those obtained from $\mathbf{M}_a'\mathbf{F}$, which is given by

$$\mathbf{M}_a'\mathbf{F} = \begin{bmatrix} 2 & 2 & 2 & 1 & 2 & 3 & 2 & 1 & 1 & 2 \\ 0 & 3 & 3 & 0 & 4 & 2 & 0 & 3 & 2 & 1 \end{bmatrix}$$

Notice that the two rows of $\mathbf{M}_a'\mathbf{F}$ are identical to the two distinct rows of $\mathbf{P}_a\mathbf{F}$. The dual scaling results from $\mathbf{P}_a\mathbf{F}$ and $\mathbf{M}_a'\mathbf{F}$ are identical,

$$\eta^2 = 0.2204, \quad \delta = 100\%, \quad \mathbf{x} = \begin{bmatrix} 2.13 \\ -0.43 \\ -0.43 \\ 2.13 \\ -0.71 \\ 0.43 \\ 2.13 \\ -1.07 \\ -0.71 \\ 0.71 \end{bmatrix}.$$

Scores for the subjects can be calculated by $\mathbf{y} = \mathbf{D}_r^{-1}\mathbf{F}\mathbf{x}/\rho$, which is

$$\mathbf{y} = \begin{bmatrix} 4.54 \\ 1.06 \\ -0.30 \\ 0.51 \\ -0.51 \\ 0.70 \\ -1.57 \\ -0.51 \\ -1.32 \\ -0.30 \\ -1.57 \\ -0.76 \end{bmatrix}.$$

As you see, these scores are proportional to those obtained by forced classification (see Table 16.5, A). Similarly, to maximize the effects of B and AB, $\mathbf{P}_b\mathbf{F}$ and $\mathbf{P}_{ab}\mathbf{F}$ can be subjected to dual scaling. Again, the dimension of each of these matrices in the current example is 12×10, while the matrices $\mathbf{M}_b'\mathbf{F}$ and $\mathbf{M}_{ab}'\mathbf{F}$ are each 2×10, as seen here.

$$\mathbf{M}_b'\mathbf{F} = \begin{bmatrix} 2 & 4 & 0 & 1 & 3 & 2 & 1 & 2 & 2 & 1 \\ 0 & 1 & 5 & 0 & 3 & 3 & 1 & 2 & 1 & 2 \end{bmatrix},$$

$$\mathbf{M}_{ab}'\mathbf{F} = \begin{bmatrix} 2 & 1 & 3 & 1 & 3 & 2 & 1 & 3 & 0 & 2 \\ 0 & 4 & 2 & 0 & 3 & 3 & 1 & 1 & 3 & 1 \end{bmatrix}.$$

If you are familiar with the analysis of variance, you can use basis matrices $\mathbf{K}_a$, $\mathbf{K}_b$, and $\mathbf{K}_{ab}$ for $\mathbf{M}_a$, $\mathbf{M}_b$, and $\mathbf{M}_{ab}$, or for $\mathbf{P}_a$, $\mathbf{P}_b$, and $\mathbf{P}_{ab}$. Applications of basis matrices to multiway data matrices are discussed in Nishisato (1972, 1980a) and Lawrence (1985). In comparing their choices of transformation matrices **P**, **M**, and **K**, the projection operator is conceptually the most attractive. However, as we have already seen, projectors are computationally not very economical. The simplest one is the use of the design matrix **M**. To capture interaction effects, however, the design matrix must be created in a special way which is different from the traditional ANOVA design matrix for interactions. An example of this was provided earlier.

As is the case with ANOVA, proportional designs can be handled in the same way as orthogonal designs. In other words, the effects of ANOVA factors can be captured independently without being confounded by other factors.

16.2 NONORTHOGONAL CLASSIFICATION

When the design is nonorthogonal (i.e., frequencies in A_1, A_2, B_1, and B_2 are not equal), the design matrix **M** may not necessarily maximize a particular effect. In nonorthogonal designs, it is well known in ANOVA that the observation vector **y** cannot be expressed in terms of such an additive fashion as Equation 16.3. This is because factors such as A, B, and AB interaction are correlated, that is, confounded with one another. Thus, in order to isolate the factors, you must resort to an appropriately constructed projection operator **P** or basis matrix **K**.

Let us look at the construction of projection operators for the nonorthogonal case, following Takeuchi, Yanai, and Mukherjee (1982). Consider a small two-factor crossed experiment, in which cell frequencies are 2 for (A_1, B_1), 1 for (A_1, B_2), 1 for (A_2, B_1), and 3 for (A_2, B_2). This is a nonorthogonal case. The design matrix **M** is given by

$$\mathbf{M} = [\mathbf{M}_m \ \ \mathbf{M}_a \ \ \mathbf{M}_b \ \ \mathbf{M}_{ab}] = \begin{bmatrix} 1 & 1 & 0 & 1 & 0 & 1 & 0 & 0 & 0 \\ 1 & 1 & 0 & 1 & 0 & 1 & 0 & 0 & 0 \\ 1 & 1 & 0 & 0 & 1 & 0 & 1 & 0 & 0 \\ 1 & 0 & 1 & 1 & 0 & 0 & 0 & 1 & 0 \\ 1 & 0 & 1 & 0 & 1 & 0 & 0 & 0 & 1 \\ 1 & 0 & 1 & 0 & 1 & 0 & 0 & 0 & 1 \\ 1 & 0 & 1 & 0 & 1 & 0 & 0 & 0 & 1 \end{bmatrix}.$$

The first column of $\mathbf{M}$ is $\mathbf{M}_m$, the next two columns are $\mathbf{M}_a$, the next two columns $\mathbf{M}_b$, and the last four columns $\mathbf{M}_{ab}$. We can define hat matrices and the corresponding projection operators, as before. In addition, we must define hat matrices for unions of two spaces. For example, the hat matrix for the union of two spaces associated with Factors A and B is

$$\mathbf{H}_{a \cup b} = (\mathbf{M}_a,\mathbf{M}_b)[(\mathbf{M}_a,\mathbf{M}_b)'(\mathbf{M}_a,\mathbf{M}_b)]^{-}(\mathbf{M}_a,\mathbf{M}_b)' \ ,$$

where the superscript minus indicates a generalized inverse. Then the corresponding projector can be defined as

$$\mathbf{P}_{a \cup b} = \mathbf{H}_{a \cup b} - \mathbf{P}_m = \mathbf{P}_a + \mathbf{P}_b - \mathbf{P}_{a \cap b} \ .$$

$\mathbf{P}_{a \cap b}$ is the projector for the intersection of the two subspaces. In the orthogonal case, $\mathbf{P}_{a \cap b} = \mathbf{O}$, that is, $\mathbf{P}_{a \cup b} = \mathbf{P}_a + \mathbf{P}_b$. Let us indicate by $\mathbf{P}_{a|b}$ the conditional projector of $\mathbf{P}_a$ given $\mathbf{P}_b$, that is, $\mathbf{P}_{a|b} = \mathbf{P}_{a \cup b} - \mathbf{P}_b$. Similarly, we can define $\mathbf{P}_{b|a}$. Using these operators, we can generate such expressions as

$$\mathbf{P}_{a \cup b} = \mathbf{P}_a + \mathbf{P}_{a|b} = \mathbf{P}_{a|b} + \mathbf{P}_b \ , \quad \mathbf{P} + \mathbf{P}_b = \mathbf{P}_{a|b} + \mathbf{P}_{b|a} + \mathbf{P}_{a \cap b}$$

With additional projectors, dual scaling of nonorthogonal multiway data becomes somewhat complicated. For instance, what do you mean by maximizing the effect of Factor A—dual scaling of $\mathbf{P}_a\mathbf{F}$, or $\mathbf{P}_{a|b}\mathbf{F}$? The situation becomes much more complex once we introduce more than two factors in the design.

In this chapter, we have looked at only the simplest case of a two-way design for only the rows of a data matrix. Once we extend it to a k-way design (k > 2), and also introduce the column design, the analysis will become increasingly more complex, not to mention in the nonorthogonal case. With designs for both rows and columns, the analysis is comparable with that of generalized forced classification to the row and column criteria. These extensions and application of forced classification to nonorthogonal designs, however, are beyond the scope of the current book.

CHAPTER 17
Additional Topics and Future Possibilities

There are a number of interesting and important topics which have neither been discussed nor examined thoroughly enough in the preceeding sections. In particular, several problematic issues related to the inherent constraints or limitations of dual scaling have yet to be addressed. As you recall, the emphasis so far has been on the potential of dual scaling in elucidating information contained in data. From the user's point of view, however, it is equally important to pay attention to controversial issues and topics that require some caution. This chapter, therefore, is devoted mainly to such issues, and it is hoped that an awareness of them will enhance our understanding of this scaling method and its future development.

17.1 PROBLEMS ASSOCIATED WITH MULTIPLE-CHOICE OPTIONS

What is the optimal number of options or alternatives for a multiple-choice question? This question has been asked so many times by so many people, but there does not seem to be a single satisfactory answer.[1] The question can be raised from several points of view. For example, how many alternatives are optimal in terms of increasing reliability, validity, discriminability, or estimating scale values? In most studies, investigators use subjective weights for options (e.g., typically 1, 2, 3, 4, 5,. . . for ordered options), and the optimal number of options is investigated. In such studies, the results were no doubt partly dependent on the subjective weights used. In spite of the strong interest in this topic, there is not much information available on the relation between the number of options and quantification of option weights. As you will recall, some of the relevant dual scaling problems were briefly mentioned in chapter 8. Now that most

[1] Some relevant articles are listed in References.

of the technical matters have been discussed, we are ready to examine those problems in detail. We will look at four of them: overquantification, the total information available for analysis, expanded dimensionality, and outliers. As you will see, these problems are interrelated. Throughout this section, we consider the $N \times m$ incidence matrix $\mathbf{F}$, where $\mathbf{F} = [\mathbf{F}_1, \mathbf{F}_2, \ldots, \mathbf{F}_j, \ldots, \mathbf{F}_n]$, $\mathbf{F}_j$ is the $N \times m_j$ incidence matrix for Item j, the sum of each row of $\mathbf{F}_j$ is 1 (i.e., $\mathbf{F}_j\mathbf{1} = \mathbf{1}$), m is the sum of m_j over n items, and N is much larger than m.

17.1.1 Overquantification

In data collection, a checklist is often prepared for recording observations. For instance, in observing children's behavior, occurrences of specific types of behavior (e.g., sulking, thumb-sucking, nail biting, screaming) are recorded by checking one of such categories as "never, sometimes, and often." The validity of such recording is often assessed by calculating the correlation between two sets of data recorded by two observers. This correlation is often used as a measure of interrater reliability. When such a checklist has many categories (options), the usual expectation is that finer discrimination by the raters (observers) leads to some reduction in the interrater reliability. This is a reasonable expectation because the raters are given added freedom in making judgments, and accumulation of slight discrepancies in judgment between the raters may cause a substantially lower value for the correlation.

This added freedom of choice works in completely the opposite way with dual scaling. When you introduce many options, there are more options to manipulate in maximizing η^2. Therefore, the more options we have, the higher the value of η^2. In an extreme case, we can make two multiple-choice items perfectly correlated by increasing the number of options for each item. To demonstrate the effects of the number of options on scaling results, let us consider a set of three continuous variables, of which the correlation matrix is given by $\mathbf{R}_c$, where

$$\mathbf{R}_c = \begin{bmatrix} 1.00 & -0.19 & 0.02 \\ -0.19 & 1.00 & 0.28 \\ 0.02 & 0.28 & 1.00 \end{bmatrix}.$$

Suppose we decide to categorize each continuous variable into a specific number of categories and subject the categorized data to dual scaling. As we have seen, we can obtain an interitem correlation matrix from each solution. Four cases will be examined: each continuous variable is discretized into 2, 4, 5, and 10 categories, and the corresponding interitem correlation matrices, based on the first solutions, are calculated, and are indicated respectively as $\mathbf{R}_2$, $\mathbf{R}_4$, $\mathbf{R}_5$, and $\mathbf{R}_{10}$:

$$\mathbf{R}_2 = \begin{bmatrix} 1.00 & 0.10 & 0.00 \\ 0.10 & 1.00 & 0.30 \\ 0.00 & 0.30 & 1.00 \end{bmatrix},$$

$$\mathbf{R}_4 = \begin{bmatrix} 1.00 & 0.31 & 0.50 \\ 0.31 & 1.00 & 0.43 \\ 0.50 & 0.43 & 1.00 \end{bmatrix},$$

$$\mathbf{R}_5 = \begin{bmatrix} 1.00 & 0.47 & 0.57 \\ 0.47 & 1.00 & 0.54 \\ 0.57 & 0.54 & 1.00 \end{bmatrix},$$

$$\mathbf{R}_{10} = \begin{bmatrix} 1.00 & 0.93 & 0.91 \\ 0.93 & 1.00 & 0.94 \\ 0.91 & 0.94 & 1.00 \end{bmatrix}.$$

There is no doubt that the number of options has a decisive influence on the outcome of quantification. In this particular example, it looks as though the dichotomous case (i.e., two-option case) results in a correlation matrix which is closest to the correlation matrix of the original continuous variables. As the number of options increases, the average interitem correlation increases rapidly to such an extent that we may no longer believe that the corresponding correlation matrix has anything to do with the original matrix.

What does this demonstration tell us? Can we say that a case of many options has in fact fabricated information which did not exist in the original data? We hope not. What then is our interpretation of the above results? They seem to imply two kinds of interpretation: one is a reminder that dual scaling is not a technique to approximate the continuous data structure, and the other is the possibility of an effective use of this overquantification phenomenon for a specific research purpose. Let us elaborate on these points.

In practice, we often encounter data sets which contain both categorical data and continuous data. The usual practice is to categorize continuous variables, and then submit the entire data set to dual scaling. This practice is fine as long as we know what dual scaling does. But if anyone expects that dual scaling results will be an approximation of data analysis of continuous variables, they are wrong. How can we present $\mathbf{R}_{10}$ as a substitute for the original matrix $\mathbf{R}_c$? This substantial discrepancy between the continuous case and the dual scaling results clearly identifies one area in which further research is needed: that is, research to develop a method that approximates continuous data structure through analysis of categorized data.

The second implication reflects the other side of the same coin. It refers to a unique property of dual scaling, that is, to enhance the visibility of differences in data, as exemplified by forced classification. Recall that in the example of Christmas party plans, we chose for forced classification the criterion pair consisting of one inexpensive and one expensive plan. The outcome was such that not only were the party plans separated into inexpensive and expensive categories, but also the subjects were cleanly divided into those who preferred inexpensive plans and those who chose expensive plans.

Let us look at another example in which two continuous variables v and w are related by $w = v^2$. Consider the following four matrices:

$$\mathbf{X}_c = \begin{bmatrix} 6 & 36 \\ 5 & 25 \\ 4 & 16 \\ 3 & 9 \\ 2 & 4 \\ 1 & 1 \\ -1 & 1 \\ -2 & 4 \\ -3 & 9 \\ -4 & 16 \\ -5 & 25 \\ -6 & 36 \end{bmatrix}, \quad \mathbf{X}_2 = \begin{bmatrix} 2 & 2 \\ 2 & 2 \\ 2 & 2 \\ 2 & 1 \\ 2 & 1 \\ 2 & 1 \\ 1 & 1 \\ 1 & 1 \\ 1 & 1 \\ 1 & 2 \\ 1 & 2 \\ 1 & 2 \end{bmatrix}, \quad \mathbf{X}_4 = \begin{bmatrix} 3 & 3 \\ 3 & 3 \\ 3 & 2 \\ 3 & 2 \\ 2 & 1 \\ 2 & 1 \\ 2 & 1 \\ 2 & 1 \\ 1 & 2 \\ 1 & 2 \\ 1 & 3 \\ 1 & 3 \end{bmatrix}, \quad \mathbf{X}_6 = \begin{bmatrix} 6 & 6 \\ 6 & 5 \\ 5 & 4 \\ 5 & 3 \\ 4 & 2 \\ 4 & 1 \\ 3 & 1 \\ 3 & 2 \\ 2 & 3 \\ 2 & 4 \\ 1 & 5 \\ 1 & 6 \end{bmatrix}.$$

$\mathbf{X}_c$ is the original matrix of two variables (v,w) with the range of v from -6 to 6 with an interval of 1. $\mathbf{X}_2$ is the coded condensed matrix obtained by dividing v and w at the respective midpoints. Similarly, $\mathbf{X}_4$ and $\mathbf{X}_6$ are data categorized into four and six, respectively. The product moment correlation between v and w, based on $\mathbf{X}_c$, is 0. If we subject the three condensed matrices to dual scaling and calculate the correlation between two item scores, we will obtain correlation coefficients 0, 1, and 1 for $\mathbf{X}_2$, $\mathbf{X}_4$, and $\mathbf{X}_6$, respectively. Do you know why? To simplify our discussion, let us indicate the values of variables v and w by their subscripts, that is, $j = v_j$, and $k = w_k$ (e.g., $v_3 = 3$, $w_9 = 9$). Then, for $\mathbf{X}_4$, dual scaling will assign identical weights to 3 and 1 (i.e., $v_3 = v_1 = a$), and a different weight, b, to 2 (i.e., $v_2 = b$), while identical weights are given to w_3 and w_2 and a different weight to w_1, that is, $w_3 = w_2 = a$, $w_1 = b$. For $\mathbf{X}_6$, dual scaling assigns weights to the options as $v_6 = v_1 = w_5 = w_6 = a$, $v_5 = v_2 = w_3 = w_4 = b$, and $v_4 = v_3 = w_2 = w_1 = c$. Thus, dual scaling of $\mathbf{X}_4$ and $\mathbf{X}_6$ will yield corresponding item-score matrices $\mathbf{Y}_4$ and $\mathbf{Y}_6$, where

$$\mathbf{Y}_4 = \begin{bmatrix} a & a \\ a & a \\ a & a \\ b & b \\ b & b \\ b & b \\ b & b \\ b & b \\ b & b \\ a & a \\ a & a \\ a & a \end{bmatrix}, \quad \mathbf{Y}_6 = \begin{bmatrix} a & a \\ a & a \\ b & b \\ b & b \\ c & c \\ c & c \\ c & c \\ c & c \\ b & b \\ b & b \\ a & a \\ a & a \end{bmatrix}.$$

Thus, it is obvious why the correlation r_{vw} is 1 for $\mathbf{X}_4$ and $\mathbf{X}_6$. As for $\mathbf{X}_2$, there is no way to make the correlation higher than 0, or lower than 0. When we look at the value of the derived correlation, overquantification makes it unnatural or spurious. But at the same time, we gain from overquantification a clear-cut demarcation for a cluster of subjects or variables. Overquantification tells us how we should group subjects and/or variables to arrive at perfect classification—a method which may be useful as an anchor point or a benchmark in research on classification.

17.1.2 Total Contributions of Options and Items

As we saw earlier, the sum of squares of Option p of Item j, SS_{jp}, is an indicator of the total amount of information that the option contributes to the entire data set. You will recall matrix $\mathbf{C}_1$, which was subjected to eigenvalue decomposition. To simplify the notation, $\mathbf{C}_1$ will be simply written $\mathbf{C}$ without the subscript 1 in the remainder of this chapter; SS_{jp} is the diagonal element of this matrix for Item j and Option p. Similarly, the sum of squares of Item j, SS_j, is the contribution of Item j to the total data set, and is the sum of SS_{jp} over p; SS_t is the total sum of squares, and is the sum of SS_j over j. These quantities are directly influenced by option frequencies and the number of options, for they are given by the following expressions:

(17.1) $$SS_{jp} = \frac{(N-f_{jp})}{nN} .$$

(17.2) $$SS_j = \sum_{p=1}^{m_j} SS_{jp} = \frac{(m_j-1)}{n} .$$

(17.3) $$SS_t = \sum_{j=1}^{n} \frac{(m_j-1)}{n} = \overline{m}-1 .$$

The first equation is not a function of the number of options, m_j. Yet it is directly related to the second equation, which is a function of m_j. Likewise, the third equation is obtained from the second equation and the average number of options is the key component of the expression. Recall that dual scaling as formulated in this book employs two constraints: that the sum of weighted responses is 0, and that the sum of squares of weighted responses is equal to f_t, which is nN if no missing responses are involved. Therefore, it would be interesting to see the total sum of squares of weighted responses of Option p of Item j, T_{jp}, the total sum of squares of responses of Item j, T_j, and the total sum of squares of weighted responses of the entire data set, T_t. They are given as follows:

(17.4) $$T_{jp} = \sum_{k=1}^{K} f_{jp}\, x_{jpk}^2 = n(N-f_{jp}) ,$$

(17.5) $$T_j = \sum_{p=1}^{m_j} T_{jp} = f_t(m_j-1) ,$$

(17.6) $$T_t = \sum_{j=1}^{n} T_j = (m-n)f_t .$$

Considering that the number of options has direct effects on these sums of squares, you might suggest standardizing the matrix **C** (i.e., $\mathbf{C}_1$) with respect to options or items. Such standardization will equalize the contribution of each option or item, but at the same time, the resultant quantification will no longer optimize η^2. In chapter 15, we saw an example of option standardization, which eliminated outliers caused by extremely small frequencies. But, apart from a possible lack of the optimization criterion, does the standardization eliminate our concern of using too many options? Definitely not. There is another problem caused by too many options: the problem of expanded dimensionality.

17.1.3 Expanded Dimensionality Due to Excessive Options

The rank of the matrix **F** is m-n+1, and the total number of nontrivial solutions is $m - n$. If we indicate by "m bar" the average number of options, then the total number of solutions can be expressed as

(17.7)
$$m-n = n(\overline{m}-1) .$$

Consider again creating categorical variables by discretizing each of n continuous variables into 2, 3, 4, and n categories. Then compare the total number of solutions from n continuous variables by principal component analysis (PCA) with those from incidence matrices of discretized variables by dual scaling, as shown in Table 17.1. This is another situation in which we can show that dual scaling does not provide an approximation to the analysis of continuous variables. Observe that PCA of n variables yields n components, but that dual scaling of infinitely discretized data (i.e., continuous data) would produce an infinite number of solutions. With respect to Table 17.1, Nishisato, Ahn and Shin (1987) presented an interesting finding that (a) when the average interitem correlation is very high (e.g., 0.995) dual scaling of k-option multiple-choice data produces (k-1) almost equally dominant squared correlation ratios; (b) as the average correlation decreases, those (k-1) dominant solutions disappear among other solutions and are no longer identifiable. Figure 17.1 shows typical examples of this finding.

TABLE 17.1
Total Number of Solutions

Data	continuous	2 options	3 options	4 options	n options
Solution	n	n	$2n$	$3n$	$n(n$-1)

The phenomenon can also be replicated under a different condition. Recall forced classification with the k-option criterion item. As the weight on the criterion increases, forced classification yields (k-1) proper solutions, associated with (k-1) dominant correlation ratios. This is the same phenomenon demonstrated previously.

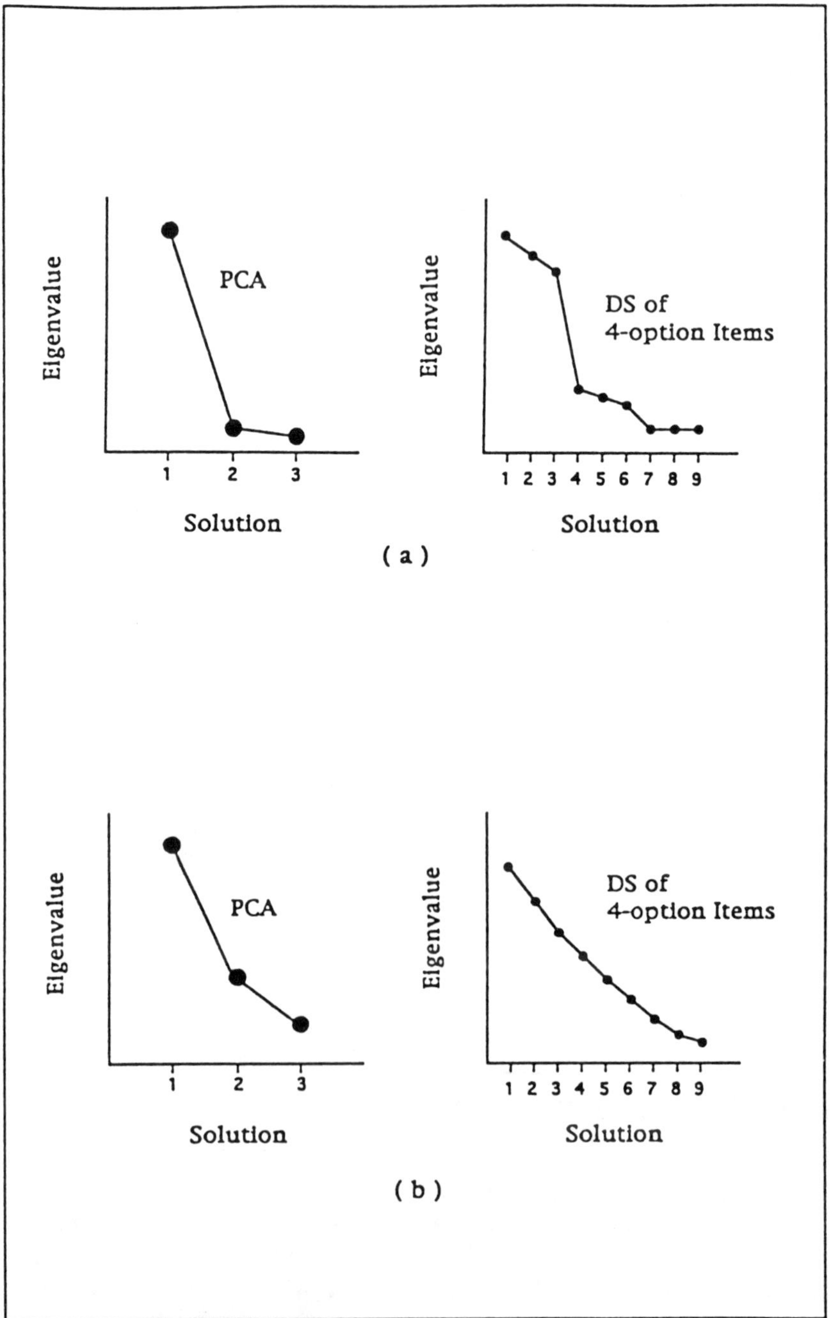

Fig. 17.1. Eigenvalues associated with highly correlated items (a), and items of moderate correlation (b).

What implications does this phenomenon have? By looking at Figure 17.1, you may wonder about two aspects. If the PCA component is explained by three dual scaling solutions, what are those two extra solutions? If the three solutions that correspond to one PCA component cannot be identified because of low interitem correlation, what can we do about it? These two questions are reasonable, considering what we must do in interpreting results. The fact that dual scaling yields many solutions shows its great potential as a data analytic tool. For instance, consider a medical checklist and forced classification of the data with "blood pressure" (high, medium, low) as the criterion. The analysis yields two dominant solutions, which capture different patterns of relations between the criterion and other variables. For example, the two patterns may be "the higher the blood pressure, the higher the scores on stress and heart ailments" and "the more extreme (high or low) the blood pressure, the higher the scores on dizziness and headache." Some data analytic methods capture only linear relations such as "the higher the blood pressure the higher the stress score." In contrast, dual scaling can identify not only such a linear relation, but also nonlinear relations such as "high headache scores come from both high blood pressure patients and low blood pressure patients." Would you not think that age is linearly related to some variables (e.g., flexibility of the body) and nonlinearly to some other variables (e.g., running time)? In dual scaling, those extra solutions caused by multiple options are means to capture all kinds of response patterns. Indeed, dual scaling looks into all possible response patterns. Since each solution picks up a specific combination of options of different items, whether the first (k-1) dual scaling solutions correspond to the first PCA solution is not as crucial as we might think. Whatever combinations of options are dominant, they should be extracted first. On the basis of this discussion, we can say that extra solutions are not per se a drawback of dual scaling, but rather a noteworthy and useful feature.

We must nevertheless be concerned with the use of too many options. Imagine, for example, having to interpret all the solutions contributed by such combinations as [strongly agree to Proposal A, strongly agree to Proposal B], [very strongly agree to A, strongly agree to B], [very strongly agree to A, moderately agree to B], [strongly agree to A, very strongly agree to B], and [moderately agree to A, moderately agree to B]. Notice that all these combinations from different solutions are telling us that they all agree to Proposals A and B. Then, why should we have to bother with all of them? In practice, many investigators wish to capture "accurate" information and use as many as five, seven, or nine options. But when such data are analyzed by dual scaling, it may be that the first few solutions tell us more or less the same thing, or that differences among solutions are not so clear as to be attributable to identifiable, unique features. Can we not use simply [disagree, neutral, agree] or [disagree, agree]? Seven or nine options for dual scaling are definitely too many to be useful, contrary to the expectation of some investigators. Two or three options are easier to analyze, and in most instances adequate for capturing important response patterns.

One final note on this topic is the reminder that we are discussing it in the context of dual scaling. If you were to pose the question on the effect of categorization of continuous variables on product-moment correlation, it would be natural to expect that the original correlation is better approximated by more categories. Notice, however, that in this case categories are typically assigned ordered values such as the integers 1,2,3 . . to calculate correlation. This is different from dual scaling.

17.1.4 Outliers

We discussed outliers caused by small frequencies of responses in Chapter 15. Ideally, we would like to have an even distribution of responses over all the response options. However, as the number of response options increases, the distribution of responses over options tends to become less even, and one or two options with very few responses will inevitably result. These options are likely to play a role in generating outliers. If your data set already has many options, you can handle the problem in two ways. First, if the options are ordered (e.g., strongly disagree, moderately disagree, neutral, moderately agree, strongly agree), options with small numbers of responses should be merged with adjacent options. Second, if the options are not ordered, discard those options with 0 or 1 response. This may sound like too drastic a measure, but in view of the present state of knowledge, it is a reasonable approach, and one that will work well.

17.2 AMOUNT OF INFORMATION EXPLAINED BY EACH SOLUTION

We often talk about the amount of information explained by individual solutions. This statistic, however, is not necessarily easy to interpret because its interpretation may depend on the types of analysis,[2] or the

---

[2] See, for example, O'Grady (1982).

kinds of data we analyze.[3] In dual scaling, we use η^2 as the optimization criterion, and we know that the sum of squared correlation ratios over all solutions is equal to the sum of the diagonal elements of the residual matrix **C**, that is, $\mathbf{C}_1$ discussed in chapter 6:

$$\sum_{k=1}^{K} \eta_k^2 = \sum_{j=1}^{m} c_{jj} = tr(\mathbf{C}), \tag{17.8}$$

where "$tr(\mathbf{C})$" reads "the trace of the matrix **C**," and it is equal to the sum of diagonal elements of **C**. As this formula indicates, $tr(\mathbf{C})$ is also equal to the sum of all squared correlation ratios. The percentage of the information explained by solution k, δ_k, is, as we have already seen, defined as

$$\delta_k = \left(\frac{100\eta_k^2}{tr(\mathbf{C})} \right) \% . \tag{17.9}$$

Note that irrespective of the data types discussed so far, η_k^2 is always bounded by 0 and 1. However, the value of $tr(\mathbf{C})$ depends on the data type. This discrepancy in $tr(\mathbf{C})$ makes it difficult to talk about δ_k without any reference to data types. Let us now look at $tr(\mathbf{C})$ for each data type.

17.2.1 Contingency Tables

Suppose that the dimension of the contingency table **F** is $n \times m$. Then the trace of the matrix **C** is given by

$$tr(\mathbf{C}) = \sum_{i=1}^{n} \sum_{j=1}^{m} \frac{f_{ij}^2}{f_{i.} f_{.j}} - 1 = \frac{\chi^2}{f_t}, \tag{17.10}$$

where χ^2 is called "chi-square," given by

$$\chi^2 = \sum_{i=1}^{n} \sum_{j=1}^{m} \frac{\left(f_{ij} - \frac{f_{i.} f_{.j}}{f_t}\right)^2}{\frac{f_{i.} f_{.j}}{f_t}} .$$

The maximal value of the chi-square statistic changes as the dimension of the contingency table changes, and it is known to be $(p-1)f_t$, where p is the smaller value of n and m. The chi-square statistic takes the minimal value

[3] See, for example, Nishisato and Sheu (1985), Nishisato (1993).

of 0 when the rows and the columns are statistically independent. Therefore, we can conclude that $tr(\mathbf{C})$ is bounded as

$$0 \leq tr(\mathbf{C}) \leq p-1 \ . \qquad (17.11)$$

When rows and columns are statistically independent, the squared correlation ratio is also 0. The maximal value of $tr(\mathbf{C})$ can be attained when there exists perfect association between rows and columns. Thus, the general expectation is that $tr(\mathbf{C})$ is a monotonically increasing function of η^2. But would this not create a problem for δ? It is often instructive to generate small numerical examples to see what will happen to the quantity in which we are interested. So let us look at the following six contingency tables, all of which have identical marginals.

$$\mathbf{F}_1 = \begin{bmatrix} 20 & 0 & 0 \\ 0 & 10 & 0 \\ 0 & 0 & 8 \end{bmatrix}, \quad \mathbf{F}_2 = \begin{bmatrix} 19 & 1 & 0 \\ 1 & 9 & 0 \\ 0 & 0 & 8 \end{bmatrix}, \quad \mathbf{F}_3 = \begin{bmatrix} 18 & 1 & 1 \\ 1 & 8 & 1 \\ 1 & 1 & 6 \end{bmatrix},$$

$$\mathbf{F}_4 = \begin{bmatrix} 16 & 3 & 1 \\ 3 & 6 & 1 \\ 1 & 1 & 6 \end{bmatrix}, \quad \mathbf{F}_5 = \begin{bmatrix} 16 & 3 & 1 \\ 3 & 4 & 3 \\ 1 & 3 & 4 \end{bmatrix}, \quad \mathbf{F}_6 = \begin{bmatrix} 14 & 3 & 3 \\ 3 & 4 & 3 \\ 3 & 3 & 2 \end{bmatrix}.$$

Since each matrix is 3×3, the maximal value of $tr(\mathbf{C})$ is 3 - 1 = 2. This value, however, changes as the distribution of cell frequencies changes. Out of these contingency tables, $\mathbf{F}_1$ has the strongest row-column association, and $\mathbf{F}_6$ the weakest. Thus, common sense would tell us that if we were to ignore the second solution, the first solution would capture the largest amount of information of $\mathbf{F}_1$ and the least amount of $\mathbf{F}_6$. Let us examine the analysis results shown in Table 17.2. Notice that $tr(\mathbf{C})$ decreases quite drastically from $\mathbf{F}_1$ to $\mathbf{F}_6$, and η^2 changes in a manner parallel to, but more gradually than, $tr(\mathbf{C})$. When we look at δ_1, we notice something quite unexpected: δ_1 *increases* as the row-column association *decreases*. Is this not contrary to what we would expect? Most investigators would say "Yes, it is totally unexpected." If we consider principal component analysis of two standardized variables, the first eigenvalue is (1 + r), where r is the positive value of the product-moment correlation between two variables and $tr(\mathbf{R})$ = 2. Thus, the higher the correlation, the greater the first eigenvalue *and* δ_1. Then why is it that δ_1 is not high in dual scaling when the row-column association is high? It is because χ^2 is comparatively larger than η^2 when the row-column association is high. What do you do if the interpretation is further complicated by two tables of different dimensions? For instance, if the dimension of the contingency table is expanded to 5×5, and all the responses fall in the main diagonal positions like $\mathbf{F}_1$, then tr(**C**) is 4, $\eta_1^2 = \eta_2^2 = \eta_3^2 = \eta_4^2 = 1.0000$, and the corresponding δ's are all equal to 0.2500. Imagine how complicated it would be to compare δ's if you were given only values of δ_1 of contingency tables of different dimensions.

The unexpected result is due to the fact that we have eliminated the contribution of the trivial solution in calculating the residual matrix **C**. Thus, one may consider replacing *tr*(**C**) by *tr*(**B'B**), discussed in chapter 6, and propose a statistic $\gamma = 100\eta_k^2/tr(\mathbf{B'B})$. See the values of γ in Tables 7.2(a) and 7.2(b). We see the expected outcome: the larger the values of η_1^2 the greater the value of γ_1. Recall, however, that γ would not be of much use, because dual scaling is a method to analyze association in the data set, that is, information in **C**, not **B'B**, and hence γ does not reflect the relative contribution of the solution in the entire set of admissible solutions.

What about comparisons of δ's within each data set rather than between different data sets? Here, too, we discover the strange behavior of δ : the smaller the row-column association, the greater the drop in value from δ_1 to δ_2. Thus in the case of contingency tables, the problem associated with δ is very complex, and further work on its use is needed. In the meantime, it is advisable not to use δ for dual scaling analysis of contingency tables.

It is opportune to have another look at the statistic "percentage homogeneity" (Ph),[4] which is given by $tr(\mathbf{C})\delta_k$, that is, $100\eta_k^2$. This statistic changes from 0 to 100 as a function of η^2, and is *free* from the effect of the matrix size. If we were to compare matrices of the same size such as the previous examples of 3×3 matrices, a more reasonable measure than δ seems to be δ* defined as

(17.12)
$$\delta* = \left(\frac{100\eta_k^2}{p-1}\right)\%$$

See δ* listed in Table 17.2(a) and (b). Instead of dividing $100\eta^2$ by *tr*(**C**), it is divided by the maximal trace value of a 3×3 contingency table. In this regard, δ* makes more sense than δ. However, as we have noted, the maximal value of δ* is 100/(*p*-1). In other words, the larger the contingency table, the smaller the value of δ* when η^2 is 1. Thus, if we wish to "standardize" δ* in terms of the dimension of the contingency table, the resultant measure is (*p*-1)δ*, which is equal to $100\eta_k^2$, the percentage homogene-

[4] Nishisato and Nishisato (1984).

ity. This is the rationale behind the statistic "percentage homogeneity." See the last columns of Table 17.2(a) and (b).

TABLE 17.2(a)
Analyses of Six Contingency Tables (Solution 1)

Data	tr(**C**)	η_1^2	δ_1	γ	δ_1*	Ph
$\mathbf{F}_1$	2.0000	1.0000	50.00	33.33	50.00	100.00
$\mathbf{F}_2$	1.7225	1.0000	58.06	33.33	50.00	100.00
$\mathbf{F}_3$	1.0600	0.6259	59.05	20.86	31.30	62.59
$\mathbf{F}_4$	0.6900	0.4829	69.98	16.09	24.15	48.29
$\mathbf{F}_5$	0.3775	0.3685	97.73	12.28	18.43	36.85
$\mathbf{F}_6$	0.1400	0.1395	99.61	4.65	6.98	13.95

TABLE 17.2(b)
Analyses of Six Contingency Tables (Solution 2)

Data	tr(**C**)	η_2^2	δ_2	γ	δ_2*	Ph
$\mathbf{F}_1$	2.0000	1.0000	50.00	33.33	50.00	100.00
$\mathbf{F}_2$	1.7225	0.7225	41.94	24.08	36.13	72.25
$\mathbf{F}_3$	1.0600	0.4341	40.95	14.47	21.71	43.41
$\mathbf{F}_4$	0.6900	0.2071	30.02	6.90	10.36	20.71
$\mathbf{F}_5$	0.3775	0.0086	2.27	0.29	0.43	0.86
$\mathbf{F}_6$	0.1400	0.0005	0.39	0.02	0.03	0.05

17.2.2 Multiple-Choice Data

This case has already been looked at in chapter 8. The trace of the matrix **C** is equal to the average number of options minus 1, that is,

$$tr(\mathbf{C}) = \sum_{j=1}^{n} \frac{m_j}{n} - 1 = \overline{m} - 1 \; . \tag{17.13}$$

Unlike in the case of contingency tables, the trace does not change as a function of the number of items,[5] n, or the consistency of the data, η^2.

[5] However, $tr(\mathbf{C})$ becomes smaller than the value indicated by (17.13) where there are some items with no variation involved (i.e., if all the subjects choose only one option of an item: everyone says "yes" to a question, or everyone fails in an item).

Thus, the statistic δ can be compared among data sets which have the same average number of options. At the same time, its dependency on the number of options contributes to another kind of problem, that is, the range of δ is a function of the average number of options, and as a consequence, δ for multiple-choice data also has an anomalous condition. More specifically, δ is bounded as

(17.14)
$$0 \leq \delta \leq \frac{100}{\overline{m}-1} .$$

Thus, δ cannot attain its absolute maximum of 1 if the average number of options is greater than 2, even when η^2 is one. The upper limits are shown in Table 17.3 as a function of the average number of options.

TABLE 17.3
Upper limits of δ

Option	2	3	5	9
maxδ	100%	50%	25%	12.5%

In this table, we assume that all of the n items have the same number of options. Notice also that the maximal value of δ is independent of the number of items, n. In chapter 8, we remarked that the coefficient of internal consistency reliability (Cronbach's α) increases as n increases, and that α becomes negative when η^2 is less than $1/n$. Yet δ is not affected by n. Why not? It is because δ is a function of the ratio of η^2 to the sum of η^2, and the correlation ratio of multiple-choice data is a function of n, thus cancelling out the influence of n. To clarify this point, recall that the trace of **C** is the average number of options minus 1 and the total number of possible solutions, that is, the rank of **C**, r(**C**), is the total number of options minus n. Therefore, the average squared correlation ratio is

(17.15)
$$\overline{\eta^2} = \frac{tr(\mathbf{C})}{r(\mathbf{C})} = \frac{\sum_{j=1}^{n} \frac{m_j}{n} - 1}{\sum_{j=1}^{n} m_j - n} = \frac{1}{n} .$$

From Equation 17.15, we can infer that δ is free from the effect of n. It is interesting to know that the average value of η^2 corresponds to the value of α being 0.

Now, what about the usefulness of the statistic δ? Unfortunately, it can be misleading, and the value of δ often disappoints many investigators.

But we have a much clearer situation here than in the case of the contingency table. Of particular importance here is the point we discussed in relation to the problem of expanded dimensionality. The information in the data is distributed over multiple-choice options, and it is natural that the average contribution of each solution is reduced as a function of the number of options. As used in chapter 8, one way to mitigate the problem of δ is to define it only with respect to "proper solutions," that is, solutions of which $\eta_k^2 \geq 1/n$ (i.e., $\alpha \geq 0$). Scaling up the value of δ has also been proposed by Benzécri (1979) and Greenacre (1987) as mentioned in chapter 8. If we require a measure for the accountability of each solution, the percentage homogeneity, discussed earlier, is a good candidate.

17.2.3 Paired Comparison and Rank-Order Data

These two data types are converted into the $N \times n$ dominance matrix $\mathbf{E}$. Although the two are different with respect to their methods of data collection and response transitivity, the corresponding dominance matrices can be treated in the same way. Indeed, if you have an $N \times n$ matrix of standardized continuous scores (i.e., N subjects and n tests), you can compare n scores of each subject and transform them to paired comparison data and rank-order data. Once you generate these two types of data from the same continuous data, they will produce identical dual scaling results.

The trace of the matrix $\mathbf{C}$ can be calculated as

$$(17.16) \qquad tr(\mathbf{C}) = \left[\frac{1}{Nn(n-1)^2}\right]\left[\frac{N(n-1)n(n+1)}{3}\right] = \frac{n+1}{3(n-1)} .$$

As n increases, $tr(\mathbf{C})$ approaches 1/3, which is the lower bound of $tr(\mathbf{C})$. As n becomes smaller, $tr(\mathbf{C})$ increases. Since n is an integer and we must have at least two stimuli, this means that $tr(\mathbf{C})$ attains its maximal value of 1 when $n = 2$. Therefore, $tr(\mathbf{C})$ is bounded as

$$(17.17) \qquad \frac{1}{3} \leq tr(\mathbf{C}) \leq 1 .$$

This means that η^2 for paired comparison and rank-order data is also bounded between 1/3 and 1. Thus, even when δ is 100%, η^2 may be smaller than 1. This is a rather unexpected phenomenon, and it calls for numerical illustrations. Let us consider the following four rank-order matrices with seven subjects and the number of stimuli changing from two to five:

$$\mathbf{F}_2 = \begin{bmatrix} 1 & 2 \\ 1 & 2 \\ 1 & 2 \\ 1 & 2 \\ 1 & 2 \\ 1 & 2 \end{bmatrix}, \quad \mathbf{F}_3 = \begin{bmatrix} 1 & 2 & 3 \\ 1 & 2 & 3 \\ 1 & 2 & 3 \\ 1 & 2 & 3 \\ 1 & 2 & 3 \\ 1 & 2 & 3 \end{bmatrix}, \quad \mathbf{F}_4 = \begin{bmatrix} 1 & 2 & 3 & 4 \\ 1 & 2 & 3 & 4 \\ 1 & 2 & 3 & 4 \\ 1 & 2 & 3 & 4 \\ 1 & 2 & 3 & 4 \\ 1 & 2 & 3 & 4 \end{bmatrix}, \quad \mathbf{F}_5 = \begin{bmatrix} 1 & 2 & 3 & 4 & 5 \\ 1 & 2 & 3 & 4 & 5 \\ 1 & 2 & 3 & 4 & 5 \\ 1 & 2 & 3 & 4 & 5 \\ 1 & 2 & 3 & 4 & 5 \\ 1 & 2 & 3 & 4 & 5 \end{bmatrix}.$$

All these matrices are exhaustively explained by one solution, as you can see from the fact that all the subjects ranked the stimuli in exactly the same manner. Let us look at the results of the analysis in Table 17.4. Despite the fact that all the cases produced only one solution and that the analysis was exhaustive, the results are influenced by the number of stimuli, n. Recall that in multiple-choice data it was δ which was constrained. In other words, even when η^2 is 1, δ is less than 100% if the average number of options is greater than 2. Now, in paired comparison and rank-order data, the situation is reversed: even when δ = 100%, η^2 cannot attain its maximum of 1 if the number of stimuli is greater than 2. Since the maximal value of $tr(\mathbf{C})$ is 1, this is a case in which δ means what our common sense tells us, that is, if δ_k is 100%, the information in the entire data set is explained by, or contained in, Solution k. One may argue that this is also true with multiple-choice data. While this is correct, in multiple-choice data a single solution may never attain a δ of 100%. In contrast to the case of incidence data, the percentage homogeneity, $100\eta^2$, is misleading and thus should not be used for paired comparison and rank-order data; δ is the meaningful statistic for dominance data.

TABLE 17.4(a)
Summary Statistics

	$\mathbf{F}_2$	$\mathbf{F}_3$	$\mathbf{F}_4$	$\mathbf{F}_5$
tr(**C**)	1.0000	0.6667	0.5556	0.5000
η^2	1.0000	0.6667	0.5556	0.5000
δ	100%	100%	100%	100%

TABLE 17.4(b)
Optimal Weights for Subjects

	Normed weights				Adjusted weights			
Subj	F_2	F_3	F_4	F_5	F_2	F_3	F_4	F_5
1	1	1	1	1	1	0.82	0.75	0.71
2	1	1	1	1	1	0.82	0.75	0.71
3	1	1	1	1	1	0.82	0.75	0.71
4	1	1	1	1	1	0.82	0.75	0.71
5	1	1	1	1	1	0.82	0.75	0.71

TABLE 17.4(c)
Optimal Weights for Stimuli

	Normed weights				Adjusted weights			
St	F_2	F_3	F_4	F_5	F_2	F_3	F_4	F_5
x_1	1	1.22	1.34	1.41	1	1	1	1
x_2	-1	0	0.45	0.71	-1	0	0.33	0.5
x_3		-1.22	-0.45	0		-1	-0.33	0
x_4			-1.34	-0.71			1	-0.5
x_5				-1.41				-1

17.2.4 Successive Categories Data

Let n and m be, respectively, the number of stimuli and the number of category boundaries. Then, recalling that successive categories data are first converted to rank-order data and then to the $N\times(n+m)$ dominance matrix, the trace of the matrix **C** can be calculated as

(17.18)
$$tr(\mathbf{C}) = \left[\frac{1}{N(n+m)(n+m-1)^2}\right]\left[\frac{N(n+m-1)(n+m)(n+m+1)}{3}\right] = \frac{n+m+1}{3(n+m-1)} .$$

The $tr(\mathbf{C})$ decreases to the limit of 1/3, as n or m or both increase. What would be the smallest values of n and m? It seems that n and m each must be at least 1. Then, when $n+m=2$, $tr(\mathbf{C})$ attains the value of 1. We therefore have the same situation as paired comparison and rank-order data, that is,

(17.19)
$$\frac{1}{3} \le tr(\mathbf{C}) \le 1 .$$

Considering all these different constraints on δ, it is difficult to compare scaling results only in terms of the value of δ. It seems that δ is informative when dominance data are analyzed, but that it can be quite misleading when it is used for incidence data. Nevertheless, δ is still useful as a supplementary source of information. More work is needed on measures of information explained by each solution.

17.3 PARTIAL ORDER CONSTRAINTS ON CATEGORY WEIGHTS

17.3.1 Contingency Tables

It is not rare to see that the rows or columns of a contingency table have ordered categories such as poor, fair, good, and excellent. In multiple-choice achievement tests, one of the response alternatives (options) to a question is the correct one or the best alternative. In addition, the teacher would know which one of the remaining is the second best, or the worst. When such data are subjected to dual scaling, there is no guarantee that the derived weights conform to the expected order. It would be strange if the weight for "good" is larger than the weight for "excellent." It would even be embarrassing if an incorrect alternative got a larger weight than the correct one. This kind of violation of the natural or expected order can and does occur in practice. What should we do about it?

TABLE 17.5
Contingency Tables with Ordered Categories

	Excellent	Good	Fair	Poor	Terrible
Treatment	x_1	x_2	x_3	x_4	x_5
1	9	5	9	13	4
2	7	3	10	20	4
3	14	13	6	7	0
4	11	15	3	5	8
5	0	2	10	30	2

There are a number of studies on this topic.[6] Out of them, the simplest one seems to be the method of successive data modification (SDM).[7] We will look at SDM with numerical examples. Consider the data reported by Bradley, Katti, and Coons (1962), as shown in Table 17.5. The SDM procedure takes the following steps:

Step 1. Specify the order constraint you wish to impose on the category weights. Let us choose the following:

[A] $$x_1 \geq x_2 \geq x_3 \geq x_4 \geq x_5$$

or, alternatively,

[B] $$x_1 - x_2 \geq 0,\ x_2 - x_3 \geq 0,\ x_3 - x_4 \geq 0,\ x_4 - x_5 \geq 0\ .$$

Step 2. Carry out dual scaling of the matrix **F**, and obtain category weights (i.e., Solution 1).

Step 3. Check if all the pairwise constraints (B) are satisfied. If they are, the current weights are our solution; go to Step 4. If they are not, identify a pair (x_p, x_q) that shows the worst (greatest) violation, and modify the corresponding columns p and q of the matrix **F** by

(17.20) $$\text{Modified } f_{ip} = \text{Modified } f_{iq} = \frac{f_{ip} + f_{iq}}{2}\ .$$

Indicate the modified matrix by **F**, and go to Step 2.

Step 4. Stop the analysis.

It may be obvious from our discussion on the principle of equivalent partitioning in chapter 14 that the modification of two columns of **F** by (17.20) is equivalent to merging columns p and q into a single column. This is an alternative to Equation 17.20, and you may choose whichever is convenient to use. The main difference between them is that (17.20) leaves

---

[6] See Bradley, Katti, and Coons (1962), Inukai (1972), Nishisato and Inukai (1972), Nishisato (1973, 1980a), Nishisato and Arri (1975), Tanaka and Asano (1978), Tanaka (1979), Tanaka and Kodake (1980), Tenenhaus (1988).

[7] Nishisato (1973); see also Nishisato (1980a).

the dimension of **F** unchanged, while the alternative method of merging reduces the number of columns by one each time the data are modified.

In our example, the standard procedure yields results:

$$\mathbf{x} = \begin{bmatrix} 1.0432 \\ 1.3047 \\ -0.5500 \\ -1.0658 \\ 0.4714 \end{bmatrix}.$$

The weights do not satisfy the order constraint (A), for if we compute the pairwise differences of (B), we obtain

$$\begin{bmatrix} x_1-x_2 \\ x_2-x_3 \\ x_3-x_4 \\ x_4-x_5 \end{bmatrix} = \begin{bmatrix} -0.2616 \\ 1.8547 \\ 0.5159 \\ -1.5372 \end{bmatrix}.$$

The worst violation occurs for the pair (4,5). Therefore, using Equation 17.20, we change the original contingency table to the new matrix **F**, where

$$\mathbf{F} = \begin{bmatrix} 9 & 5 & 9 & 8.5 & 8.5 \\ 7 & 3 & 10 & 12 & 12 \\ 14 & 3 & 6 & 3.5 & 3.5 \\ 11 & 15 & 3 & 6.5 & 6.5 \\ 0 & 2 & 10 & 16 & 16 \end{bmatrix}.$$

Notice that columns 4 and 5 are replaced with averages of the correspon-ding rows of the original matrix. Dual scaling of this matrix yields the following results:

$$\mathbf{x} = \begin{bmatrix} 1.1566 \\ 1.4062 \\ -0.5599 \\ -0.8557 \\ -0.8577 \end{bmatrix}, \quad \begin{bmatrix} x_1-x_2 \\ x_2-x_3 \\ x_3-x_4 \\ x_4-x_5 \end{bmatrix} = \begin{bmatrix} -0.2495 \\ 1.9961 \\ 0.2958 \\ 0.0000 \end{bmatrix}.$$

Out of all the pairwise comparisons, (x_1-x_2) is still negative. As you see, the previous modification resulted in identical values for x_4 and x_5. If these two columns had been merged, we would have obtained a single value, that is, -0.8557.

Let us now modify the first two columns of **F** as

$$\mathbf{F} = \begin{bmatrix} 7 & 7 & 9 & 8.5 & 8.5 \\ 5 & 5 & 10 & 12 & 12 \\ 13.5 & 13.5 & 6 & 3.5 & 3.5 \\ 13 & 13 & 3 & 6.5 & 6.5 \\ 1 & 1 & 10 & 16 & 16 \end{bmatrix} .$$

Dual scaling of this matrix yields the following solution that satisfies our order constraint (A).

$$\mathbf{x} = \begin{bmatrix} 1.2799 \\ 1.2799 \\ -0.5539 \\ -0.8609 \\ -0.8609 \end{bmatrix}, \quad \begin{bmatrix} x_1 - x_2 \\ x_2 - x_3 \\ x_3 - x_4 \\ x_4 - x_5 \end{bmatrix} = \begin{bmatrix} 0.0000 \\ 1.8338 \\ 0.3070 \\ 0.0000 \end{bmatrix} .$$

There are many other order constraints to consider. Some of them are:

1. Only the best category is known:

$$x_1 \geq (x_2, x_3, x_4, x_5) .$$

Therefore, the set of nonnegative contrasts is given by

$$\begin{bmatrix} x_1 - x_2 \\ x_1 - x_3 \\ x_1 - x_4 \\ x_1 - x_5 \end{bmatrix} \geq 0 .$$

2. The best and the second best are known. The set of nonnegative contrasts is given by

$$\begin{bmatrix} x_1 - x_2 \\ x_2 - x_3 \\ x_2 - x_4 \\ x_2 - x_5 \end{bmatrix} \geq 0 .$$

3. The best and the worst are known. Then the set of pairwise nonnegative contrasts is given by

$$\begin{bmatrix} x_1-x_2 \\ x_1-x_3 \\ x_1-x_4 \\ x_2-x_5 \\ x_3-x_5 \\ x_4-x_5 \end{bmatrix} \geq 0 .$$

Are these clear? These pairwise contrasts are a necessary and sufficient set. Thus, all the other nonnegative contrasts are redundant.

17.3.2 Multiple-Choice Data

In the case of multiple-choice data, an order constraint is imposed on the set of options of each item. Since the response patterns are such that the sum of each row of $\mathbf{F}_j$ (i.e., the incidence matrix of Item j) is 1, it may be easier to merge two columns of disordered options than to assign the average of two corresponding elements, which is either 0 or .5. If you decide to merge two columns, the elements of the merged column are either 0 or 1. Let us look at these two types of modifications, using the following example:

$$\mathbf{F} = [\mathbf{F}_1 \quad \mathbf{F}_2 \quad \mathbf{F}_3] = \begin{bmatrix} 1 & 0 & 0 & 1 & 0 & 0 & 1 & 0 & 0 \\ 0 & 1 & 0 & 0 & 1 & 0 & 0 & 1 & 0 \\ 0 & 0 & 1 & 1 & 0 & 0 & 0 & 0 & 1 \\ 0 & 1 & 0 & 0 & 0 & 1 & 0 & 0 & 1 \\ 0 & 0 & 1 & 0 & 1 & 0 & 0 & 0 & 1 \\ 1 & 0 & 0 & 1 & 0 & 0 & 0 & 1 & 0 \\ 1 & 0 & 0 & 0 & 1 & 0 & 1 & 0 & 0 \end{bmatrix} .$$

Consider the following order constraints:

$$x_1 \leq x_2 \leq x_3, \quad x_4 \leq x_5 \leq x_6, \quad x_7 \leq x_8 \leq x_9 .$$

The corresponding nonnegative contrasts are

$$\begin{bmatrix} x_2-x_1 \\ x_3-x_2 \end{bmatrix} \geq 0, \quad \begin{bmatrix} x_5-x_4 \\ x_6-x_5 \end{bmatrix} \geq 0, \quad \begin{bmatrix} x_8-x_7 \\ x_9-x_8 \end{bmatrix} \geq 0 .$$

Dual scaling of the data matrix yields

$$\begin{bmatrix} x_1 \\ x_2 \\ x_3 \end{bmatrix} = \begin{bmatrix} -1.2380 \\ 1.1971 \\ 0.5605 \end{bmatrix}, \quad \begin{bmatrix} x_4 \\ x_5 \\ x_6 \end{bmatrix} = \begin{bmatrix} -0.6770 \\ 0.0202 \\ 1.9701 \end{bmatrix}, \quad \begin{bmatrix} x_7 \\ x_8 \\ x_9 \end{bmatrix} = \begin{bmatrix} -1.3454 \\ -0.2996 \\ 1.0967 \end{bmatrix}.$$

Our constraints indicate that the weights within each item must be in the increasing order. Then, we see only one violation of the constraint, that is, the order of the second and third options of Item 1. Thus, we must modify $\mathbf{F}_1$. Let us examine the two means of modification, one to merge columns 2 and 3 and the other to replace columns 2 and 3 with the corresponding row averages. The two modified submatrices are indicated by $\mathbf{F}_{1(1)}$ and $\mathbf{F}_{1(2)}$,

$$\mathbf{F}_{1(1)} = \begin{bmatrix} 1 & 0 \\ 0 & 1 \\ 0 & 1 \\ 0 & 1 \\ 0 & 1 \\ 1 & 0 \\ 1 & 0 \end{bmatrix}, \quad \mathbf{F}_{1(2)} = \begin{bmatrix} 1 & 0 & 0 \\ 0 & 0.5 & 0.5 \\ 0 & 0.5 & 0.5 \\ 0 & 0.5 & 0.5 \\ 0 & 0.5 & 0.5 \\ 1 & 0 & 0 \\ 1 & 0 & 0 \end{bmatrix}.$$

We substitute one of these matrices for $\mathbf{F}_1$ in $\mathbf{F}$, and carry out the analysis. The two matrices yield weights for Item 1 respectively as

$$\begin{bmatrix} x_1 \\ x_2 \end{bmatrix} = \begin{bmatrix} -1.2589 \\ 0.9441 \end{bmatrix}, \quad \begin{bmatrix} x_1 \\ x_2 \\ x_3 \end{bmatrix} = \begin{bmatrix} -1.2589 \\ 0.9441 \\ 0.9441 \end{bmatrix},$$

and option weights for Items 2 and 3,

$$\begin{bmatrix} x_4 \\ x_5 \\ x_6 \end{bmatrix} = \begin{bmatrix} -0.6246 \\ 0.0124 \\ 1.8365 \end{bmatrix}, \quad \begin{bmatrix} x_7 \\ x_8 \\ x_9 \end{bmatrix} = \begin{bmatrix} -1.3575 \\ -0.4020 \\ 1.1730 \end{bmatrix}.$$

We can now see that all the constraints are satisfied. As you see, when the order of two options is reversed, the solution is obtained by equating the two option weights. This is the best we can do in order to maximize η^2 under these constraints. In the relevant literature, you often see the phrase "weak order" to indicate the relation "greater than or equal to." If the order relation is strictly "greater than," it is referred to as a "strong order."

The SDM method was used by Nishisato and Arri (1975) to determine an initial approximation of the solution in their separable programming approach. Tanaka and Kodake (1980) stated that the SDM method "has a possibility to pool excessively and to obtain the initial approximation not sufficiently near the true value" (p. 44). Yet the method has so far provided optimal solutions for a number of cases examined. However, readers are referred to the Tanaka-Kodake method and the Tenenhaus method for potentially more efficient algorithms than the simple SDM.

17.4 HORSESHOE EFFECTS

There is a well-known study by Guttman (1950), in which he demonstrated that Guttman's perfect unidimensional scale can be decomposed into orthogonal components of various polynomial degrees, linear, quadratic, cubic, quartic, and so forth. For instance, consider the following subject-by--item data matrix **A**, in which subjects are arranged in the descending order of their scores (i.e., the number of 1s), and the items in the ascending order of difficulty:

$$\mathbf{A} = \begin{bmatrix} 1 & 1 & 1 & 1 & 1 & 1 & 1 \\ 1 & 1 & 1 & 1 & 1 & 1 & 0 \\ 1 & 1 & 1 & 1 & 1 & 0 & 0 \\ 1 & 1 & 1 & 1 & 0 & 0 & 0 \\ 1 & 1 & 1 & 0 & 0 & 0 & 0 \\ 1 & 1 & 0 & 0 & 0 & 0 & 0 \\ 1 & 0 & 0 & 0 & 0 & 0 & 0 \end{bmatrix}.$$

Suppose that **A** is subjected to principal component analysis, and six components are extracted. If you plot their scores of each component as a function of subjects' scores, it is obvious that the first component is a linear function of the observed scores (i.e., the number of 1s), the second component a quadratic function, the third component a cubic function, and so forth, revealing polynomial functions of successive degrees. Consider now plotting the subjects in terms of the first two components, that is, two components as two axes of the graph. You will see an arc, or "horseshoe."

This phenomenon occurs in dual scaling, too. When you plot solution 1 against solution 2 from a contingency table, it is not uncommon that the graph shows a horseshoe. It is easy to generate the phenomenon when binary data are analyzed. Since a horseshoe often appears in cases where we suspect a single dominant solution, it is often regarded as an artifact.

A simple method for handling a horseshoe, therefore, is to adopt only the first solution.

Hill (1979; see also Hill and Gauch, 1980) devised a method of piecewise centering to get rid of the horseshoe, and called it "detrended correspondence analysis." There are other methods available to deal with horseshoe effects. Interested readers are referred to a compendium by van Rijckevorsel (1987).

From a practical point of view, the question here is whether or not the horseshoe phenomenon has to be eliminated from analysis. Is it appropriate to use piecewise centering (detrended correspondence analysis) to linearize the two-dimensional plot? If linearization is the purpose of the transformation, why should we not use only the first solution that is a linear function of, for example, the popularity of political candidates, or the ability of students? Since dual scaling yields orthogonal solutions in the form of $\mathbf{x}_p'\mathbf{D}_c\mathbf{x}_q$ being zero ($p \neq q$), it is natural to expect from the case of a dominant linear component of preference or ability that the orthogonal polynomials will be fitted to the residual data matrix.

17.5 PARTIALLY OPTIMAL SCALING OF MULTIPLE-CHOICE DATA

Partially optimal scaling is not a new topic, but in the light of recent work on analysis of categorical data the old topic may be viewed from new perspectives.

17.5.1 Item Quantification

Let x_{pj} be the weight for Option p of Item j. Then we can postulate two models for x_{pj}[8]:

(a) Crossed Design. When subjects are asked to answer all the multiple--choice questions in terms of a common set of options, x_{pj} can be decomposed as follows:

$$x_{pj} = \mu + \alpha_j + \beta_p + \gamma_{jp} ,$$

where μ is the mean, α_j the contribution of Item j, β_p the contribution of Option p, and γ_{pj} the contribution of the "Item $j \times$ Option p" interaction.

[8] See Nishisato and Torii (1971), Nishisato (1980a).

(b) *Nested Design*. When each item is answered by a different set of options, x_{pj} can be decomposed as follows:

$$x_{pj} = \mu + \alpha_j + \beta_{p(j)} ,$$

where $\beta_{p(j)}$ is the contribution of Option p within Item j.

Up to now, this book treated x_{pj} as a single entity without any reference to its composition. The approach taken so far is reasonable because in dual scaling $\alpha_j = 0$ for all the items. However, the decompositions just mentioned shed some light on hidden aspects of quantification. For instance, Nishisato and Torii have shown that the interaction γ_{pj} in the crossed design is related to the unreliability of items, and hence is not a component that we would wish to maximize. The term μ applies to all the items, and thus is not a component of interest from the scaling point of view. Moreover, these decompositions led to the idea of partially optimal scaling.

Nishisato and Inukai (1972)[9] proposed *partially optimal scaling* (POS), in which quantification is carried out only with respect to items, while keeping the contribution of options constant. This case applies to such a situation in which the investigator employs a priori, or subjective, weights for all the options of items. In practice, it is not unusual for investigators to adopt such weights as 1, 2, 3 for options such as "disagree," "neutral," and "agree," respectively, and to analyze data, typically without further quantification. From the scaling point of view, such a priori weighting for options still leaves room for the quantification of items. To describe the procedure of POS, let us define the following notation:

$\mathbf{A}^* = (a_{ij}^*)$ = the $N \times n$ (subject-by-item) data matrix, expressed in terms of scores, preassigned to the options,

$a_{ij} = a_{ij}^* - \Sigma a_{ij}^*/N$ = the deviation score Subject i from the item mean of Subject i and Item j,

$\mathbf{A} = (a_{ij})$ = the $N \times n$ matrix of deviation scores,

$\mathbf{D}_a$ = the $n \times n$ diagonal matrix of the column sums of squared deviation scores,

$\mathbf{D}_N$ = the $N \times N$ diagonal matrix of the numbers of responses from N subjects = $n N \mathbf{I}$, if there is no missing response,

$\mathbf{x}$ = an $n \times 1$ vector of weights for n items,

$\mathbf{a}$ = the $n \times 1$ vector of column totals of $\mathbf{A} = \mathbf{0}$,

f_t = the total number of responses in $\mathbf{A} = nN$, if there is no missing response.

---

[9] See Inukai (1972) for computational results.

By the definition of **a**, the origin of derived measurement is 0, that is, $\mathbf{a}'\mathbf{x} = 0$. The squared correlation ratio of the weighted responses is therefore given by

$$\eta^2 = \frac{\mathbf{x}'\mathbf{A}'\mathbf{D}_N^{-1}\mathbf{A}\mathbf{x}}{\mathbf{x}'\mathbf{D}_a\mathbf{x}} = \frac{\mathbf{w}'\mathbf{B}'\mathbf{B}\mathbf{w}}{\mathbf{w}'\mathbf{w}},$$

where $\mathbf{B} = \mathbf{D}_N^{-1/2}\mathbf{A}\,\mathbf{D}_a^{-1/2}$, and $\mathbf{w} = \mathbf{D}_a^{1/2}\mathbf{x}$. This expression is comparable to the one used in dual scaling, except that the former does not contain the second trivial solution. Thus, our problem is to carry out the eigenvalue decomposition of $\mathbf{B}'\mathbf{B}$.

This formulation is in essence that of principal component analysis. However, there is an important implication of the framework of POS: conditional analysis, given option weights. This point is revisited in the concluding remarks of this book.

17.5.2 Quantification of Both Items and Options

There are occasions in which some items have preassigned option weights and others do not. It is convenient to know how to combine partially optimal scaling and dual scaling, that is, item weighting and option weighting. This procedure is due to Inukai (1972) and is also reported in Nishisato and Inukai (1972). Let us define the following notation:

$\mathbf{G}^* = (g_{ij}^*)$ = the $N\times(h+m-t)$ response matrix in mixed format, where the first h columns represent h items with preassigned option weights as elements and the remaining $(m-t)$ columns represent the responses to the $(m-t)$ options of $(n-h)$ items, coded as response patterns of 1's and 0's,

$g_{ij} = g_{ij}^* - \Sigma g_{ij}^*/N$, if g_{ij}^* are preassigned values,

$g_{ij} = g_{ij}^*$, otherwise, thus leaving g_{ij} to be binary numbers,

$\mathbf{G} = (g_{ij})$ = the $N\times(h+m-t)$ matrix of deviation scores and binary numbers,

$\mathbf{g}$ = the $(h+m-t)\times 1$ vector of column totals of $\mathbf{G}$,

$\mathbf{D}_g = \mathrm{diag}(\Sigma g_{ij}^2)$ = the diagonal matrix of order $(h+m-t)$,

$\mathbf{D}_N$ = the $N\times N$ diagonal matrix of the numbers of responses = $n\mathbf{NI}$, if there is no missing response,

$\mathbf{x}$ = a vector of weights for the h items and the $(m-t)$ options

$c_k = 0$, if the kth column of $\mathbf{G}^*$ has a preassigned value,

$c_k = 1$, otherwise,

$\mathbf{c}$ = the $(h+m-t)\times 1$ vector of c_k,

f_t = the total number of scores in $\mathbf{G}^*$ = nN, if there is no missing response. Using the condition that $\mathbf{g}'\mathbf{x} = 0$, the squared correlation ratio can be expressed as

$$\eta^2 = \frac{\mathbf{x}'\mathbf{G}'D_N^{-1}\mathbf{G}\mathbf{x}}{\mathbf{x}'\mathbf{D}_g\mathbf{x}} = \frac{\mathbf{w}'\mathbf{B}'\mathbf{B}\mathbf{w}}{\mathbf{w}'\mathbf{w}},$$

where $\mathbf{w} = \mathbf{D}_g^{1/2}\mathbf{x}$, and $\mathbf{B} = \mathbf{D}_N^{-1/2}\mathbf{G}\ \mathbf{D}_g^{-1/2}$. It is interesting to note two aspects of this formulation. First, the eigenequation of $\mathbf{B}'\mathbf{B}$ has two trivial solutions: (a) $\mathbf{w} = \mathbf{0}$, and (b) $\mathbf{w} = \mathbf{D}_g^{1/2}\mathbf{c}$, that is, $\mathbf{x} = \mathbf{c}$. The eigenvalue associated with the second trivial solution is

$$\eta^2 = \frac{\mathbf{c}'\mathbf{G}'\mathbf{D}_N^{-1}\mathbf{G}\mathbf{c}}{\mathbf{c}'\mathbf{D}_g\mathbf{c}} = \frac{N(n-h)^2}{nN(n-h)} = 1 - \frac{h}{n}.$$

Thus, as expected, when all the responses are in the response-pattern format (i.e., $h = 0$), the eigenvalue associated with the second trivial solution is 1, and this value becomes 0 under partially optimal scaling (i.e., $h = n$, all the responses have preassigned scores). In general, the eigenvalue associated with the second trivial solution for mixed data lies between 0 and 1. To eliminate the second trivial solution from $\mathbf{B}'\mathbf{B}$, we obtain the residual matrix $\mathbf{C}_1$ as follows:

$$\mathbf{C}_1 = \mathbf{B}'\mathbf{B} - \left(1 - \frac{h}{n}\right)\frac{\mathbf{D}_g^{1/2}\mathbf{c}\mathbf{c}'\mathbf{D}_g^{1/2}}{\mathbf{c}'\mathbf{D}_g\mathbf{c}} = \mathbf{B}'\mathbf{B} - \frac{\mathbf{D}_g^{1/2}\mathbf{c}\mathbf{c}'\mathbf{D}_g^{1/2}}{f_t}.$$

17.5.3 Merits and Demerits of Two Quantifications

Now that all the necessary formulas are presented, the main question is the merits and demerits of option weighting (i.e., dual scaling) and item weighting (partially optimal scaling). It is obvious that more degrees of freedom are available for option weighting than for item weighting, and we can therefore expect that the maximized squared correlation ratio from option weighting, η_o^2, is generally greater than that from item weighting, η_{it}^2. More precisely,

$$\eta_o^2 \geq \eta_{it}^2 .$$

The two correlation ratios become equal under two conditions: (a) when the preassigned option weights of each item under item weighting are proportional to the corresponding optimal option weights, obtained under option weighting, and (b) when all the items are *dichotomous*. Since the condition stated under (a) is rarely met in practice, it can be said that as far as the maximization of η^2 is concerned, option weighting (dual scaling) is preferable to item weighting.

On the other hand, there is a good reason why we might instead prefer item weighting to option weighting. It is the problem of data size. Imagine analyzing 100 items with five options per item. Under item weighting, the data matrix is $N\times100$, while the data size becomes $N\times500$ under option weighting. There are many situations in which those multiple-choice options are ordered categories, and many researchers know from experience that integer weighting (e.g., 1, 2, 3, 4, 5 for five categories) does a fine job in data analysis. In this example, item weighting provides 100 solutions, assuming N is much larger than the number of items or options, while option weighting yields 400 solutions. This discrepancy in number between the two methods of quantification widens as the number of multiple-choice options increases. Then, as was discussed earlier, it would be necessary to deal with those problems associated with the large number of options in option weighting.

As for the mixed case of item weighting and option weighting, there is not much information available. It offers, however, an interesting framework in which the merits and demerits of the two quantification methods may be pursued by changing the proportion of the preassigned part of the data to the entire data set.

The *dichotomous* case provides the starting point, from which the two methods depart. As such, it is possible to explore many characteristics of both methods of scaling, using dichotomous data. Interested readers are referred to Inukai (1972), Nishisato and Inukai (1972), Nishisato (1980a), and Yamada and Nishisato (1993).

17.6 CONCLUDING REMARKS

Since the time when Maskelyne dismissed his assistant Kinnebrook at Greenwich in 1796,[10] the problem of individual differences has caught the attention of many researchers. However, as the tide of probabilistic reasoning swept over the scientific community and the nobility partook of

[10] Because "Kinnebrook observed the time of stellar transits almost a second later than he (Maskelyne) did" (Boring, 1957, p. 134).

games of "chance," the old, descriptive approach to statistics was for the most part put aside with the eventual emergence of inferential statistics. Individuals were put together to produce an "average" person as a unit of analysis, even though nobody in the group looked like or behaved like the average person. Over a long period of time, however, some psychologists and educational researchers have shown a keen interest in individuals, with the development of mental tests, personality inventories, and achievement tests. In statistics, too, people started to talk about "exploratory analysis" and "data analysis." Individual differences have resurfaced as a legitimate topic for statistical investigation.

Dual scaling is one of the techniques to analyze individual differences. From this book, the reader will know that dual scaling is mainly descriptive, exploratory, multivariate, and a method of nonlinear quantification. These characteristics are in contrast to many popular statistical procedures that are inferential, confirmatory, and linear. To step down from the traditional well-developed inferential statistical procedures to a more mundane sample-oriented procedure of dual scaling may not be appealing to some. Yet when we realize that data analysis must start with the understanding of our sample data, we suddenly see the potential of dual scaling as a technique of data analysis.

Since Richardson and Kuder (1933) proposed what Horst (1935) called the method of reciprocal averages, there have been a number of important contributions to this type of quantification method, as discussed earlier. It is easy to identify a few outstanding studies in those early days, for instance, Hirschfeld (1935), Fisher (1940), Guttman (1941, 1946), and Maung (1941). Then, starting in 1950, Hayashi established his school of quantification theory in Japan, and he and his colleagues published a number of relevant articles under the name of "Hayashi's theory of quantification." In the early 1960s, another strong evolutionary movement in data analysis took root, led by Benzécri in France, and he and his school have had a decisive impact on statistics under the name of "analyse des correspondances factorielles." In late 1960s, the Leiden school in the Netherlands, led by Jan de Leeuw, emerged to become a driving force of modern psychometrics under the rubric of homogeneity analysis. Around the same time, research on quantification methods began in Toronto, first under the name of optimal scaling and then dual scaling.

There have been many outstanding individual contributions outside these groups. Over the past 60 years, many developments have been made, and helped to clarify the basic idea of this quantification method. Edwin Diday (1991) once stated in personal conversation that most of the theoretical development of analyse des correspondances had been completed in France some 15 years before, and that a new approach to data analysis, called symbolic data analysis, would replace it in the future. He is probably right, but, as readers of this book may have noted, dual scaling has not reached the apex of its development. There are still many research problems of importance and interest for further work. There is no doubt that the method will further improve in its usefulness and statistical rigor.

Nishisato concluded his 1980 book by stating, "Through constant interplay between empirical and theoretical work one can perhaps expand the scope of its applications in effective and useful ways. Inferential aspects of dual scaling deserve immediate attention by researchers" (p.205). Aitken (1982) "heartily endorsed" those final words. It is this inferential aspect of dual scaling that has yet to be explored and investigated before it can be put into practice.

As a promising avenue to inferential study, the contingency table provides an ideal starting point. There are a number of studies in which the relationship between correspondence analysis (dual scaling) and loglinear analysis is explored.[11] In a recent review article, Goodman (1991) summarized the current knowledge on the relationship, and various studies address the implications of introducing an inferential framework to dual scaling. Millones (1991) identified partially optimal scaling, discussed in Section 17.5, as a general model for quantification, and explored dual scaling in relation to a number of association models. Thus, at least for contingency tables and multiple-choice data, research has provided a starting point for an inferential framework for dual scaling.

During the past three decades, item response theory (IRT)[12] has steadily increased in popularity as a research topic. IRT offers another possibility that may be instrumental in generating an inferential framework for dual scaling. The use of a logistic function in IRT and loglinear analysis also suggests a possibility for dual scaling to be formulated through information theory.[13] Once successfully formulated by the information-theoretic terms, dual scaling may produce results that can be trans-

---

[11] See, for instance, Choulakian (1988), Goodman (1985, 1986, 1991), van der Heijden, Falguerolles, and de Leeuw (1989), and van der Heijden and de Leeuw (1985), and Tsujitani (1988a, b).

[12] See, for instance, Birnbaum (1968), Crocker and Algina (1986), Hambleton (1983), Lord (1980), Lord and Novick (1968), Traub and Wolfe (1981).

[13] See, for example, Garner (1962), Khinchin (1957), Kullback (1959), McGill (1954), Reza (1961), Shannon and Weaver (1949), Watanabe (1969), and Wiener (1948).

lated into a simpler interpretive framework. This idea, however, has yet to be worked out.

What has been presented here suggests some important problems for future work: its graphical method must be much further developed, perhaps in relation to cluster analysis[14]; an expert system for interpreting the results as well as deciding on the depth and breadth of analysis needs to be developed; more meaningful statistics (e.g., measures of quality of graphical display and of information explained) should be derived; and an inferential framework (e.g., a confidence interval or region for each weight and each score) is urgently needed. To some people, the approach of dual scaling has been exhaustively investigated, and hence lacks novelty; to most others, however, it is a technique that offers immense possibilities for applications. Each time it is applied to a new problem, interesting ideas often emerge for further clarification of the method.

Finally, it should be reiterated that the current book has covered the basic characteristics of this quantification method and the work done mainly in Toronto, and that there are too many other outstanding and relevant studies carried out elsewhere to mention, for instance, those conducted in the United States, Canada, the Netherlands, Japan, France, the United Kingdom, Germany, Russia, South Africa, Australia, Italy, and Belgium. In this regard, too, it is evident that the technique has had significant impact on data analysis. Yet at the same time, the method has not been exhaustively studied, and it leaves a great deal of room for further development.

---

[14] See, for example, Anderberg (1973), Everitt (1974), Hartigan (1975), Jain and Dubes (1988), Jardine and Sibson (1971), Mucha (1992), and Sneath and Sokal (1973).

Appendix
The DUAL3 Statistical Software Series

There are a number of computer programs available for quantifying categorical data.[1] Considering the scope and focus of this book, however, the DUAL3 Statistical Software Series package is the one that offers the same kinds of analysis covered here. For this reason, a brief overview of the DUAL3 package is provided in this section.

Data Types

- Incidence data — contingency/frequency tables, multiple-choice data, sorting data.
- Dominance data — paired comparison data, rank-order data, successive categories data.

Modes of Analysis

- Standard dual scaling (correspondence analysis, multiple correspondence analysis).
- Forced classification with multiple-choice data, sorting data, paired comparison data, rank-order data.
- Multiway analysis via forced classification.
- Graphical display of symmetric scaling — linear and planar — for contingency/frequency tables, paired comparison data, rank-order data, successive categories data.

Operational Features

- Dynamic memory allocation.
- Easy-to-use, fully menu-driven and interactive.
- Output to printer, disk, or screen.

---

[1]See, for example, Green (1992) for references.

Hardware Requirements

- IBM machines, IBM compatibles.
- 512K RAM.
- Monitor and printer.

Prior to the use of the program, a file of the data to be analyzed must be created. Once the program starts, the following basic questions are asked:

- Title of the study .
- Name of the data file.
- Number of rows.
- Number of columns.
- Number of solutions to extract.
- Name of the output file for graphical display.
- Name for the entire output.
- Location where the output should be directed: screen, printer, or disk.

Depending on the data type, additional questions may be asked — for example, the number of options of each item (multiple-choice data), the criterion for forced classification (multiple-choice data, sorting data), the pair of criterion variables for forced classification (paired comparison data, rank-order data), and the number of categories (successive categories data).

The package was written in BASIC by Ira Nishisato and Shizuhiko Nishisato, together with the *DUAL3 Users' Guide* and *An Introduction to Dual Scaling*. Inquiries should be directed to MicroStats, 9 St. George's Road, Islington, Ontario, Canada M9A 3S9.

References

Ahn, H. (1991). Effects of missing responses in multiple choice data on dual scaling. Unpublished doctoral dissertation, University of Toronto, Canada.

Aitkin, M. (1982). Review of S. Nishisato, "Analysis of categorical data: Dual scaling and its applications". *Royal Statistical Society Journal, Series A, 145*, 513-516.

Anderberg, M.R. (1973). *Cluster analysis for applications*. New York: Academic.

Bahadur, R.R. (1961). A representation of the joint distribution of responses to n dichotomous items. In H. Solomon (ed.), *Studies in item analysis and prediction* (pp.158—176). Stanford: Stanford University Press.

Benali, H., & Escofier, B. (1987). Stabilité de l'analyse factorielle des correspondances multiples en cas de données manquantes et de modalités a faibles effectifs [Stability of multiple correspondence analysis with missing responses and weak association]. *Revue de Statistiques Appliquées* (France), *35*, 41-52.

Bendig, A.W. (1954). Reliability and the number of rating categories. *Journal of Applied Psychology, 38*, 38-40.

Benson, P.H. (1971). How many scales and how many categories shall we use in consumer research? *Journal of Marketing, 35*, 59-61.

Benzécri, J.-P. (1979). Sur le calcul des taux d'inertie dans l'analyse d'un questionnaire [On the calculation of the total inertial in the analysis of a questionnaire]. *Cahiers de l'Analyse des Données* (France), *4*, 377-378.

Benzécri, J.-P. (1982). *Histoire et préhistoire de l'analyse des données* [History and prehistory of data analysis]. Paris, France: Dunod.

Benzécri, J.-P. (1992) *Correspondence Analysis Handbook*. New York: Marcel Dekker.

Benzécri, J.-P., & Benzécri, F. (1980). *L'analyse des correspondances: Exposé elementaire* [Correspondence analysis: Elementary exposition]. Paris, France: Dunod.

Benzécri, J.-P. et al. (1973). *L'Analyse des Données: II. L'Analyse des Correspondances* [Data analysis: II. Correspondence analysis]. Paris, France: Dunod.

Birnbaum, A. (1968). Some latent trait models and their use in inferring an examinee's ability. In F.M. Lord & M.R.Novick: *Statistical theories of mental test scores*. Reading, MA.: Addison—Wesley.

Bock, H.H. (Ed.) (1988). *Classification and related methods of data analysis.* Amsterdam, the Netherlands: North-Holland.

Bock, R.D. (1960). *Methods and applications of optimal scaling*. (Research Memorandum No. 25). The University of North Carolina Psychometric Laboratory, Chapel Hill, NC.

Bock, R.D., & Jones, L.V. (1968). *The measurement and prediction of judgement and choice.* San Francisco: Holden-Day.

Boring, E.G. (1957). *A history of experimental psychology*. New York: Appleton-Century-Crofts.

Bouroche, J.M. (1977). *Analyse des données en marketing* [Data analysis in marketing]. Paris, France: Masson.

Bouroche, J.M., & Saporta, G. (1980). *L'Analyse des données* [Data analysis]. Paris, France: Presses Universitaires de France.

Bradley, R.A., Katti, S.K., & Coons, I.J. (1962). Optimal scaling for ordered categories. *Psychometrika*, *27*, 355-374.

Bradu, D., & Gabriel, K. R. (1978). The biplot as a diagnostic tool for models of two-way tables. *Technometrics*, *20*, 47-68.

Cailliez, F., & Pagès, J. P. (1976). *Introduction à l'analyse des données* [Introduction to data analysis]. Paris, France: SMASH.

Carroll, J.D., Green, P.E., & Schaffer, C.M. (1986). Interpoint distance comparisons in correspondence analysis. *Journal of Marketing Research*, *23*, 271-280.

Carroll, J.D., Green, P.E., & Schaffer, C.M. (1987). Comparing interpoint distances in correspondence analysis: A clarification. *Journal of Marketing Research*, *24*, 445-450.

Carroll, J.D., Green, P.E., & Schaffer, C.M. (1989). Reply to Greenacre's commentary on the Carroll—Green—Schaffer scaling of two-way correspondence analysis solutions. *Journal of Marketing Research*, *26*, 366-368.

Chan, D. O. (1978). Treatment of missing data by optimal scaling. Unpublished master's thesis, University of Toronto, Canada.

Chapney, H., & Marshall, H. (1939). Optimal refinement of the rating scale. *Journal of Applied Psychology*, *23*, 323-331.

Chipuer, H.M. (1986). Dual scaling of several sets of categorical data. Unpublished master's thesis, University of Toronto, Canada.

Choulakian, V. (1988). Exploratory analysis of contingency tables by loglinear formulation and generalizations of correspondence analysis. *Psychometrika*, *53*, 235-250.

Coombs, C.H. (1964). *A theory of data*. New York: Wiley.

Coppi, R., & Bolasco, S. (Eds.) (1989). *Multiway data analysis.* Amsterdam, the Netherlands: North-Holland.

Covert, D. (1990). Consumer perceptions of the similarity of 51 car models: An application of dual scaling to sorting data. Unpublished manuscript.

Cox, E.P, III. (1980). The optimal number of response alternatives for a scale: A review. *Journal of Marketing Research, 18,* 407-422.

Crocker, L., & Algina, J. (1986). *Introduction to classical and modern test theory.* New York: Holt, Rinehart & Winston.

Cronbach, L.J. (1951). Coefficient alpha and internal structure of tests. *Psychometrika, 16,* 297-334.

Day, D.A. (1989). Investigating the validity of the English and French versions of the Myers-Briggs Type Indicator. Unpublished doctoral dissertation, University of Toronto, Canada.

de Leeuw, J. (1973, 1984). *Canonical analysis of categorical data.* Psychological Institute, University of Leiden, the Netherlands, Report RN 007-68 (reprinted as a book and published in 1984 from DSWO Press, Leiden) .

de Leeuw, J. (1983). On the prehistory of correspondence analysis. *Statistica Neerlandica, 37,* 161-164.

Dempster, A.P. (1969). *Elements of continuous multivariate analysis.* Reading, MA: Addison-Wesley.

Dempster, A.P., Laird, N.M., & Rubin, D.B. (1977). Maximum likelihood estimation from incomplete data via the EM algorithm (with discussion). *Journal of Royal Statistical Society, Series B, 39,* 1-38.

Diday, E. (1991). Personal communication.

Eckart, C., & Young, G. (1936). The approximation of one matrix by another of lower rank. *Psychometrika, 1,* 211-218.

Everitt, B.S. (1974). *Cluster analysis.* New York: Wiley.

Fisher, R. A. (1940). The precision of discriminant function. *Annals of Eugenics, 10,* 422-429.

Fisher, R. A. (1948). *Statistical mehthods for research workers* (10th edition). London: Oliver & Boyd.

Ford, B.L. (1983). An overview of hot deck procedures. In W.G. Madow, I. Olkin, & D.B. Rubin (Eds.), *Incomplete data in sample surveys: Vol.2. Theory and annotated bibliography.* New York: Academic.

Franke, G.R. (1985). Evaluating measures through data quantification: Applying dual scaling to an advertising copytest. *Journal of Business Research, 13,* 61-69.

Gabriel, K.R. (1971). The biplot graphical display of matrices with applications to principal component analysis. *Biometrika, 58,* 453-467.

Garmize, L.M., & Rychlak, J.F. (1964). Role-play validation of a socio-cultural theory of symbolism. *Journal of Consulting Psychology, 28*, 107-115.

Garner, W.R. (1960). Rating scales, discriminability, and information transmission. *Psychological Review*, *67*, 343-352.

Garner, W.R. (1962). *Uncertainty and structure as psychological concepts*. New York: Wiley.

Gaul, W., & Schader, M. (Eds.). (1986). *Classification as a tool of research.* Amsterdam, the Netherlands: North-Holland.

Gaul, W., & Schader, M. (Eds.). (1988). *Data, expert knowledge and decisions.* Heidelberg, Germany: Springer-Verlag.

Gessaroli, M. E. (1985). A Monte Carlo investigation of the Type I error rates of three multivariate tests applied to categorical data. Unpublished doctoral dissertation, University of Toronto, Canada.

Gibson, L. (1993). An investigation of the generalized forced classification procedure and its application to diminishing outlier effects. Unpublished master's thesis, Universityi of Toronto, Canada.

Gifi, A. (1990). *Nonlinear multivariate analysis*. New York: Wiley.

Goldstein, H. (1987). The choice of constraints in correspondence analysis. *Psychometrika*, *52*, 207-215.

Goodman, L.A. (1985). The analysis of cross-classified data having ordered and/or unordered categories: Association models, correlation models, and asymmetry models for contingency tables with or without missing entries. *Annals of Statistics*, *13*, 10-69.

Goodman, L.A. (1986). Some useful extensions of the usual correspondence approach and the usual loglinear model approach in the analysis of contingency tables. *International Statistical Review*, *54*, 243-309.

Goodman, L.A. (1991). Measures, models, and graphical displays in the analysis of cross-classified data. *Journal of the Americal Statistical Association, 86*, 1085-1111.

Green, P.E. (1992). Review of MCA+ and CORANA: Correspondence analysis. *Journal of Marketing Research, 14*, 478-480.

Green, P.E., & Rao, V.R. (1970). Rating scales and information recovery-How many scales and response categories to use? *Journal of Marketing*, *34*, 33-39.

Greenacre, M. J. (1984). *Theory and applications of correspondence analysis*. London: Academic.

Greenacre, M. J. (1987). *Measuring total variation and its components in multiple correspondence analysis*. Murray Hill, NJ, AT&T Bell Laboratories Statistical Research Report.

Greenacre, M.J. (1989). The Carroll-Green-Schaffer scaling in correspondence analysis: A thoretical and empirical appraisal. *Journal of Marketing Research*, *26*, 358-365.

Guest, L. (1962). A comparison of two-choice and four-choice questions. *Journal of Advertising Research*, *2*, 32-34.

Guilford, J.P. (1954). *Psychometric methods* (2nd edition). New York: McGraw-Hill.

Guttman, L. (1941). The quantification of a class of attributes: A theory and method of scale construction. In The Committee on Social Adjustment (Ed.), *The prediction of personal adjustment* (pp.319—348). New York: Social Science Research Council.

Guttman, L. (1946). An approach for quantifying paired comparisions and rank order. *Annals of Mathematical Statistics*, *17*, 144-163.

Guttman, L. (1950). The principal components of scale analysis. In S.A. Stouffer, L. Guttman, E.A. Suchman, P.F. Lazarsfeld, S. A. Star, and J.A. Clausen (Eds), *Measurement and prediction*. Princeton: Princeton University Press.

Guttman, L. (1956). Best possible systematic estimates of communalities. *Psychometrika*, *21*, 273-285.

Halpine, S. (1991). Phase angle differences in visual evoked potentials: Do they affect estimates of the signal? Unpublished doctoral dissertation, University of Toronto, Canada.

Hambleton, R.K. (1983). *Applications of item response theory*. Vancouver, Canada: Educational Research Institute of British Columbia.

Hartigan, J.A. (1975). *Clustering algorithms*. New York: Wiley.

Hayashi, C. (1950). On the quantification of qualitative data from the mathematico-statistical point of view. *Annals of the Institute of Statistical Mathematics* (Japan), *2*, 35-47.

Hayashi, C. (1977). *Data kaiseki no kangaekata* [Foundations of Data Analysis]. Tokyo: Toyo Kenzaishimpo-sha.

Healy, M.J.R., & Goldstein, H. (1976). An approach to the scaling of categorized attributes. *Biometrika*, *63*, 219-229.

Heiser, W.J. (1981). Unfolding analysis of proximity data. Unpublished doctoral dissertation, University of Leiden, the Netherlands.

Heuer, J. (1979). *Selbsmord bei Kindern und Jugendlichen* [Suicide by children and youths]. Stuttgart, Germany: Ernst Klett Verlag.

Hill, M.O. (1979). DECORANA — A FORTRAN program for detrended correspondence analysis and reciprocal averaging. *Ecology and Systematics Technical Report*, Cornell University.

Hill, M.O., & Gauch, H.G. (1980). Detrended correspondence analysis, an improved ordination technique. *Vegitatio, 42*, 47-58.

Hirschfeld, H. O. (1935). A connection between correlation and contingency. *Cambridge Philosophical Society Proceedings*, *31*, 520-524.

Hoffman, D.L. & Franke, G.R. (1986). Correspondence analysis: Graphical representation of categorical data in marketing research. *Journal of Marketing Research*, *23*, 213-217.

Hollingshead, A.B. (1949). *Elmtown's youth: The impact of social classes on adolescents*. New York: Wiley.

Horst, P. (1935). Measuring complex attitudes. *Journal of Social Psychology*, *6*, 369-374.

Hotelling, H. (1933). Analysis of a complex of statistical variables into principal components. *Journal of Educational Psychology*, *24*, 417-441, 498-520.

Inukai, Y. (1972). Optimal versus partially optimal scaling of polychotomous items. Unpublished master's thesis, University of Toronto, Canada.

Iwatsubo, S. (1987). *Suryokaho no kiso* [Foundations of quantification methods]. Tokyo, Japan: Asakura Shoten.

Jackson, D.N., & Helmes, E. (1979). Basic structure content scaling. *Applied Psychological Measurement*, *3*, 313-325.

Jacoby, J., & Matell, M.S. (1971). Three-point scales are good enough. *Journal of Marketing Research*, *8*, 495-500.

Jain, A.K., & Dubes, R.C. (1988). *Algorithms for clustering data*. Englewood Cliffs, NJ: Prentice—Hall.

Jardine, N., & Sibson, R. (1971). *Mathematical taxonomy*. New York: Wiley.

Khinchin, A.I. (1957). *Mathematical foundations of information theory*. New York: Dover.

Kiers, H.A.L. (1989). *Three-way methods for the analysis of qualitative and quantitative two-way data*. Leiden, the Netherlands: DSWO Press.

Komazawa, T. (1978). *Tagenteki data bunseki no kiso* [Foundations of multiway data analysis] . Tokyo, Japan: Asakura Shoten.

Komazawa, T. (1982). *Suryoka riron to data shori* [Quantification theory and data analysis] . Tokyo: Asakura Shoten.

Komorita, S.S., & Graham, W.K. (1965). Number of scale points and the reliability of scales. *Educational and Psychological Measurement*, *25*, 987-995.

Koster, J.T.A. (1989). *Mathematical aspects of multiple correspondence analysis for ordinal variables*. Leiden, the Netherlands: DSWO Press.

Kullback, S. (1959). *Information theory and statistics*. New York: Wiley.

Lancaster, H. O. (1958). The structure of bivariate distributions. *Annals of Mathematical Statistics*, *29*, 719-736.

Lawrence, D.R. (1985). Dual scaling of multidimensional data structures: An extended comparison of three methods. Unpublished doctoral dissertation, University of Toronto, Canada.

Lebart, L., Morineau, A. & Tabard, N. (1977). *Techniques de la Description Statistique: Méthodes et Logiciels pour l'Analyse des Grands Tableaux* [Techniques of statistical description: Methods and softwares for analysis of large tables of data]. Paris: Dunod.

Lebart, L., Morineau, A., & Warwick, K.M. (1984). *Multivariate Descriptive Statistical Analysis*. New York: Wiley.

Lehmann, D.R., & Hulbert, J. (1972). Are three-point scales always good enough? *Journal of Marketing Research, 9*, 444-446.

Lingoes, J.C. (1964). Simultaneous linear regression: An IBM 7090 program for analyzing metric/nonmetric or linear/nonlinear data. *Behavioral Science, 9*, 87-88.

Lissitz, R.W., & Green, S.B. (1975). Effect of the number of scale points on reliability : A Monte Carlo approach. *Journal of Applied Psychology, 60*, 10-13.

Lord, F.M. (1958). Some relations between Guttman's principal components of scale analysis and other psychometric theory. *Psychometrika, 23*, 291-296.

Lord, F.M. (1980). *Applications of item response theory to practical testing problems*. Hillsdale, NJ: Lawrence Erlbaum Associates.

Lord, F.M., & Novick, M.R. (1968). *Statistical theories of mental test scores*. Reading, MA: Addison-Wesley.

MacNeill, I.B., & Umphrey, G.J. (Eds.). (1987). *Foundations of statistical inference*. Dordrecht: D. Reidel Publishing Company.

Matell, M.S., & Jacoby, J. (1971). Is there an optimal number of alternatives for Likert scale items? Study I: Reliability and validity. *Educational and Psychological Measurement, 31*, 657-674.

Matell, M.S., & Jacoby, J. (1972). Is there an optimal number of alternatives for Likert scale items? *Journal of Applied Psychology, 56*, 506-509.

Maung, K. (1941). Measurement of association in contingency tables with special reference to the pigmentation of hair and eye colours of Scottish school children. *Annals of Eugenics, 11*, 189-223.

McDonald, R. P. (1983). Alternative weights and invariant parameters in optimal scaling. *Psychometrika, 48*, 377-392.

McDonald, R.P., Torii, Y., & Nishisato, S. (1979). Some results on proper eigenvalues and eigenvectors with applications to optimal scaling. *Psychometrika, 44*, 211-227.

McGill, W.J. (1954). Multivariate information transmission. *Psychometrika, 19*, 97-116.

Meulman, J. (1982). *Homogeneity analysis of incomplete data*. Leiden, the Netherlands: DSWO Press.

Meulman, J. (1986). *A distance approach to nonlinear multivariate analysis*. Leiden, the Netherlands: DSWO Press.

Millones, O. (1991). Dual scaling in the framework of the association and correlation models under maximum likelihood. Unpublished doctoral dissertation, University of Toronto, Canada.

Mosier, C.I. (1946). Machine methods in scaling by reciprocal averages. *Proceedings, Research Forum* (pp. 35—39). Edicath, NY: International Business Corporation.

Mucha, H.-J. (1992). *Clusteranalyse mit mikrocomputern* [Cluster analysis with microcomputers]. Berlin, Germany: Akademie Verlag.

Nishisato, S. (1971). Transform factor analysis: A sketchy presentation of a general approach. *Japanese Psychological Research, 13*, 155-166.

Nishisato, S. (1972a). Analysis of variance of categorical data through selective scaling. *Proceedings of the 20th International Congress of Psychology*. (p. 279). Tokyo, Japan: Science Council of Japan.

Nishisato, S. (1972b). Analysis of variance through optimal scaling. *Proceedings of the First Canadian Conference in Applied Statistics "STATISTICS '71 CANADA"* (pp.306-317). Montreal: Sir George Williams University Press.

Nishisato, S. (1973). *Optimal scaling of partially ordered categories*. Paper presented at the Annual Meeting of the Psychometric Society, University of Chicago, IL.

Nishisato, S. (1975). *Oyo shinri shakudo-ho: Shitsuteki data no bunseki to kaishaku* [Applied Psychological Scaling: Analysis and Interpretations of Qualitative Data]. Tokyo, Japan: Seishin Shobo.

Nishisato, S. (1976). *Optimal scaling as applied to different forms of data*. Measurement and Evaluation of Categorical Data Technical Report No. 4, Department of Measurement and Evaluation, the Ontario Institute for Studies in Education, Toronto, Canada.

Nishisato, S. (1977). Shakudo-ho to sono kanren ryoiki ni okeru saikin no dohkoh: Bunken ni yoru review [Recent developments in scaling and related areas: A bibliographic review] (1) . *Japanese Journal of Behaviormetrics, 4*, 74-95. (in Japanese with English abstract) .

Nishisato, S. (1978a). Optimal scaling of paired comparison and rank order data: An alternative to Guttman's formulation. *Psychometrika, 43*, 263-271.

Nishisato, S. (1978b). Keiryo shinrigaku: Sono dohkoh [Psychometrics: Its frontiers]. *Mathematical Sciences, 183*, 69-73.

Nishisato, S. (1978c). Shakudo-ho to sono kanren ryoiki: Tajigen shakudoho [Recent developments in scaling and related areas: Multidimensional scaling] (2) . *Japanese Journal of Behaviormetrics, 5*, 37-55. (in Japanese with English abstract) .

Nishisato, S. (1979a). Sohtsui shakudo-ho to sono rekishi [Dual scaling and its historical development]. *Mathematical Sciences, 190*, 69-73.

Nishisato, S. (1979b). Dual scaling and its variants. In R.E. Traub (Ed.), *Analysis of test data (New directions for testing and measurement quarterly sourcebook)* (pp.1—12). San Francisco: Jossey-Bass.

Nishisato, S. (1980a). *Analysis of Categorical Data: Dual Scaling and its Applications*. (Mathematical Exposition series No. 24) Toronto, Canada: University of Toronto Press.

Nishisato, S. (1980b). Dual scaling of successive categories data. *Japanese Psychological Research, 22*, 134-143.

Nishisato, S. (1981). Mathematical expositions of dual scaling. In Y.P. Chaubey & T.D. Dwivedi (Eds.). *Topics in applied statistics*. (pp.629—640). Montreal, Canada: Concordia University Press.

Nishisato, S. (1982) *Shitsuteki data no suryoka: Dual scaling to sono oyo* [Quantifying qualitative data: Dual scaling and its applications]. (Statistics Library Series) Tokyo, Japan: Asakura Shoten (revised in 1984).

Nishisato, S. (1984a). Dual scaling by reciprocal medians. *Proceedings of the 32nd Scientific Conference of the Italian Statistical Society*. Sorrento, 141-147.

Nishisato, S. (1984b). Forced classification: A simple application of a quantification technique. *Psychometrika, 49*, 25-36.

Nishisato, S. (1984c). Unpublished workshop handout.

Nishisato, S. (1985). Methods for handling outlier responses in dual scaling. *Proceedings of the 49th Annual Meeting of the Japanese Psychological Association*, Tokyo, 501.

Nishisato, S. (1986a). Classification with a variety of categorical data. In W. Gaul & M. Schader (Eds.) , *Classification as a tool of research*. (pp. 353—359). Amsterdam, the Netherlands: Elsevier Science (North-Holland).

Nishisato, S. (1986b). Generalized forced classification for quantifying categorical data. In E. Diday, Y. Escoufier, L. Lebart, J. Pages, Y. Schektman, & R. Tomassone (Eds.) , *Data analysis and informatics, IV* (pp.351—362). Amsterdam, the Netherlands: North-Holland.

Nishisato, S. (1986c). Multidimensional analysis of successive categories. In J. de Leeuw, W. Heiser, J. Meulman, & F. Critchley (Eds.) , *Multidimensional data analysis* (pp. 249—250). Leiden, the Netherlands: DSWO Press.

Nishisato, S. (1986d). *Quantification of categorical data: A bibliography 1975-1986*. Toronto, Canada: MicroStats.

Nishisato, S. (1987). Robust techniques for quantifying categorical data. In I. B. MacNeil & G. J. Umphrey (Eds.) , *Foundations of statistical inference* (pp. 209—217). Dordrecht: D. Reidel Publishing.

Nishisato, S. (1988a). Assessing quality of joint graphical display in correspondence analysis and dual scaling. In E. Diday, Y. Escoufier, L. Lebart, J. Pages, Y. Schektman, & R Tomassone (Eds.) , *Data analysis and informatics, V* (pp. 409—416). Amsterdam, the Netherlands: North-Holland.

Nishisato, S. (1988b). Dual scaling: Its development and comparisons with other methods. In *Proceedings of the Annual Meeting of the German Society of Operations Research*, Berlin, Germany, 376-389.

Nishisato, S. (1988c). *Effects of coding on dual scaling*. Paper presented at the Annual Meeting of the Psychometric Society, University of California, Los Angeles.

Nishisato, S. (1988d). Forced classification procedure of dual scaling: Its mathematical properties. In H. H. Bock (Ed.) , *Classification and related methods of data analysis* (pp. 523—532). Amsterdam, the Netherlands: North-Holland.

Nishisato, S. (1988e). Market segmentation by dual scaling through generalized forced classification. In W. Gaul & M. Schader (Eds.) , *Data, expert knowledge and decisions* (pp. 268—278). Berlin, Germany: Springer-Verlag.

Nishisato, S. (1990). Dual scaling of designed experiments. In M. Schader & W. Gaul (Eds.) , *Knowledge, data and computer-assisted decisions*. (NATO ASI Series F: Computer and Systems Science Vol. 61, pp. 115—125). Berlin, Germany: Springer-Verlag.

Nishisato, S. (1991). Standardizing multidimensional space for dual scaling. In the *Proceedings of the 20th Annual Meeting of the German Operations Research Society*, Hohenheim University, 584-591.

Nishisato, S. (in press). Graphical representation of quantified categorical data: Its inherent problems. *Journal of Statistical Planning and Inference.*

Nishisato, S. (in press). On quantifying different types of categorical data. To appear in *Psychometrika, 58*, No. 4.

Nishisato, S. (in press). Optimization and data structure: Seven faces of dual scaling. *Annals of Operations Research.*

Nishisato, S., & Ahn, H. (in press). When not to analyze data: Decision making on missing responses in dual scaling. *Annals of Operations Research.*

Nishisato, S., Ahn, H., & Shin, A. (1987). *Dual scaling of a large data set: Some problems*. Paper presented at the Fifth European Meeting of the Psychometric Society, University of Twente, Enschede, the Netherlands, June 24—26.

Nishisato, S., & Arri, P.S. (1975). Nonlinear programming approach to optimal scaling of partially ordered categories. *Psychometrika, 40,* 525-548.

Nishisato, S., & Gaul, W. (1989). Marketing data analysis by dual scaling. *International Journal of Research in Marketing, 5,* 151-170.

Nishisato, S., & Gaul, W. (1990). An approach to marketing data analysis: The forced classification procedure of dual scaling. *Journal of Marketing Research, 27,* 354-360.

Nishisato, S., & Inukai, Y. (1972). Partially optimal scaling of items with ordered categories. *Japanese Psychological Research, 14,* 109-119.

Nishisato, S., & Kolić, M. (1992). *From dual scaling to clustering: An alternative to joint graphical display*. Paper presented at the Annual Meeting of the Psychometric Society, Ohio State University, Columbus, OH, July 9—12.

Nishisato, S., & Lawrence, D.R. (1981). *Dual scaling of multidimensional tables: A comparative study*. Paper presented at the Annual Meeting of the Psychometric Society, University of North Carolina, Chapel Hill, NC, May 27—29.

Nishisato, S., & Lawrence, D.R. (1982). *Dual scaling of multidimensional tables: Proportional and nonorthogonal cases*. Paper presented at the Annual Meeting of the Psychometric Society, Université du Québec, Montreal, Canada, May 30—June 2.

Nishisato, S., & Lawrence, D.R. (1984). *Dual scaling of multidimensional tables: Further results*. Paper presented at the Annual Meeting of the Psychometric Society, University of California, Santa Barbara, CA, June 25—27.

Nishisato, S., & Lawrence, D.R. (1989). Dual scaling of multiway data matrices: Several variants. In R. Coppi & S. Bolasco (Eds.) , *Multiway data analysis* (pp.317—326). Amsterdam, the Netherlands: North-Holland.

Nishisato, S. & Levine, R. (1975). *Optimal scaling of omitted responses*. Paper presented at the Annual Meeting of the Psychometric Society, University of Iowa, Iowa City, IA, April 24—26.

Nishisato, S., & Nishisato, I. (1983). *DUAL3 statistical software series*. Toronto, Canada: MicroStats.

Nishisato, S., & Nishisato, I. (1984). *An introduction to dual scaling*. Toronto, Canada: MicroStats.

Nishisato, S., & Nishisato, I. (1986). *DUAL3 users' guide*. Toronto, Canada: MicroStats.

Nishisato, S., & Sheu, W.J. (1980). Piecewise method of reciprocal averages for dual scaling of multiple-choice data. *Psychometrika, 45*, 467-478.

Nishisato, S., & Sheu, W.J. (1984). A note on dual scaling of successive categories data. *Psychometrika, 49*, 493-500.

Nishisato, S., & Sheu, W.J. (1985). A note on measures of consistency and explained variance for quantification of categorical data. Unpublished manuscript.

Nishisato, S., & Torii, Y. (1971). *Multifactor model of optimal scaling*. Paper presented at the Annual Meeting of the Psychometric Society, Washington University, St. Louis, MO.

Nishisato, S., & Yamauchi, H. (1974). Principal components of deviation scores and standard scores. *Japanese Psychological Research, 16*, 162-170.

O'Grady, K.E. (1982). Measures of explained variance: Cautions and limitations. *Psychological Bulletin, 92*, 766-777.

Orchard, T., & Woodbury, M.A. (1972). A missing information principle: Theory and applications. In *Proceedings of the Sixth Berkeley Symposium on Mathematical Statistics and Probability, Vol. 1* (pp. 697—715). Berkeley: University of California Press.

Poon, W.P. (1977). Transformations of data matrices in optimal scaling. Unpublished master's thesis, University of Toronto, Canada.

Ramsay, J.O. (1973). The effect of number of categories in rating scales on precision of estimation of scale values. *Psychometrika, 37*, 513-532.

Rao, C.R. (1965). *Linear statistical inference and its applications*. New York: Wiley.

Reza, A.M. (1961). *An introduction to information theory*. New York: McGraw—Hill.

Reitman, R.M., & Davison, L.A. (1974). *Clinical neuropsychology*. New York: Winston.

Richardson, M., & Kuder, G.F. (1933). Making a rating scale that measures. *Personnel Journal, 12*, 36-40.

Sachs, J. (1984). Approximating an underlying continuous variable with dual scaling. Unpublished master's thesis, University of Toronto, Canada.

Saito, T. (1980). *Tajigen shakudo-ho* [Multidimensional scaling]). Tokyo, Japan: Asakura Shoten. (in Japanese)

Saporta, G. (1979). *Theories et méthodes de la statistique* [Theory and methods of statistics]. Paris, France: Technip.

Saporta, G. (1980). *About some remarkable properties of generalized canonical analysis*. Paper presented at the European Meeting of the Psychometric Society, University of Groningen, Groningen, the Netherlands, July.

Schader, M., & Gaul, W. (Eds.). (1990). *Knowledge, data and computer-assisted decisions*. Berlin, Germany: Springer-Verlag.

Shannon, C.E., & Weaver, W. (1949). *Mathematical theory of communication*. Urbana: University of Illinois Press.

Sheu, W.J. (1982). Development of new algorithms in dual scaling in dual scaling for multi-way classification. Unpublished doctoral dissertation, University of Toronto, Canada.

Sneath, P.H.A., & Sokal, R.R. (1973). *Numerical taxonomy*. San Francisco: W.H. Freeman.

Stebbins, C. L. (1950). *Variation and evolution*. New York: Columbia University Press.

Stewart, F.M. (1973). *Introduction to linear algebra*. Princeton: Van Nostrand.

Symonds, P.M. (1924). On the loss of reliability in ratings due to coarseness of the scale. *Journal of Experimental Psychology*, *7*, 456-461.

Tago, A. (1966). *Atama no taiso* [Exercises for the Brain] (Vol. 1). Tokyo, Japan: Kobunsha.

Takane, Y. (1980a). Analysis of categorizing behavior. *Behaviormetrika*, *8*, 75-86.

Takane, Y. (1980b). *Tajigen shakudo ho* [Multidimensional scaling]. Tokyo, Japan: University of Tokyo Press.

Takane, Y., & Shibayama, T. (1991). Principal component analysis with external information on both subjects and variables. *Psychometrika*, *56*, 97-120.

Takeuchi, K., Yanai, H., & Mukherjee, B.N. (1982). *The foundations of multivariate analysis: A unified approach by means of projection onto linear subspaces*. New Delhi, India: Wiley Eastern.

Tanaka, Y. (1979). Optimal scaling of arbitrarily ordered categories. *Annals of the Institute of Statistical Mathematics*, *31*, 115-124.

Tanaka, Y., & Asano, C. (1978). A generalized method of optimal scaling for partially ordered categories. In *Proceedings of the Third Symposium on Computational Statistics* (pp.324—330). Berlin: Physica-Verlag.

Tanaka, Y., & Kodake, K. (1980). Computational aspects of optimal scaling for ordered categories. *Behaviormetrika*, *7*, 35-46.

Tenenhaus, M. (1982). Review of S. Nishisato: "Analysis of categorical data: Dual scaling and its applications." *Psychometrika*, *47*, 120-121.

Tenenhaus, M. (1988). Canonical analysis of two convex polyhedral cones and applications. *Psychometrika*, *53*, 503-524.

Tenenhaus, M., & Young, F.W. (1985). An analysis and synthesis of multiple correspondence analysis, optimal scaling, dual scaling, homogeneity analysis and other methods for quantifying multivariate categorical data. *Psychometrika, 50*, 91-119.

Torgerson, W.S. (1958). *Theory and methods of scaling*. New York: Wiley.

Torii, Y. (1977). Some generalizations of optimal scaling. Unpublished doctoral dissertation, University of Toronto, Canada.

Traub, R.E. (Ed.) (1979). *New directions in testing and meaurement*. San Francisco: Jossey-Bass.

Traub, R.E., & Wolfe, R.G. (1981). Latent trait theories and assessment of educational achievement. In D.C. Berliner (Ed.), *Review of Research in Education 9*. Washington: American Educational Research Association.

Tsujitani, M. (1988a). Graphical analysis of association in cross-classification having ordered categories. *Behaviometrics, 24*, 41-53.

Tsujitani, M. (1988b). Optimal scaling for association models when category scores have a natural ordering. *Statistics & Probability Letters, 6*, 175-180.

Tversky, A. (1964). On the optimal number of alternatives at a choice point. *Journal of Mathematical Psychology, 1*, 386-391.

van Buuren, S., & Heiser, W.J. (1989). Clustering N objects into K groups under optimal scaling of variables. *Psychometrika, 54*, 699-706.

van Buuren, S., & van Rijckevorsel, J.L.A. (1992). Imputation of missing categorical data by maximizing internal consistency. *Psychometrika, 57*, 567-580.

van der Burg, E. (1988). *Nonlinear canonical correlation and some related techniques*. Leiden, the Netherlands: DSWO Press.

van der Heijden, P.G.M. (1987). *Correspondence analysis of longitudinal categorical data*. Leiden, the Netherlands: DSWO Press.

van der Heijden, P.G.M., & de Leeuw, J. (1985). Correspondence analysis used complementary to loglinear analysis. *Psychometrika, 50*, 429-447.

van der Heijden, P.G.M, Falguerolles, A., & de Leeuw, J. (1989). A combined approach to contingency table analysis using correspondence analysis and log-linear analysis. *Applied Statistics, 38*, 249-292.

van Rijckevorsel, J.L.A. (1987). *The application of fuzzy coding and horseshoe in multiple correspondence analysis*. Leiden, the Netherlands: DSWO Press.

van Rijckevorsel, J.L.A., & de Leeuw, J. (Eds.). (1988). *Component and correspondence analysis: Dimension reduction by functional approximation*. New York: Wiley.

Watanabe, S. (1969). *Knowing and guessing: A quantitative study of inference and information*. New York: Wiley.

Weingarden, P., & Nishisato, S. (1986). Can a method of rank ordering reproduce paired comparisons? An analysis by dual scaling (correspondence analysis) . *Canadian Journal of Marketing Research. 5*, 151-170.

Wiener, N. (1948). *Cybernetics*. New York: Wiley.

Willet, M. (1986). An investigation of the effects of forced classification of successive categories data, using dual scaling. Unpublished master's thesis, University of Toronto, Canada.

Xu, L. (1993). Ties in rank-order data and dual scaling. Unpublished master's thesis, University of Toronto, Canada.

Yamada, F., & Nishisato, S. (in press). Sohtsui shakudoho ni kansuru ikutsuka no tokusei [Several mathematical properties of dual scaling as applied to dichotomous item-category data]. *Japanese Journal of Behaviormetrics, 20*.

Yanai, H., & Takane, Y. (1977). *Tahenryo kaiseki ho* [Multivariate Analysis] . Tokyo, Japan: Asakura Shoten.

Young, F.W. (1984). Scaling. *Annual Review of Psychology*, *35*, 55-81.

Author Index

A,B

C

D,E,F

G

H

I,J,K

L,M,N

O,P,R

S,T,U

V,W,X,Y

Subject Index

A

B,C

D

E,F

G

H,I

L

M

N,O

P

R

S

T,U,V,W

Elements of Dual Scaling:

An Introduction to Practical Data Analysis